2
Plant Cell Monographs

Series Editor: David G. Robinson

Plant Cell Monographs

Recently Published Titles

Plant Endocytosis
Volume Editors:
Šamaj, J., Baluška, F., Menzel, D.
Vol. 1, 2005

Somatic Embryogenesis
Volume Editors: Mujib, A., Šamaj, J.
Vol. 2, 2006

Somatic Embryogenesis

Volume Editors:
A. Mujib · J. Šamaj

With 53 Figures and 21 Tables

Springer

Volume Editors:

Dr. A. Mujib
Department of Botany
Hamdard University
Hamdard Nagar
110062 New Delhi
India

Dr. Jozef Šamaj
Institute of Cellular and Molecular Botany
Rheinische Friedrich-Wilhelms-University Bonn
Department of Plant Cell Biology
Kirschallee 1
D-53115 Bonn
Germany

Series Editor:

Professor Dr. David G. Robinson
Ruprecht-Karls-University of Heidelberg
Heidelberger Institute for Plant Sciences (HIP)
Department Cell Biology
Im Neuenheimer Feld 230
D-69120 Heidelberg
Germany

ISSN 1861-1370
ISBN-10 3-642-42473-2 Springer Berlin Heidelberg New York
ISBN-13 978-3-642-42473-1 Springer Berlin Heidelberg New York
DOI 10.1007/b136454

Springer is a part of Springer Science+Business Media

springer.com

Softcover re-print of the Hardcover 1st edition 2006

Editor: Dr. Christina Eckey, Heidelberg
Desk Editor: Anette Lindqvist, Heidelberg
Cover design: *Design & Production* GmbH, Heidelberg
Typesetting and Production: LE-TEX Jelonek, Schmidt & Vöckler GbR, Leipzig

Printed on acid-free paper 149/3152 YL – 5 4 3 2 1 0

Editors

A. Mujib is senior lecturer at the Department of Botany, Hamdard University, New Delhi, India. He received his Ph.D. degree from the University of Kalyani (India) working on in vitro studies of bulbous ornamentals. Later he was appointed to a position in the Department of Chemical Engineering, Indian Institute of Technology (IIT), Kharagpur. He teaches molecular genetics, cytogenetics, and plant biotechnology. His current research interests include somatic embryogenesis and gene expression, analysis of somaclonal variants with a focus on transposons, and in vitro mutagenesis using cell culture as a system. His scientific work has resulted in more than 55 publications including research papers, review articles, and book chapters.

Jozef Šamaj received his Ph.D. degree from the Comenius University in Bratislava, Slovakia. He completed three post-doctoral programmes supported by Eurosilva, the Alexander von Humboldt Foundation, and the EU Marie Curie Programme in well-recognized laboratories in Toulouse, Bonn, and Vienna. His scientific work has focused on the cell biology of somatic embryogenesis, lignification in tree species, arabinogalactan proteins, the cytoskeleton, and signalling proteins. Jozef Šamaj has co-authored more than 65 research papers, reviews, and book chapters. Currently he is a senior lecturer and associate researcher at the Institute of Cellular and Molecular Botany in Bonn, Germany. His main research interest centers on the role of stress-induced MAP-kinase cascades in relation to vesicular trafficking and the cytoskeleton during stress responses in *Arabidopsis thaliana*.

Preface

Somatic embryogenesis is an unique process in plants in which embryos are formed in culture from somatic cells. Although the phenomenon is known since 1950s, there is a continuously growing interest up to now. In recent years, information is being accumulated rapidly in the field of *in vitro* embryogenesis. Even though internet and scientific journals are quite efficient in disseminating current knowledge, the need of regular updated information in a structured way, i.e in this book, is always desirable for a variety of reasons. Thus, in the present book, a sincere effort has been attempted to collect and compile the recent advances made in somatic embryogenesis research. The book contains 22 chapters which will refresh our view on various important aspects of *in vitro* embryogenesis. The origin, development and other structural/morphological details of embryos have been discussed in chapters 2 and 7. Some chapters (1, 10 and 15) are particularly devoted to biochemical, physiological and cytological aspects of embryogenesis. The book also contains topics on the role of genotype, environment, plant hormones, arabinogalactan-proteins and cytoskeleton as well as brief focus on genetic transformation in relation to embryogenesis (chapters 3, 4, 8, 12, 16, 17 and 19). Chapter 7 describes why somatic plant cells start to form embryos in culture. Besides, a good attention has been paid to subjects that deal with protein markers, antibodies (chapter 14) and differential gene expression during embryogenesis (chapter 20). As regards to somatic embryogenesis, this *in vitro* system in many ways, mimics zygotic embryogenesis, so a thoughtful discussion on this subject is wisely incorporated (chapters 5 and 18). In current years, several genes were found to be associated with embryogenesis, therefore chapters 5 and 11 are discussing the role of genes such as *Baby Boom, Leafy Cotyledons* etc. Different genome wide expression technologies analyzing "embryo specific" genes have been described critically in chapter 6. Haploid embryogenesis, a major tool in crop improvement program has been discussed in chapter 21. Last but not least, embryo culture and applications of *in vitro* embryogenesis in Arabidopsis are discussed in chapter 22.

We hope that this latest book on embryogenesis will be a very useful source of information and references to a wide range of audience especially to researchers, graduate students, teachers and many others, working in the areas of agriculture, botany, plant biotechnology, horticulture and related fields.

A. Mujib is grateful to S. Banerjee, Associate Editor, CSIRO publishing, Australia, who has gone through some of the papers, and to several present laboratory research students for their timely help. Lastly, we both are indebted to our authors for their valuable contributions.

New Delhi, India A. Mujib
Bonn, Germany J. Šamaj

Contents

Plant Cell Monogr (2)
A. Mujib · J. Šamaj: Somatic Embryogenesis
DOI 10.1007/7089_021/Published online: 20 October 2005

Storage Proteins and Peroxidase Activity During Zygotic and Somatic Embryogenesis of Firs (*Abies* sp.)

A. Kormut'ák (✉) · B. Vooková

Institute of Plant Genetics and Biotechnology, Slovak Academy of Sciences, Akademicka 2, P.O. Box 39A, 950 07 Nitra, Slovakia
andrej.kormutak@savba.sk

Abstract Somatic embryogenesis was initiated from immature embryos of *Abies concolor* (Gord. et Glend), *A. numidica* De Lann. and *A. cilicica* Carr., *A. alba* Mill. as well as in hybrid fir *A. cilicica* × *A. nordmanniana*. Soluble protein profiles and peroxidase activity were compared in developing zygotic and somatic embryos of silver fir (*A. alba* Mill.). On the basis of sodium dodecyl sulfate polyacrylamide gel electrophoresis of soluble proteins a high degree of homology was established between the two types of embryos. A higher peroxidase activity was registered throughout zygotic embryogenesis than during somatic embryo development but the opposite was true at the stage of mature embryos. Isoperoxidase composition reflected more efficiently the developmental stages of zygotic embryogenesis than those of somatic embryogenesis.

1 Introduction

Somatic embryogenesis has become a major tool in the study of plant embryology, as it is possible in culture to manipulate cells of many plant species to produce somatic embryos in a process that is remarkably similar to zygotic embryogenesis (Thorpe 2000).

Induction of somatic embryogenesis in the genus *Abies* has been demonstrated in five pure species: *A. alba* (Hristoforoglu et al. 1995; Schuller et al. 2000), *A. nordmanniana* (Nørgaard and Krogstrup 1991, 1995), *A. balsamea* (Guevin et al. 1994), *A. fraseri* (Guevin and Kirby 1997; Rajbhandari and Stomp 1997) and *A. cephalonica* (Krajňáková and Häggman 1997).

In our laboratory, embryogenic cultures of hybrid firs have been derived from immature *A. alba* × *A. alba*, *A. alba* × *A. nordmanniana* (Gajdošová et al. 1995), *A. alba* × *A. cephalonica*, *A. alba* × *A. numidica* (Salajová et al. 1996) and mature *A. alba* × *A. cephalonica* zygotic embryos (Salajová and Salaj 2003/2004).

2
Somatic Embryogenesis of *Abies* sp.

2.1
Initiation of Embryogenic Tissue from Immature Zygotic Embryos

Embryogenic tissue was induced from immature embryos of *A. concolor* (Gord. et Glend), *A. numidica* De Lann. and *A. cilicica* Carr. derived from self-pollination as well as in hybrid fir from interspecific crosses *A. cilicica* × *A. nordmanniana*. Immature seeds were surface-sterilized for 10 min in 10% H_2O_2. Endosperms containing embryos (from July) or embryos after excision from the megagametophyte (from August) were plated on Schenk and Hildebrandt (SH) initiation medium (Schenk and Hildebrandt 1972) with 1 $mg\,l^{-1}$ benzylaminopurine and 2% sucrose. The medium was solidified with 0.3% Phytagel. The cultures were kept in the dark at 21–23 °C. After 4–8 weeks of explant cultivation, white, mucilaginous extrusions were observed from the micropylar end of the megagametophyte. Early zygotic embryos in megagametophytes, collected in early July, produced more readily embryogenic cultures. Embryogenic tissue of *A. concolor* was induced in 5.6% of explants, and of *A. numidica* in 6.8% of explants (Vooková and Kormut'ák 2004). In *A. cilicica*, the initiation of embryogenic tissue frequency ranged between 5.4 and 63.5%, and in *A. cilicica* × *A. nordmanniana* between 3.0 and 27.6% (Vooková and Kormut'ák 2003). For *Abies*, the cytokinin as a sole growth regulator was sufficient to induce somatic embryogenesis in immature (Schuller et al. 1989; Nørgaard and Krogstrup 1991) as well as in mature (Hristoforoglu et al. 1995) embryo explants.

2.2
Proliferation of Embryogenic Cultures

Embryogenic tissue proliferated on SH initiation medium with supplement of 0.05% L-glutamine and 0.1% casein hydrolysate and were subcultured every 3 weeks. More than 90% of the responding explants developed embryogenic tissue within 1 month of culture. The embryogenic cultures in *Abies* sp. regardless of their different origin exhibited the common morphological features. It was found in our previous experiments (Hřib et al. 1997) that embryogenic tissue of *A. alba* shows many similarities with habituated non-organogenic sugar beet callus (Gaspar et al. 1988).

A. numidica embryogenic culture was used as a model for characterization of cell lines (Vooková and Kormut'ák 2002a). Embryogenic cell lines have been divided into two groups on the basis of morphology and growth characteristics of somatic embryos according to Mo et al. (1996). The cell line representing group B with undeveloped somatic embryos was stimulated to undergo maturation by treatment with plant growth regulators.

2.3 Somatic Embryo Maturation and Germination

Somatic embryo maturation of *Abies* species and hybrid was achieved on modified Murashige and Skoog (MS) medium (Murashige and Skoog 1962) supplemented with 4% maltose, 10% polyethyleneglycol 4000 (PEG-4000), $10\,\mathrm{mg\,l^{-1}}$ abscisic acid (ABA) and $500\,\mathrm{mg\,l^{-1}}$ L-glutamine and casein hydrolysate.

Maturation of fir somatic embryos is promoted by ABA. ABA plays an important role in conifer embryogenesis. It inhibits cleavage polyembryony, allowing embryo singulation, its further development and maturation (Boulay et al. 1988). The production of cotyledonary somatic embryos in *A. cilicica* and *A. cilicica* × *A. nordmanniana* was influenced by ABA. The addition of $20\,\mathrm{mg\,l^{-1}}$ ABA into the maturation medium was the most effective for maturation (Vooková and Kormuťák 2003).

The literature data indicated that carbohydrates as a source of carbon or osmotica influenced somatic embryogenesis in *Abies*. Lactose and sorbitol favoured *A. alba* somatic embryo maturation up to an early cotyledonary stage (Schuller et al. 2000). Maltose gave a better maturation response and the addition of PEG-4000 to the medium promoted the maturation of somatic embryos in *A. nordmanniana* (Nørgaard 1997) and *A. alba* × *A. numidica* (Salaj et al. 2004).

In *A. numidica*, the effect of subculture period and the concentration of PEG and maltose was confirmed on maturation of somatic embryo (Vooková and Kormuťák 2002b). The maturation was promoted by PEG-4000, at 7.5 to 10%. Maltose (3 to 6%) significantly enhanced the yield of mature embryos. It seems that choice of the basal medium for somatic embryo maturation is also important. Embryogenic tissues of *A. cilicica*, *A. numidica*, *A. concolor* and *A. cilicica* × *A. nordmanniana* hybrid were cultured on SH, Gresshoff and Doy (GD; Gresshoff and Doy 1972) and modified MS media. The tendency for better maturation on SH and MS media was common for all cultures tested (Table 1). GD medium was not suitable because maturation was slow and achieved only the precotyledonary stage of development (Vooková and Kormuťák 2003, 2004). Exogenously applied myo-inositol ($100\,\mathrm{mg\,l^{-1}}$) influenced somatic embryogenesis of *A. numidica* although this process occurred on media with and without this compound (Vooková et al. 2001).

Prior to germination, isolated mature somatic embryos with four to six cotyledons were subjected to partial drying in the dark at 21–23 °C for 3 weeks. Mature somatic embryos were placed in small Petri dishes (60-mm diameter). The Petri dish was open and placed on moist filter paper in a bigger Petri dish (90-mm diameter), which was sealed with Parafilm. Then desiccated mature somatic embryos were transfered to a germination medium and cultured in the light (16-h photoperiod).

Table 1 Numbers (± standard error, *SE*) of cotyledonary somatic embryos of *Abies* species and hybrid (per gram of embryogenic tissue) matured on Schenk and Hildebrandt (*SH*), Murashige and Skoog (*MS*) and Gresshoff and Doy (*GD*) media, and germination frequency of somatic embryos on SH medium

Species/hybrid	SH	MS	GD	Germination (%)	SE ±
A. cilicica	6±1.5	16±1.9	0	74.99	6.81
A. numidica	16±4.9	26±2.9	1±0.7	85.45	4.11
A. concolor	–	61±7.5	0	71.10	5.22
A. cilicica × *A. nordmanniana*	3±1.3	45±6.6	0	83.61	11.40

Media for germination are routinely used with sucrose in 2% concentration, and with (Nørgaard 1997) or without (Salajová et al. 1996; Guevin and Kirby 1997) activated charcoal. In our experiment (Vooková and Kormuťák 2001) no significant differences were detected between MS and SH media. The addition of 1% activated charcoal or 0.05 mg l^{-1} indole-3-butyric acid into both media had a positive influence on *A. numidica* embryo germination. A high rooting percentage (85%) was recorded on half SH medium with 1% sucrose and activated charcoal. It seems that this medium is widely applicable. We have used it successfully for germination of other *Abies* sp. and hybrid (Table 1). With increased sucrose concentration the germination was reduced.

3
Storage Proteins of Conifer Seeds

Comparative study of zygotic and somatic embryogenesis in conifers has shown that except for morphological similarity there exists a high degree of biochemical homology between zygotic and somatic embryos of conifers, especially with respect to their storage proteins (Hakman et al. 1990). Because of their accumulation during embryo development, the latter were reported to be excellent markers for comparison of zygotic and somatic embryo programmes (Flinn et al. 1993). On the basis of similarities of the protein molecular weight, the somatic embryos of *Picea glauca* (Flinn et al. 1991; Misra et al. 1993), *Picea abies* (Hakman 1993; Hakman et al. 1990) and *Pinus strobus* (Klimaszewska et al. 2004) were shown to contain the same storage proteins as the corresponding zygotic embryos. The greater biochemical similarity of somatic embryos to their zygotic counterparts is believed to improve the conversion of somatic embryos to plants (Klimaszewska et al. 2004). According to Cyrr et al. (1991) the criteria for obtaining high-quality somatic embryos include both the formation of storage reserves that are analogous to those of seed embryos and the absence of precocious germination. The authors presented evidence suggesting that differences between the performance of *Picea glauca* somatic and seed embryos during germination and early growth could be attributed to the differences in the kinetics of storage reserve utilization.

As far as the nature of conifer seed storage proteins is concerned, both insoluble crystalloids and soluble matrix proteins were identified (Misra and Green 1990). Insoluble proteins have molecular masses in their non-reduced form of 57 kDa, whereas in reduced form they migrate as three distinct groups of proteins in the molecular mass range of 42 kDa, 34.5–35 kDa and 22.5–23 kDa. The soluble fraction involves proteins in the molecular mass range of 27–30 kDa. In two of the three *Picea* species analysed the 34.5-kDa protein band was absent, indicating interspecific variation in quality of storage reserves (Misra and Green 1990). In *Pinus strobus*, the most abundant were the buffer-insoluble 11S globulins of molecular mass 59.6 kDa, which dissociate under reduced conditions to 38.2 – 40.0 and 22.5–23.5-kDa range polypeptides, and buffer soluble 7S vicilin-like proteins of molecular mass 46.0–49.0 kDa, which did not separate under reduced conditions. Other relatively abundant soluble proteins were in the ranges of 25–27 and 27–29 kDa (Klimaszewska et al. 2004). The *Abies* species lack 55 kDa $\alpha\beta$-dimer legumin-like proteins in their seeds and were reported to deviate conspicuously from *Cedrus*, *Larix*, *Picea* and *Pseudotsuga*. Other proteins are present in *Abies* seeds like in the remaining *Pinaceae*. Their soluble fraction involves 43-, 28- and 16-kDa proteins (Jensen and Lixue 1991).

Our data derived from comparison of the sodium dodecyl sulfate polyacrylamide gel electrophoresis (SDS-PAGE) protein profiles of both zygotic and somatic embryos of silver fir (*A. alba* Mill.) indicate the presence of some additional proteins which meet the criteria of storage reserves. Their origin was traced from the cub-like embryo stage until germinating embryos during zygotic embryogenesis and from the non-embryogenic callus until regenerated emblings during somatic embryogenesis.

Figure 1 illustrates the dynamics of soluble protein synthesis during silver fir zygotic embryo development. At least ten major components along with numerous minor protein bands may be distinguished in the SDS-PAGE profile of mature embryos. The approximate molecular masses of the major proteins correspond to 55, 46, 40, 36, 30, 26, 24, 22, 18 and 14 kDa, respectively (Fig. 1, lane F with arrows). Their presence in embryos may be traced already at the precotyledonary stage (lane B). In particular, it is true of the 55- and 46-kDa proteins, which represent the prominent components of the soluble protein profile of young zygotic embryos. The only exception is the 24-kDa protein, whose synthesis seems to begin at the advanced cotyledonary stage only (lane D). We infer, this protein belongs to the category of Lea proteins that are synthesized during late embryogenesis and which are believed to prevent embryos from damage from desiccation and from precocious germination during somatic embryo development (Dong and Dunstan 2000; Zimmerman 1993). Within the context of a continuous synthesis of an

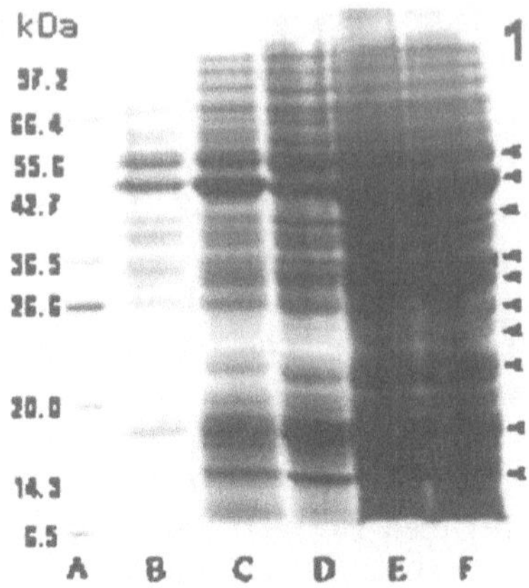

Fig. 1 Sodium dodecyl sulfate polyacrylamide gel electrophoresis (*SDS-PAGE*) profiles of soluble proteins during zygotic embryo development. *A* molecular size marker, *B* pre-cotyledonary embryos, *C* early cotyledonary embryos, *D* advanced cotyledonary embryos, *E* morphologically differentiated embryos, *F* physiologically mature embryos

overwhelming majority of proteins during embryo development an abrupt increase in the amount of the 46-, 40-, 36-, 30-, 22- and 18-kDa proteins at the advanced cotyledonary stage was rather conspicuous (lane D). All these proteins dominate the soluble protein profile of mature embryos. Their identity as seed storage reserves was inferred from degradation of individual proteins during seed germination. Figure 2 illustrates that SDS-PAGE protein profiles of zygotic embryos are identical during the first 48 h of seed imbibition (lanes H, I). Profound changes appear only when the radicle emerges from a seed coat. The 24-kDa protein is depleted completely at this stage, while the proteins of 46, 36, 26 and 22 kDa are consumed only partially (lane J). During advanced germination (lane K) and at the seedling stage (lane L) the degradation of 46- and 36-kDa proteins is completed. The depletion of the 26- and 22-kDa proteins is also considerable but not complete. Their synthesis seems to be resumed at the seedling stage along with a strengthened synthesis of the 55-kDa protein and de novo synthesis of the 19-kDa protein (lane L). On the basis of the abundance criterion and degradation kinetics during germination, it seems reasonable to ascribe the storage reserve function to the 46-, 36-, 26-, 24- and 22-kDa proteins in silver fir zygotic embryos. This figure is very similar to that found for *Picea abies*, where three major seed storage proteins of 42, 33 and 22 kDa were distinguished by Stabel et al. (1990). Hakman et al. (1990) have in addition included among *Picea abies* storage proteins a 28-kDa protein. An essentially similar situation was also found in *Picea glauca* zygotic embryos with 43-, 33-, 22-, 18- and 16-kDa proteins dominating the SDS-PAGE profile and with less abundant 28- and 24-kDa proteins (Flinn et al. 1993). Recently, Klimaszewska et al. (2004) reported seed storage proteins in zygotic embryos of *Pinus strobus* involving soluble proteins with

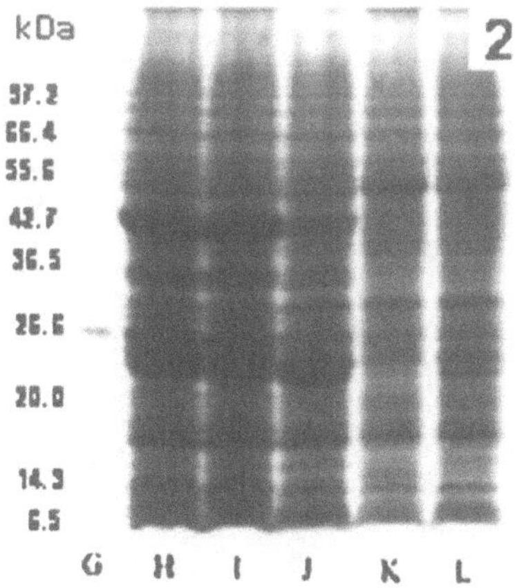

Fig. 2 SDS-PAGE profiles of soluble proteins during germination of seeds. *G* molecular size marker, *H* dormant embryos after 24-h imbibition, *I* dormant embryos after 48-h imbibition, *J* beginning of seed germination, *K* advanced seed germination, *L* seedlings

little deviating molecular mass ranges of 46.0–49.0, 38.2–40.0, 25–27, 27–29 and 22.5–23.5 kDa. With special reference to major storage proteins detected in *A. alba*, they seem to fall into these classes of proteins as well. According to Gifford (1988) and Gifford and Tolley (1989) this suggests that storage proteins may be conserved among the conifers, although the relative amount of different proteins differ among the species.

4
SDS-PAGE Protein Profile of *A. Alba* Somatic Embryos

As far as somatic embryos of silver fir are concerned, their SDS-PAGE protein profiles were comparable with the corresponding profiles of zygotic embryos. Among the proteins detected, the most abundant were those with molecular masses of 53, 46, 40, 36, 30, 28, 24, 20 and 18 kDa, respectively (Fig. 3). As an exception, the presence of the 53-kDa protein in somatic embryos may be mentioned instead of the 55-kDa protein detected in zygotic embryos. Also, the 14-kDa protein of somatic embryos was expressed less than the corresponding fraction of zygotic embryos.

An overwhelming majority of abundant proteins may be traced in developing somatic embryos. They are already weakly expressed in embryogenic callus (lane C) and become very distinct at the globular, torpedo and cotyledonary stages of somatic embryos (lanes D–F). The 46-kDa protein is an exception in this respect, exhibiting the highest concentration at the cotyledonary stage only (lane F). However, during desiccation of mature somatic embryos this protein is depleted preferentially (lane G). The same is also true

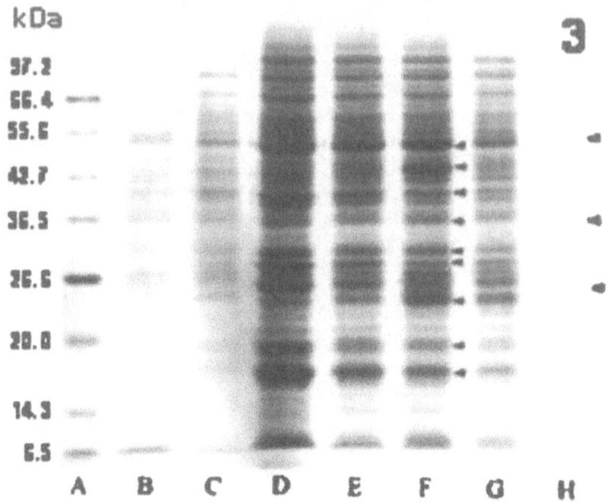

Fig. 3 SDS-PAGE profiles of soluble proteins during somatic embryogenesis. *A* molecular size marker, *B* non-embryogenic callus, *C* embryogenic callus, *D* globular stage, *E* torpedo stage, *F* cotyledonary stage, *G* mature somatic embryos after desiccation, *H* regenerated emblings

of the rest of the soluble proteins when at the stage of regenerated emblings nearly all proteins were consumed. Detectable amounts were found only in the case of 53-, 36- and 24-kDa proteins (lane H with arrows). Owing to the buffer-soluble nature of these proteins we assume they represent the soluble matrix proteins as quoted by Misra and Green (1990).

5 Peroxidase Activity in Developing Zygotic and Somatic Embryos of *A. Alba*

In contrast to soluble proteins the differences between zygotic and somatic embryos of silver fir in peroxidase activity are more evident. The enzyme was found to exhibit 3 times higher activity in mature somatic embryos than in dormant zygotic embryos (Table 2). No activity was detected in precotyledonary zygotic embryos. Starting with the early cotyledonary stage, a decline in peroxidase activity was registered throughout zygotic embryogenesis, and the situation was similar during somatic embryogenesis. However, peroxidase activity changed abruptly during two stages of somatic embryogenesis. The first stage was the transition of non-embryogenic to embryogenic callus, accompanied by a conspicuous decline in specific enzyme activity. The higher peroxidase activity in non-embryogenic callus is due to increased levels of phenolic substances in this tissue, some of which serve as substrates in peroxidase-catalysed reactions (Hrubcová et al. 1994). The second stage was that of regenerated emblings, which had 7 times higher peroxidase

Table 2 Changes in peroxidase activity during zygotic and somatic embryogenesis of silver fir (From Kormutak et al. 2003, with permission of Versalius University Medical Publisher in Cracow)

Developmental stage	Specific activity
Zygotic embryogenesis	
Precotyledonary	0
Early cotyledonary	0.25 ± 0.00
Advanced cotyledonary	0.20 ± 0.00
Morphologically difffenet embryos	0.16 ± 0.00
Physiologically mature embryos	0.08 ± 0.00
Somatic embryogenesis	
Non-embryogenic callus	3.15 ± 0.050
Embryogenic callus	0.93 ± 0.015
Precotyledonary	0.11 ± 0.005
Cotyledonary	0.11
Mature embryos	0.24
Regenerated emblings	1.83 ± 0.080

activity than mature somatic embryos. Obviously, this increase in enzyme activity is a part of metabolic events underlying embryo germination and progressive embling development. The higher metabolic potential of mature somatic embryos than that of mature zygotic embryos may probably be ascribed to the different levels of dormancy which seem to be lower in somatic embryos.

The changes outlined in peroxidase activity during zygotic embryogenesis were also paralleled by the changes in isoenzyme composition. The only exception were embryos at the precotyledonary stage lacking peroxidase activity but containing as many as seven to eight isoenzymes (Fig. 4, lane A). Starting with the early cotyledonary stage until mature embryos, the number of isoperoxidases followed closely the tendencies in peroxidase activity. The early and advanced cotyledonary embryos accordingly possessed the highest number of isoperoxidases visualized in the gels as eight intensively stained bands (lanes B, C). Also, morphologically differentiated and physiologically mature zygotic embryos with lowered peroxidase activity possessed very similar isoenzyme profiles consisting of five isoperoxidases (lanes D, E).

Like in zygotic embryos, a close correlation between peroxidase acitity and its isoenzyme composition has been observed during somatic embryogenesis as well. As shown in Fig. 5, the high enzyme activity of both embryogenic callus (lane A) and regenerated emblings (lane D) is also reflected by the enriched isoenzyme profiles involving nine to ten isoperoxidases as compared with six to seven isoperoxidases detected in precotyledonary (lane B) and cotyledonary (lane C) embryos. However, as a molecular marker, this enzyme

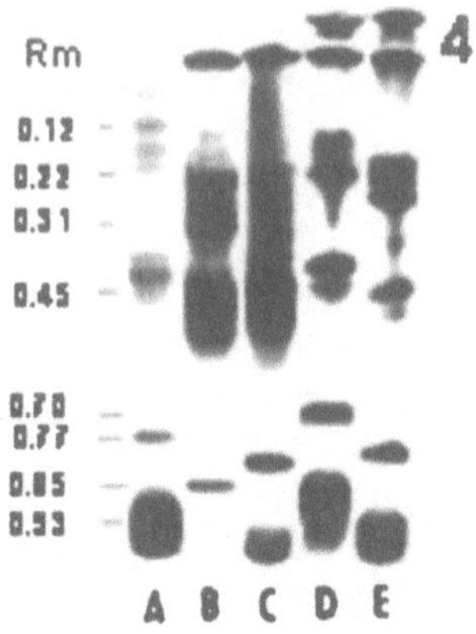

Fig. 4 Isoperoxidase composition of developing zygotic embryos. *A* precotyledonary embryos, *B* early cotyledonary embryos, *C* advanced cotyledonary embryos, *D* morphologically differentiated embryos, *E* physiologically mature embryos

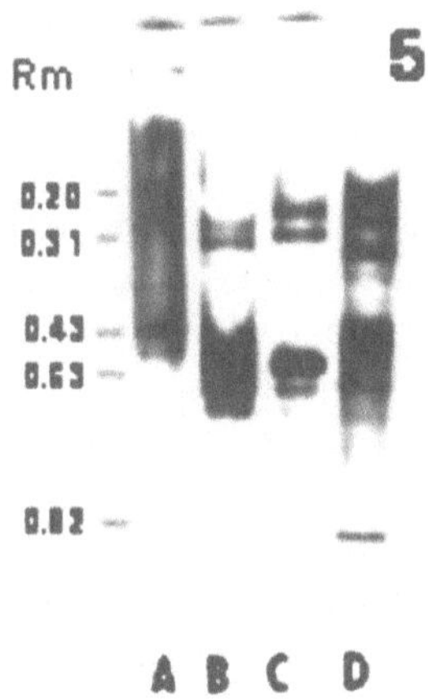

Fig. 5 Isoperoxidase composition of developing somatic embryos. *A* embryogenic callus, *B* torpedo stage, *C* cotyledonary stage, *D* regenerated emblings

seems to be more indicative of individual stages of zygotic embryogenesis than those of somatic embryogenesis.

6 Conclusions and Future Prospects

Emblings of *A. concolor*, *A. numidica*, *A. cilicica*, *A. alba* and *A. cilicica* × *A. nordmanniana* hybrid firs have been regenerated from immature zygotic embryos via somatic embryogenesis.

In spite of the postulated divergency of *Abies* storage proteins from other *Pinaceae*, the data presented indicate a high degree of similarity between soluble protein profiles of silver fir and the corresponding profiles of *Picea* and *Pinus* sp. Most probably, these proteins represent the soluble matrix proteins. The question whether insoluble proteins as the main constituent of the conifer storage reserves share the genus- or species-specific features remains to be answered. Additional experiments which will help to resolve this point are highly desirable. A high degree of homology has also been observed between zygotic and somatic embryogenesis of silver fir with respect to the major storage proteins represented by ten or nine fractions, respectively. The only difference observed so far was related to the dynamics of the 46-kDa protein synthesis. As the main component of the soluble protein profile this protein seems to be synthesized continuously during zygotic embryogenesis starting with the precotyledonary stage of embryo development. In contrast, during somatic embryo development its synthesis becomes conspicuous at the cotyledonary stage only. A remarkable feature of the somatic embryo soluble protein dynamics is their nearly complete depletion in regenerated emblings. This aspect of *Abies* somatic embryo development needs to be verified as well. The metabolic potential as revealed by peroxidase activity seems to be higher in developing zygotic embryos than in somatic ones; however, zygotic embryos after reaching maturity become enzymatically more quiescent than somatic embryos. Isoperoxidase composition can be discriminated more clearly between individual stages of zygotic embryo development than in somatic embryogenesis. In order to find out efficient molecular markers of *Abies* embryogeny, additional isoenzyme systems have to be involved in future comparative studies.

Acknowledgements Financial support of the work from the Slovak Grant Agency VEGA, project no. 2/3192/24 is highly appreciated.

References

Boulay MP, Gupta PK, Krogstrup P, Durzan DJ (1988) Development of somatic embryos from cell suspension cultures of Norway spruce (*Picea abies* karst). Plant Cell Rep 7:134–137

Cyrr DR, Webster FB, Roberts DR (1991) Biochemical events during germination and early growth of somatic embryos and seed of interior spruce (*Picea glauca engelmanii* complex). Seed Sci Res 1:91–97

Dong JZ, Dunstan DI (2000) Molecular biology of somatic embryogenesis in conifers. In: Jain SM, Minocha SC (eds) Molecular biology of woody plants. vol 1. Kluwer, Dordrecht, p 51–87

Flinn BS, Roberts DR, Taylor JEP (1991) Evaluation of somatic embryos of interior spruce. Characterization and developmental regulation of storage proteins. Physiol Plant 82:624–632

Flinn BS, Roberts DR, Newton CH, Webster FB, Taylor JEP (1993) Storage protein gene expression in zygotic and somatic embryos of interior spruce. Physiol Plant 89:719–730

Gajdošová A, Vooková B, Kormut'ák A, Libiaková G, Doležel J (1995) Induction, protein composition and DNA ploidy level of embryogenic calli of silver fir and its hybrids. Biol Plant 37:169–176

Gaspar T, Kevers C, Penel C, Crèvecoeur M, Greppin H (1988) Biochemical characterization of normal and habituated sugarbeet calli. Relationship with anatomy, habituation and organogenesis. Potsdamer Forsch 57:21–30

Gifford DJ (1988) An electrophoretic analysis of the seed protein from *Pinus monticola* and eight other species of pine. Can J Bot 66:1808–1812

Gifford DJ, Tolley MC (1989) The seed proteins of white spruce and their mobilization following germination. Physiol Plant 77:254–261

Gresshoff PM, Doy CH (1972) Development and differentiation of haploid *Lycopersicum esculentum* (tomato). Planta 107:161–170

Guevin TG, Kirby EG (1997) Induction of embryogenesis in cultured mature zygotic embryos of *Abies fraseri* (Pursh) Poir. Plant Cell Tissue Org Cult 49:219–222

Guevin TG, Micah V, Kirby EG (1994) Somatic embryogenesis in cultured mature zygotic embryos of *Abies balsamea*. Plant Cell Tissue Org Cult 37:205–208

Hakman I (1993) Embryology in Norway spruce (*Picea abies*). An analysis of the composition of seed storage proteins and deposition of storage reserves during seed development and somatic embryogenesis. Physiol Plant 87:148–159

Hakman I, Stabel P, Engström P, Eriksson T (1990) Storage protein accumulation during zygotic and somatic embryo development in *Picea abies* (Norway spruce). Physiol Plant 80:441–445

Hřib J, Vooková B, Kormut'ák A (1997) Biochemical differences between normal callus and embryogenic suspensor mass of silver fir. Biol Plant 39:507–513

Hristoforoglu K, Schmidt J, Bolhar-Nordenkamp H (1995) Development and germination of *Abies alba* somatic embryos. Plant Cell Tissue Org Cult 40:277–284

Hrubcová M, Cvikrová M, Eder J (1994) Peroxidase activities and content of phenolic acids in embryogenic and non-embryogenic alfalfa cell suspension cultures. Biol Plant 39:175–182

Jensen U, Lixue Ch (1991) *Abies* seed protein profile divergent from other *Pinaceae*. Taxon 40:435–440

Kormut'ák A, Salaj T, Matúšová R, Vooková B (2003) Biochemistry of zygotic and somatic embryogenesis in silver fir (*Abies alba* Mill.). Acta Biol Cracoviensia, Ser Botan 45:59–62

Klimaszewska K, Morency F, Jones-Overton C, Cooke J (2004) Accumulation pattern and identification of seed storage proteins in zygotic embryos of *Pinus strobus* and in somatic embryos from different maturation treatments. Physiol Plant 121:682–690

Krajňáková J, Häggman H (1997) Regeneration of *Abies cephalonica* through somatic embryogenesis. In: Identification and control of phase changes in rejuvenation. COST 822, 4th Meeting of WG3, Nitra, Slovakia, 15–19 October 1997, pp 102–104

Misra S, Green MJ (1990) Developmental gene expression in conifer embryogenesis and germination. I. Seed proteins and protein body composition of mature embryo and the megagametophyte of white spruce (*Picea glauca* [Moench] Voss.). Plant Sci 68:163–173

Misra S, Attree SM, Leal J, Fowke LC (1993) Effect of abscisic acid, osmoticum, and desiccation on synthesis of storage proteins during the development of white spruce somatic embryos. Ann Bot 71:11–22

Mo LH, Egertsdotter U, von Arnold S (1996) Secretion of specific extracellular proteins by somatic embryos of *Picea abies* is depended on embryo morphology. Ann Bot 77:143–152

Murashige T, Skoog F (1962) A revised medium for rapid growth and bioassays with tobacco tissue cultures. Physiol Plant 15:473–497

Nørgaard JV (1997) Somatic embryo maturation and plant regeneration in *Abies nordmanniana* LK. Plant Sci 124:211–221

Nørgaard JV, Krogstrup P (1991) Cytokinin induced somatic embryogenesis from immature embryos of *Abies nordmanniana* LK. Plant Cell Rep 9:509–513

Nørgaard JV, Krogstrup P (1995) Somatic embryogenesis in *Abies* spp. In: Jain SM, Gupta PK, Newton RJ (eds) Somatic embryogenesis in woody plants, vol 3. Gymnosperms. Kluwer, Dordrecht, pp 341–355

Rajbhandari N, Stomp A-E (1997) Embryogenic callus induction in Fraser fir. Hort Sci 32:737–738

Roth R, Schmidt J (1998) Induction of embryogenic suspensor masses from cotyledons of zygotic seedlings of *Abies alba*. In: Symp. Zum Gedenken an die 100 Wiederkehr der Begründung der Gewebekultur durch Gottlieb Haberlandt, Vienna, 8–9 October 1998, p 45

Salajová T, Salaj J (2001) Somatic embryogenesis and plantlet regeneration from cotyledon explants isolated from emblings and seedlings of hybrid firs. J Plant Physiol 158:747–755

Salajová T, Salaj J (2003/2004) Somatic embryo formation on mature *Abies alba* × *Abies cephalonica* zygotic embryo explants. Biol Plant 47:7–11

Salajová T, Salaj J, Jásik J, Kormut'ák A, Hakman I (1996) Embryogenic culture initiation and somatic embryo development in hybrid firs (*Abies alba* × *Abies cephalonica* and *Abies alba* × *Abies numidica*). Plant Cell Rep 15:527–530

Salaj T, Matúšová R, Salaj J (2004) The effect of carbohydrates and polyethylene glycol on somatic embryo maturation in hybrid fir *Abies alba* × *A. numidica*. Acta Biol Cracov Bot 46:159–167

Schenk RU, Hildebrandt AC (1972) Medium and techniques for induction and growth of monocotyledonous and dicotyledonous plant cell cultures. Can J For Res 50:199–204

Schuller A, Reuther G, Geier T (1989) Somatic embryogenesis from seed explants of *Abies alba*. Plant Cell Tissue Org Cult 17:53–58

Schuller A, Kirchner-Ness R, Reuther G (2000) Interaction of plant growth regulators and organic C and N components in the formation and maturation of *Abies alba* somatic embryos. Plant Cell Tiss Org Cult 60:23–31

Stabel P, Eriksson T, Engström P (1990) Changes in protein synthesis upon cytokinin-mediated adventitious bud induction and during seedling development in Norway spruce, *Picea abies*. Plant Physiol 92:1174–1183

Thorpe TA (2000) Somatic embryogenesis: morphogenesis, physiology, biochemistry and molecular biology. Korean J Plant Tissue Cult 27:245–258

Vooková B, Kormut'ák A (2001) Effect of sucrose concentration, charcoal, and indole-3-butyric acid on germination of *Abies numidica* somatic embryos. Biol Plant 44:181–184

Vooková B, Kormut'ák A (2002a) Effect of plant growth regulators treatment on proliferation and somatic embryo development of Algerian fir (*Abies numidica* De Lann.). Folia Oecolog 29:141–146

Vooková B, Kormut'ák A (2002b) Some futures of somatic embryo maturation of Algerian fir. In Vitro Cell Dev Biol 38:549–561

Vooková B, Kormut'ák A (2003) Plantlet regeneration in *Abies cilicica* and *Abies cilicica* × *Abies nordmanniana* hybrid via somatic embryogenesis. Turk J Bot 27:71–76
Vooková B, Kormut'ák A (2004) Propagation of some *Abies* species by somatic embryogenesis. Acta Univ Latv Biol 676:257–260
Vooková B, Kormut'ák A, Hřib J (2001) Effect of myo-inositol on somatic embryogenesis of *Abies numidica*. J Appl Bot 75:46–49
Vooková B, Matúšová R, Kormut'ák A (2003) Secondary somatic embryogenesis in *Abies numidica* De Lann. Biol Plant 46:513–517
Zimmerman JL (1993) Somatic embryogenesis: A model for early development of higher plants. Plant Cell 5:1411–1423

Plant Cell Monogr (2)
A. Mujib · J. Šamaj: Somatic Embryogenesis
DOI 10.1007/7089_028/Published online: 2 December 2005

Origin, Development and Structure of Somatic Embryos in Selected Bulbous Ornamentals: BAP as Inducer

A. Mujib[1] (✉) · S. Banerjee[2] · P. D. Ghosh[3]

[1]Department of Botany, Hamdard University, 110062 New Delhi, India
amujib3@yahoo.co.in

[2]CSIRO Publishing, 3066 Melbourne, VIC, Australia

[3]Department of Botany, Kalyani University, 741 235 Kalyani, West Bengal, India

Abstract Somatic embryogenesis in three important ornamentals is discussed in this chapter. Direct somatic embryo development on the explant tissues (bulb-scale) was noticed in *Hippeastrum hybridum* and *Eucharis grandiflora*, both of which are members of Amaryllidaceae. At the time of initiation the embryos were small, water droplet-like, opaque structures, and such development was entirely restricted to the outer rows of scales only. In *Crinum asiaticum*, callus-mediated (indirect) embryo formation was observed on the flower-bud callus, whereas the bulb-scale callus was largely non-embryogenic. Plant growth regulator, such as 6-benzylaminopurine (BAP), frequently induced somatic embryos in *Hippeastrum* and *Eucharis*, and the addition of naphthaleneacetic acid (NAA) further increased the frequency of somatic embryo production. Unlike in many other plant systems, 2,4-D had little or no role in inducing somatic embryogenesis in *Hippeastrum* and *Eucharis*. The somatic embryos eventually gave rise to individual plantlets, though occasionally exhibiting dormancy. Histological and scanning electron microscopic observations during the stages of embryo development are presented. The embryo-derived plants had normal chromosome numbers. Besides the importance of true-to-type propagation, somatic embryogenesis offers a system of studying the various facets of non-zygotic embryo development.

1
Introduction

In plant cell and tissue culture, non-zygotic or somatic cells are induced to form embryos by a complex process of cell divisions, eventually developing into complete plants and thus demonstrating the phenomenon of totipotency. The developmental pathway of non-zygotic embryogenesis is very similar to zygotic embryogenesis. This unique occurrence has many useful applications in tissue culture and micropropagation, and is practised in a number of plant groups, including bulbous ornamentals. These groups of plants reproduce vegetatively and the process of multiplication is very slow. Besides the advantage of rapid propagation, somatic embryogenesis has several basic applications to agriculture and biotechnology. Since the initial reports (Reinert 1958; Steward et al. 1958), various facts about somatic embryogenesis have unfolded in the literature (Meinke 1995). Data on the origin, development, embryo structure and other morphological information have been regularly accumulated for a wide range of plants (Yeung 1995). The physiology, biochemistry and lately molecular information on embryogenesis are relatively new (Schmidt et al. 1997; Perry et al. 1999). Among the important inducers of somatic embryogenesis, the role of nutrient media, inorganic reduced nitrogen and other additives have been noted, but the role of plant growth regulators (PGR) such as auxin in embryogenesis has always been emphasized in plant systems (Davletova et al. 2001; Pasternak et al. 2002). However, there is relatively less information on the effect of cytokinin (Mujib et al. 1998). Cytological studies to assess the nature of "callus-to-embryo" regenerated plants have also been conducted in some plants. In the present chapter, the origin of somatic embryos, their development and related ontogeny are described in three bulbous ornamentals. The role of PGR, especially 6-benzylaminopurine (BAP) and naphthaleneacetic acid (NAA), is summarized for this plant group. Chromosomal analysis is presented to determine the status of the regenerated plants.

2 Origin, Development and Structure of Somatic Embryos

Bulb-scale explants (basal plate with leaf base) showed an initial swelling upon culture, became green and within 30–40 days of culture somatic embryos originated on the outer walls of the expanded scale-leaves (Fig. 1a). At an early stage, they appeared as tiny water droplets and gradually became enlarged and turned opaque. Morphologically the embryos were globular at the time of initiation, and sometimes had a swollen base with apical notches; however, a wide variation in morphology was noted. The incidence of such embryo formation increased progressively on the outer scales. A light microscopic study of leaf cross sections showed that embryos originated from the outer epidermal or adjacent mesophyll cell layers (Fig. 1c). The asynchronously developed embryos matured and developed, and germinated into individual plantlets while they were still attached to the mother explants (Fig. 1b). The developed embryos always had roots at the basal end.

The origin of embryos and their further progression and germination are identical in *Eucharis grandiflora* (Figs. 1d,e). Outer rows of scales are the active zone on which frequent, asynchronous embryos were initiated. Scanning electron microscopic (SEM) observations reveal a range of embryo types that are very delicately attached to the mother explants (Figs. 2a,b). The embryos of *Crinum asiaticum* were shiny, globular structures clustered together on a common callus matrix (Fig. 3a) which turned into bipolar structures later. At the cotyledonary stages, embryos had large furrows at the apex; some of these were broad and elongated. SEM analyses show embryos with a lateral

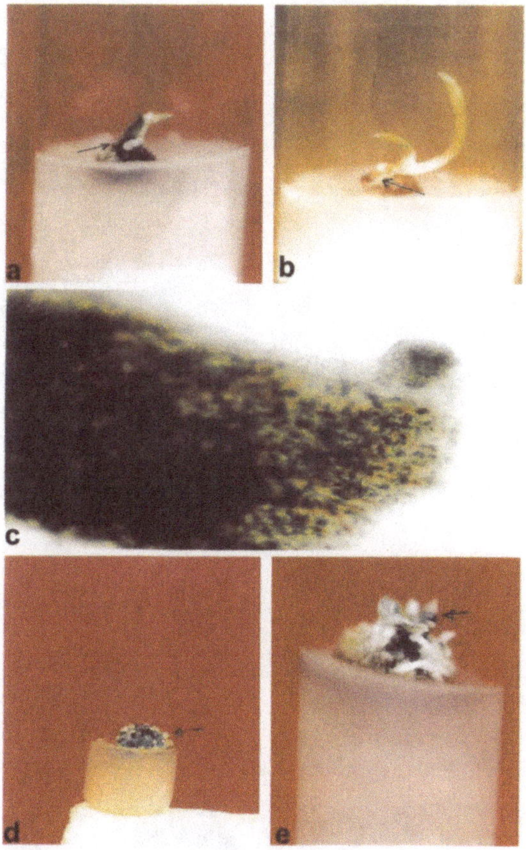

Fig. 1 Somatic embryogenesis in *Hippeastrum hybridum* and *Eucharis grandiflora*. **a** Formation of somatic embryos (*arrow*) on a bulb-scale explant (×10). **b** Germinated somatic embryo attached to mother explant, root (*arrow*) at the base (×1). **c** Leaf cross section showing an embryo originates from the subepidermal region (×80). **d,e** Somatic embryos (*arrows*) at different stages in *E. gradiflora* (×10)

notch (Fig. 3b) that creeps along the callus surface. Embryos were easily detachable from the underlying callus surface. Embryos with shoot and root axes were also clearly visible under histological preparations (Figs. 3c,d) of fully developed embryos.

2.1 The Role of 2,4-D in Embryogenesis

The relationship of 2,4-D (2,4-dichlorophenoxyacetic acid) with embryogenesis has been demonstrated in a large number of plants. In *C. asiaticum*, it was evident that 2,4-D was essential for callus induction and growth, and also equally effective in somatic embryogenesis. The somatic embryo was in-

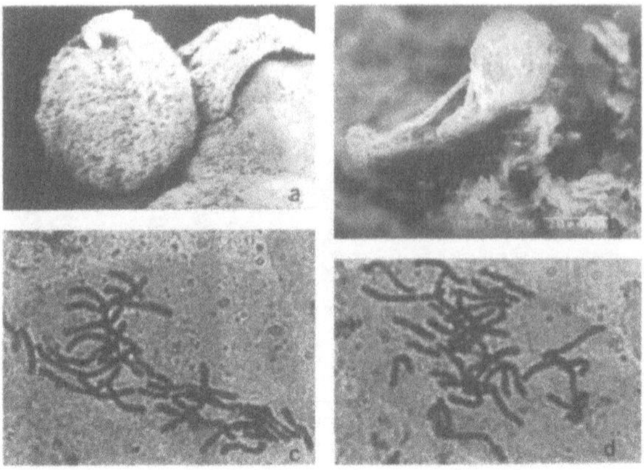

Fig. 2 Scanning and cytological preparations. **a** Scanning electron micrograph of a globular embryo in *H. hybridum* (×10). **b** Scanning micrograph of an embryo with swollen base in *E. grandiflora* (×75). **c,d** Two metaphase plates of regenerated root tips in *H. hybridum* (×500)

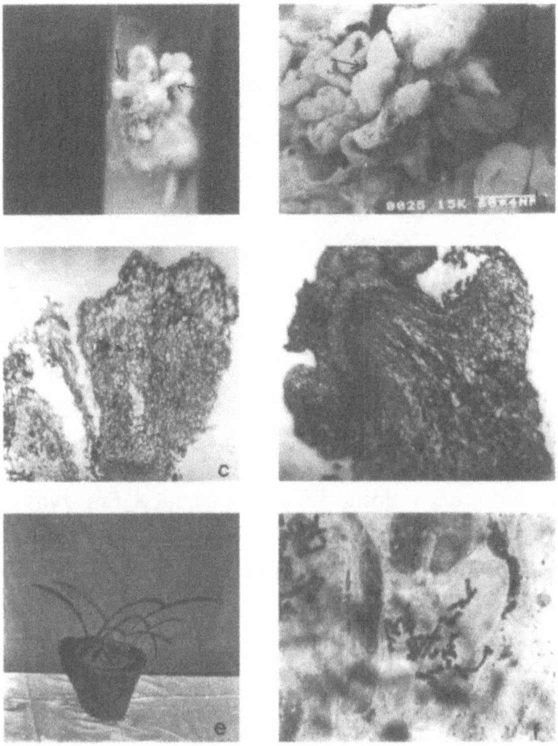

Fig. 3 Somatic embryogenesis and plant regeneration in *C. asiaticum*. **a** Somatic embryos from flower bud callus (×10). **b,c** Lateral notch development (*arrow*) and differentiation of shoot meristem (×40, ×225). **d** Longitudinal section of a fully developed embryo (×225). **e** Regenerated plant in pot. **f** A diploid cell with 33 chromosomes (×500)

duced in the 2.36–9.05 μm range. The same requirement, i.e. the presence of 2,4-D, was earlier found to be necessary in other plant systems during the early developmental stages (perhaps up to the globular stage) (Vasil and Vasil 1982; Tabei et al. 1991; Gray et al. 1993). In most cases, the developmental morphology of the embryo was noticed up to the cotyledonary stages. The incidence of plant regeneration was not significantly higher, as the induced somatic embryos did not germinate at all or germinated at a very low frequency. In contrast, *Hippeastrum* and *Eucharis* were entirely unresponsive to the 2,4-D signal.

2.2 Cytokinin and Embryogenesis

Except for some sporadic reports (Malik and Saxena 1992; Iantcheva et al. 1999), the role of cytokinin in embryogenesis is relatively less. However, unlike other systems, the concept is little different in these plant groups. Successful embryogenesis was noticed in BAP-added media and the number was quite high in these three ornamentals. Table 1 shows that lower concentrations of BAP (0.44 and 2.22 μm) induce somatic embryos in *Eucharis.* BAP is also very effective in *Hippeastrum hybridum* and *C. asiaticum.* However, addition of NAA (ineffective when used singly) in BAP-supplemented media accelerated the frequency of embryogenesis (Table 1) and embryo numbers in culture (Tables 2 and 3). In BAP-added medium or with NAA the embryo germinated into a plantlet in the same media without other treatments being required during the maturation and germination time. The entire process

Table 1 Growth regulators and embryo formation on bulb-scale explant in *E. grandiflora*

Growth regulator (μM) 2,4-D	BAP	NAA	Number of embryos/explant (Mean ± SD)
0	0	0	0
1.23	0	0	0
2.26	0	0	0
4.52	0	0	0
0	0.44	0	4.8 ± 0.58
0	2.22	0	4.0 ± 0.89
0	0	2.68	0
0	0	5.37	0
0	2.22	2.68	2.4 ± 0.51
0	2.22	5.37	9.6 ± 0.81
0	2.22	11.74	5.2 ± 0.37
0	4.40	2.68	10.2 ± 1.02
0	8.9	2.68	1.8 ± 0.37

Table 2 Preferred growth regulators for the induction of embryos from bulb-scale explant on two different culture media in *H. hybridum*

Growth regulator (μM)		Number of embryos/explant (Mean ± SD)	
NAA	BAP	MS	KC
0	2.22	4.44 ± 1.01	3.20 ± 1.16
0	4.40	3.40 ± 0.80	2.80 ± 0.74
2.68	2.22	2.20 ± 0.74	2.60 ± 0.48
2.68	4.40	7.20 ± 0.74	6.60 ± 1.01
2.68	8.90	3.20 ± 0.89	2.60 ± 0.80
5.32	2.22	6.00 ± 0.63	4.40 ± 1.62
10.60	2.22	3.80 ± 1.32	3.40 ± 1.01

Table 3 Callusing and embryogenesis in *C. asiaticum*. Embryo number and germination from flower-bud callus, cultured on MS medium supplemented with different growth regulators

Growth regulator (μM)	Callus intensity		Callusing %	Embryogenesis %	No. of embryos/ callus mass	Germination (Mean ± SD)
	Bulb scale	Flower bud				
2,4-D (2.26)	++	+++	63	28	1.75 ± 0.43	0.25 ± 0
2,4-D (9.05)	+++	++	85	60	3.50 ± 0.50	0
BAP (2.22)	-	-	18	20	2.50 ± 1.11	0.75 ± 0
BAP (2.22)+ NAA (5.32)	+++	++	48	43	2.75 ± 0.82	0.75 ± 0
BAP (4.40)+ NAA (2.68)	+	+	72	56	4.20 ± 0.82	1.0 ± 0.4
Kn (2.32)+ NAA (5.37)	++	++	42	13	1.25 ± 0.43	0.25 ± 0

Each treatment had 4–8 replicas; +, ++, +++, - represent poor, moderate and prolific callusing and no response, respectively; Kn=kinetin

(embryo to plantlet) took about 3–4 months, thereby suggesting a kind of dormancy of unknown nature. Since the somatic embryos do not have any seed coat, the inhibitor(s) or physiological barrier(s) noted during quiescence must definitely lie within the embryo itself. The application of thidiazuron (TDZ), a new compound tried as a plant growth regulator with "cytokinin-like activity", was found to be effective in some legumes (Murthy et al. 1995). Cytokinin-induced embryogenesis was, however, reported in a number of plant species when zygotic embryos (immature/mature) were cultured in vitro (Maheswaran and Williams 1984; Norgaard and Krogstrup 1991). In cytokinin-added media, auxin plays a dual (stimulatory and inhibitory) role in embryogenesis (Kysely and Jacobsen 1990; Mo and Von Arnold 1991). The precise role of cytokinin in inducing somatic embryos is yet to be found; high endogenous cytokinin in cultured tissues that decreases embryogenic potential was observed earlier (Wenck et al. 1988).

2.3
Differential Behaviour of Callus

In *C. asiaticum*, the initiation of callus was observed within a week from the cut edges of the explant. Both bulb-scale and flower bud induced callus; the extent, efficiency and nature, however, varied markedly and were dependent on explant type and media composition. The bulb-scale calli were soft, yellow with characteristic red pigment and fast growing, sometimes with gummy exudates, while flower-bud calli were pale yellow without pigment and slow growing. For callus induction and growth, the flower bud required a lower level of 2,4-D (2.26 μm) in comparison to the bulb-scale, in which a higher 2,4-D level was active. The addition of coconut water (10–15%, v/v) to 2,4-D supplemented media further improved callusing. Importantly, only the flower-bud calli developed somatic embryos and the bulb-scale calli were totally non-embryogenic. Although the reason for differential behaviour of calli originating from different tissue explants is not clearly known, various endogenous plant growth regulator levels may be responsible for such dual responses (Wernicke and Milkovits 1986; Mujib et al. 1996). The bulb-scale calli showed high shoot regenerating ability via the organogenic process.

2.4
Direct and Indirect Embryogenesis

In *H. hybridum*, the embryo originated directly on explants (bulb-scale) without any intervening callus phase. The epidermal cells and some sub-epidermal cells were committed or programmed to be pre-embryonic cells. The number of embryos on the explant also increased with time. The embryonic signal (the nature is unknown) is accumulated, expressed and restricted to the outer rows of scales. The embryogenic trigger was primarily noticed in BAP and/or with NAA supplemented media. Interestingly, media containing 2,4-D were entirely unresponsive in the embryogenic programme. The embryos were regularly developed on Knudson C (KC) medium—a "less rich" medium compared with MS medium which contains higher levels of inorganic salts, especially nitrogen.

Direct embryogenesis on explants was again observed in *E. grandiflora*. The developmental pathway is also identical and importantly both plants are members of the same family, Amaryllidaceae. In *C. asiaticum*, indirect embryogenesis was observed to be the mode where somatic embryos were developed from previously induced meristematic callus cells. Although the embryogenesis percentage is moderate to high (up to 60%), only a few cells in the callus cluster produced embryos (maximum four embryos/callus mass) (Fig. 3a, Table 3). In *Crinum*, however, there was no visible sign of direct embryogenesis.

3 Cytological Preparation

Squash preparation of regenerated root tip cells of *H. hybridum* revealed that the number of triploid cells, $3n$ $(2n) = 33$, was prevalent (Figs. 2c,d; Table 4); however, alterations of basic chromosome number were also noticed in the form of hyperploid and hypoploid cells at the same frequency (1.2–2.1%). In indirect embryogenesis (callus mediated) there is always a risk of somaclonal variation; however, cytological analysis of *C. asiaticum* (Fig. 3e) confirmed the expected mother's ploidy, i.e. $3n = 33$ (Fig. 3f) in the regenerants. The regenerated plant also exhibited a very low frequency of hyperploid and hypoploid cells. Besides numerical alteration of the chromosome, embryo-regenerated cells showed several abnormalities like the presence of laggards, bridges, micronuclei, etc. The karyotypic change (i.e. in number and structure of chromosome) is not uncommon in regenerated plants; however, variation occurring in pre-existing somatic cells, particularly in vegetatively propagated plants, was earlier established in many genera (Skene and Barlass 1983; Van Aartrijk and Vander Linde 1986; Mujib et al. 2000).

Table 4 Ploidy status of the regenerated plants; values are expressed as mean and SD

Regenerated plant	Triploidy %	Hyperploidy %	Hypoploidy %
H. hybridum	69.9 ± 1.82	1.82 ± 0.38	1.74 ± 2.0
C. asiaticum	68.16 ± 6.04	3.14 ± 0.38	1.98 ± 0.17

4 Conclusion

The phenomenon of totipotency of plant cells has been reconfirmed in the event of embryogenesis and plant regeneration, which is well established in a wide array of plant species. This non-zygotic embryogenesis is regarded as a unique system for studying the whole process of differentiation from a single cell. In this process, several exogenous triggers have so far been recognized in the literature (Pedroso and Pais 1995b; Li and Demarly 1995; Mordhorst et al. 1997). Plant growth regulators, especially auxins, are most frequently used to initiate the embryogenic signal. The role of genotypes and genetically induced embryogenesis has also been well documented in plant cell culture (Metheson et al. 1990; Barbulova et al. 2002). After the establishment of important external and internal inducers, focus has been recently shifted to understand the mechanism of gene regulation during this developmental process. Several model systems, including *Arabidopsis*, carrot and *Medicago*, have been clinically investigated and a variety of genes have been identified as having important regulatory roles, although the functions of isolated genes and expressed proteins are not always fully understood. However, the continuous accumulation of data on structural, biochemical and physiological aspects of in vitro somatic embryogenesis with currently employed molecular genetics techniques will provide a gamut of information on the biology of somatic embryogenesis that is as yet unknown. Moreover, the advantage of the embryogenic system, which is capable of producing unlimited embryos/propagules within a short period of time, and the development of transgenics possibilities may add fuel to continue research strongly in this direction.

References

Barbulova A, Iantcheva A, Zhiponova M, Vlahova M, Atanasov A (2002) Establishment of embryogenic potential of economically important Bulgarian alfalfa cultivars (*Medicago sativa* L.). Biotechnol Biotechnol Equip 16:55–63

Davletova S, Meszaros T, Miskolczi P, Oberschall A, Torok K, Magar Z, Dudits D, Deak M (2001) Auxin and heat shock activation of a novel member of the calmodulin-like domain protein kinase gene family in cultured alfalfa cells. J Exp Bot 52:215–221

Gray DJ, McColley DW, Compton ME (1993) High frequency embryogenesis from quiescent seed cotyledons of *Cucumis melo* cultivars. J Am Soc Hortic Sci 118:425–432

Iantcheva A, Barbulova A, Vlahova M, Kondorosi E, Elliott M, Atanassov A (1999) Regeneration of diploid annual medics via direct somatic embryogenesis promoted by thidiazuron and benzylaminopurine. Plant Cell Rep 18:904–910

Kysely W, Jacobsen HJ (1990) Somatic embryogenesis from pea embryos and shoot apices. Plant Cell Tissue Organ Cult 20:7–14

Li XQ, Demarly Y (1995) Characterization of factors affecting regeneration frequency of *Medicago lupulina* L. Euphytica 86:143–148

Maheswaran G, Williams EG (1984) Direct somatic embryoid formation on immature embryos of *Trifolium repens*, *T. pratense* and *Medicago sativa* and rapid clonal propagation of *T. repens*. Ann Bot 54:201–211

Malik K, Saxena PK (1992) Regeneration in *Phaseolus vulgaris* L.: High frequency induction of direct shoot formation in intact seedlings by BAP and TDZ. Planta 186:384–388

Meinke DW(1995) Molecular genetics of plant embryogenesis. Annu Rev Plant Physiol Plant Mol Biol 46:369–394

Metheson SL, Nowak J, Maclean N (1990) Selection of regenerative genotypes from highly productive cultivars of alfalfa. Euphytica 45:105–112

Mo LH, Von Arnold S (1991) Origin and development of embryogenic cultures from seedlings of Norway spruce. J Plant Physiol 138:223–230

Mordhorst AP, Toonen MAJ, DeVries SC (1997) Plant embryogenesis. Crit Rev Plant Sci 16:535–576

Mujib A, Bandyopadhyay S, Jana BK, Ghosh PD (1996) Growth regulator involvement and somatic embryogenesis in *Crinum asiaticum*. Indian J Plant Physiol 1:84–86
Mujib A, Bandyopadhyay S, Jana BK, Ghosh PD (1998) Direct somatic embryogenesis and in vitro plant regeneration in *Hippeastrum hybridum*. Plant Tissue Cult 8:19–25
Mujib A, Bandyopadhyay S, Ghosh PD (2000) Tissue culture derived plantlet variation in *Caladium bicolor* L., an important ornamental. Plant Tissue Cult 10:149–155
Murthy B, Murch S, Saxena PK (1995) Thidiazuron-induced somatic embryogenesis in intact seedlings of peanut (*Arachis hypogea*): endogenous growth regulator levels and significance of cotyledons. Physiol Plant 94:268–276
Norgaard JV, Krogstrup P (1991) Cytokinin-induced somatic embryogenesis from immature embryos of *Abies nordmanniana* Lk. Plant Cell Rep 9:509–513
Pasternak TP, Prinsen E, Ayaydin F, Miskolczi P, Potters G, Asard H, Onckelen HAV, Dudits D, Feher A (2002) The role of auxin, pH and stress in the activation of embryogenic cell division in leaf protoplast derived cells of alfalfa. Plant Physiol 129:1807–1819
Pedroso MC, Pais S (1995) Factors controlling somatic embryogenesis. Plant Cell Tissue Organ Cult 43:147–154
Perry SE, Lehti MD, Fernandez DE (1999) The MADs domain protein AGAMOUS-like 15 accumulates in embryonic tissues with diverse origins. Plant Physiol 120:121–130
Reinert J (1958) Morphogenese und ihre Kontrolle an Gewebekulturen aus Karotten. Naturwissenschaften 45:344–345
Schmidt ED, Guzzo F, Toonen MA, deVries SC (1997) A leucine-rich repeat containing receptor-like kinase marks somatic plant cells competent to form embryos. Development 124:2049–2062
Skene KGM, Barlass M (1983) Studies on the fragmented shoot apex of grapevine. IV. Separation of phenotypes in periclinal chimera in vitro. J Exp Bot 34:1271–1280
Steward FC, Mapes MO, Mears K (1958) Growth and organized development of cultured cells. II. Organization in cultures grown from freely suspended cells. Am J Bot 45:705–708
Tabei Y, Kanno T, Nishio T (1991) Regulation of organogenesis and somatic embryogenesis by auxin in melon, *Cucumis melo* L. Plant Cell Rep 10:225–229
Van Aartrijk J, Vander Linde PCG (1986) In vitro propagation of flower bud crops. In: Zimmerman RH, Griesbach RJ, Hammerschlag FA, Lawson RJ (eds) Tissue culture as a plant production system for horticultural crops. Kluwer, Dordrecht, pp 317–337
Vasil V, Vasil IK (1982) Characterization of an embryogenic cell suspension culture derived from cultured inflorescences of *Pennisetum americanum* (Pearl millet, Graminae). Am J Bot 699:1441–1449
Wenck AR, Conger BV, Trigiano RN, Sams CE (1988) Inhibition of somatic embryogenesis in orchardgrass by endogenous cytokinins. Plant Physiol 88:990–992
Wernicke W, Milkovits L (1986) Development gradient in wheat leaves. Responses of leaf segments in different genotypes cultured in vitro. J Plant Physiol 115:49–58
Yeung EC (1995) Structural and developmental patterns in somatic embryogenesis. In: Thorpe TA (ed) In vitro embryogenesis in plants. Kluwer, Dordrecht, pp 205–248

Plant Cell Monogr (2)
A. Mujib · J. Šamaj: Somatic Embryogenesis
DOI 10.1007/7089_035/Published online: 2 December 2005

Environmental Design Considerations for Somatic Embryogenesis

Takanori Hoshino (✉) · Joel L. Cuello

Department of Agricultural and Biosystems Engineering, The University of Arizona, 507 Shantz Building, Tucson, AZ 85721-0038, USA
takanori@email.arizona.edu, cuelloj@email.arizona.edu

Abstract In addition to the biomolecular, physiological, and biochemical aspects of somatic embryogenesis, careful design of environmental conditions is necessary to ensure the successful induction and development of somatic embryos for different plant species. A dissolved oxygen concentration, for instance, below 10% generally inhibits the differentiation of somatic embryos, while the same is promoted at 40, 80, or 100%, depending on the plant species. Certain plant species also exhibit inhibition of somatic embryo differentiation at high dissolved oxygen concentration, such as at 80%. Cell density influences somatic embryogenesis by changing the concentrations of conditioning factors released by plant cells and embryos into the culture medium. High initial cell density, in general, results in inhibition of somatic embryo differentiation on account of inhibitory compounds released by cells into the culture medium. Partial medium replacement has been employed to rectify this situation. In terms of the general influence of light, red light promotes and blue light inhibits the induction of somatic embryos. Blue light, however, generally promotes the development of somatic embryos.

1 Introduction

Investigation of the various critical aspects of somatic embryogenesis is necessary in order to establish protocols for the successful induction and development of somatic embryos for different plant species. Recent studies, for instance, have focused on the biomolecular (Takahara et al. 2004), physiological (Godbole et al. 2004; Konradova et al. 2002), and biochemical (Sharma et al. 2004; Ramarosandratana and Stade 2004) aspects of somatic embryogenesis. The environmental factors that interact with and influence somatic embryogenesis constitute another critical aspect that needs careful consideration. This is especially true since it is a given that certain environmental factors need to be controlled and regulated for the important practical applications of somatic embryogenesis, i.e., artificial seed technology and automated plant mass production using bioreactors (Onishi et al. 1994).

Only a handful of studies, however, have addressed the optimization of specific environmental factors for somatic embryogenesis. Examples include those that investigated the effects on somatic embryogenesis of cell density and conditioning factors (CFs; Bellincampi and Morpurgo 1987, 1989; Vries et al. 1988; Osuga and Komamine 1994; Higashi et al. 1998), dissolved oxygen (DO) concentration (Kessell and Carr 1972; Jay et al. 1992; Archambault et al. 1994; Shimazu and Kurata 1997), medium pH (Hofmann et al. 2004), nutrient and plant hormone composition in the medium (Jimenez 2001), as well as humidity (Meskaoui and Tremblay 1999; Bomal and Tremblay 1999). This chapter underscores the effects on somatic embryogenesis of three critical environmental factors: (1) DO concentration; (2) cell density; and (3) light quality and intensity.

2 Dissolved Oxygen Concentration

The effects of DO concentration on somatic embryogenesis are mainly twofold: influencing the biomass of undifferentiated cells; and influencing the development or differentiation of somatic embryos. Archambault et al. (1994) reported that the biomass (0.7–9.7 g of dry weight per liter) of undifferentiated cells of transformed California poppy (*Eschscholtzia californica*) at high DO (60% of air saturation) exceeded that of the control (0.2–10 g of dry weight per liter) at a DO of 20%. By contrast, a low DO (5–10%) yielded a lower biomass (0.2–3.3 g of dry weight per liter) compared with that of the control. Jay et al. (1992) reported that the stationary phase of the dry-mass curve of the undifferentiated cells of carrot (*Daucus carota* L.) occurred after 10 days of culturing for 100% DO, while that for 10% DO occurred with a 3-day delay. There was no significant difference in the final dry mass, approximately 4.5 g of dry weight per liter, for 100 and 10% DO levels. Jay et al. (1992) concluded that the results had a nutritional basis. They showed that while glucose uptake commenced after 4 days of culturing for 100% DO, glucose uptake started after 6 days of culturing for 10% DO. Also, complete consumption of glucose (defined as less than 2 g L^{-1}) in the medium occurred on day 10 for 100% DO, while it took another 3 days (on day 13) for 10% DO for the glucose to be completely consumed. The foregoing results indicated that high DO concentration generally resulted in higher biomass of undifferentiated cells. It should be noted, however, that inhibitory effects at 40% relative oxygen partial pressure in bioreactors were observed by Hohe et al. (1999) on cell proliferation of florist cyclamen (*Cyclamen persicum*) relative to the effects at 20 and 30%. A reduction of up to two thirds in yield in a packed cell volume and a decrease of more than 50% in growth rate in one genotype were observed.

In terms of the effects of DO on the development or differentiation of somatic embryos from embryonic callus cells, Kessell and Carr (1972) reported that lower than 16% DO was quite detrimental to the production of carrot somatic embryos. Jay et al. (1992) reported that carrot somatic-embryo production was inhibited by approximately 75% at 10% DO compared with that at 100% DO. They also found that the DO level supplied during cell proliferation did not affect cell differentiation. Archambault et al. (1994) reported that cell differentiation of transformed *E. californica* cells was slow at 60% DO and was inhibited at low (5–10%) DO. Feria et al. (2003) also reported that the total number of somatic embryos that were induced from embryogenic cells of coffee (*Coffea arabica* cv.) was greater (71 072 somatic embryos per liter) at 80% DO than that (36 941 somatic embryos per liter) at 50% DO. Meanwhile, Shimazu and Kurata (1999) reported that the total number of somatic embryos that differentiated from carrot embryogenic cells was not affected at 4–40% DO. They also found that the development of carrot somatic embryos into the torpedo-shaped or heart-shaped stage was enhanced at 20–40% DO, while the same was completely inhibited at less than 7% DO. Also, they found that increasing DO from 4 to 7% increased the sugar consumption by the somatic embryos. By contrast, no significant difference in sugar consumption was observed when DO was varied from 20 to40%. Feria et al. (2003) reported that the development of coffee somatic embryos into the torpedo-shaped stage was enhanced at 50% DO, and was inhibited at 80% DO. Thus, different levels of DO were required to enhance torpedo-shaped differentiation.

3 Cell Density

The predominant effects of cell density on somatic embryogenesis appear to be indirect, rather than direct. The adjustment of cell density influences somatic embryogenesis through the following: (1) change in the concentrations of the CFs which plant cells and embryos release into the culture medium; (2) change in the amount of nutrients or gas which individual plant cells or embryos can consume; and (3) physical stress caused by increasing the physical contact among plant cells and embryos when cell density is increased. In studies that investigated the effects of cell density on somatic embryogenesis, it was established that the change in the concentrations of the CFs which plant cells and embryos release into the culture medium was the most significant aspect of manipulating cell density (Osuga and Komamine 1994; Osuga et al. 1993, 1997; Bellincampi and Morpurgo 1987, 1989; Higashi et al. 1998).

Bellincampi and Morpurgo (1987, 1989) investigated the effects of CFs released from plant cells of carrot (*D. carota* L.) into cell suspension culture medium, and determined that at least two different CFs were released from carrot cells into the culture medium. In the first study (Bellincampi and Morpurgo 1987), they concluded that (1) the first CF increased growth by cell division activity, and significantly enhanced the plating efficiency (defined as the ratio of the number of proliferating colonies to the number of initial plating units) of carrot cells, (2) the CF was physically and chemically very stable, being resistant to boiling and to acid or alkaline pH, and was strongly hydrophilic, and (3) the CF had a low molecular mass of 700 Da. Their results also suggested the species-unspecificity of the CF.

In the second study (Bellincampi and Morpurgo 1989), evidence was provided for the presence of a second growth-stimulating CF. But while the plating efficiency as influenced by the first CF was completely dependent on the initial cell density, the plating efficiencies as influenced by the second CF after 20 days of growth remained very similar for different initial cell densities. They suggested that the second CF had relatively low hydrophilicity and, thus, diffused slowly, and might have also been unstable.

Sung and Okimoto (1981) explored the relationship between cell density and embryo differentiation of carrot (*D. carota* L.). In their study, they found that a globular embryo was induced even under low concentrations of exogenous auxin (in this case 2,4-dichlorophenoxyacetic acid) at low cell density (2×10^4 cells mL^{-1}). Differentiation into torpedo-shaped embryos, however, was completely inhibited under that condition. By contrast, the differentiation of somatic embryos was strongly inhibited at high initial cell density (4×10^6 cells mL^{-1}). This fact indicated that an inhibitory CF was released from carrot cells during cell proliferation, and differentiation of somatic embryos was repressed when a high concentration of inhibitory CF was brought about by high cell density. This result was also supported by the results obtained by Fridborg and Eriksson (1975). They found that the differentiation of carrot somatic embryos was stimulated by the addition of activated charcoal, and that the differentiation was observed even in the presence of 1 mg L^{-1} α-naphthalene acetic acid, which would typically inhibit differentiation. They suggested that inhibitory compounds were removed by the activated charcoal.

Osuga et al. (1993, 1994) reported that cell density did not affect the development of carrot embryogenic cell clusters into globular or heart-shaped embryos. They also found that the total numbers of somatic embryos obtained at different initial cell densities were statistically similar when initial cell densities ranged from 0.5 to 2.0×10^3 cell clusters per milliliter. No torpedo-shaped embryo formation, however, was observed when the cell density exceeded 1.0×10^3 cell clusters per milliliter. Previous studies reported that the rate of somatic embryo development was enhanced when cell density was high (Halperin 1967; Hari 1980). Osuga (1993, 1994) concluded in his study, however, that such enhancement at high cell density was caused by stimulation of growth of single cells (or very small cell clusters) into embryogenic cell clusters by cell division. This conclusion agreed with the results obtained by Bellincampi and Morpurgo (1987, 1989).

Osuga et al. (1997) also found an enhancement in the development of embryogenic carrot cell clusters into globular embryos at high cell density with partial replacement of the medium. They also confirmed that this was not caused by either physical stress or the enrichment of nutrients by replacement of the medium. They found that a greater number of globular embryos was obtained with partial medium replacement compared with entire medium replacement. Thus, they concluded that both inhibitory CF and promotive CF were released during cell proliferation. The results of Higashi et al. (1998) supported the inhibitory effects of high cell density (defined as greater than 1.0-mL packed cell volume per liter of medium). They found that the inhibitory effects caused by physical stress and by the change in the amount of available nutrients were not as critical as the negative effects of the inhibitory CF, which was released during cell proliferation and had a molecular mass of less than 3500 Da. Interestingly, Osuga et al. (1993, 1994) also reported that when globular embryos were cultured at different embryo densities, their results showed that the rate of torpedo-shaped embryo formation decreased linearly as embryo density increased from approximately 100 embryos per milliliter to 500 embryos per milliliter.

4
Light Quality and Intensity

That light affects somatic embryogenesis has been known for over 30 years through the pioneering studies by Ammirato and Steward (1971) on the effects of light on the growth of somatic embryos of hemlock water-parsnip (*Sium suave*) cells and by Halperin (1966) and Ammirato and Steward (1971) on the effects of light on the morphological characteristics of carrot somatic embryos. Of the critical environmental factors, however, light is the one whose effects on somatic embryogenesis have been the least investigated. Indeed, there is a paucity of published literature on the subject. What is more, three major issues make it difficult to analyze the specific effects of light quality and intensity on somatic embryogenesis in existing literature. These include (1) the different definitions of light quality used in the available studies, (2) the problematic spectral noises generated by the conventional experimental lighting systems, consisting of fluorescent tubes and light filters, used in such studies, and (3) the different light intensities applied to embryogenesis, which makes difficult the isolation of the morphological effects from the photosynthetic effects. Further studies are clearly needed to analyze and determine the specific effects of light environments on somatic embryogenesis.

Micheler and Lineberger (1987) explored the effects of light quality on carrot somatic embryos by examining the effects of four blue light (480 ± 100 nm), green light (540 ± 50 nm), red light (660 ± 70 nm), and white light, with light intensities ranging from 5 to 50 μmol m^{-2} s^{-1}. When cell cultures were exposed under red or green light, a similar number of somatic embryos, approximately 9000 embryos per milliliter, was obtained after 14 days of culturing. By contrast, significant inhibitions were observed under blue light, resulting in approximately 3000 embryos per milliliter, and especially under white light, fewer than 2000 embryos per milliliter. Also, they showed that the effects of red light and green light did not change with different light intensities, while the negative effects of blue light and white light increased as the light intensity rose from 5 to 30 μmol m^{-2} s^{-1}. Indeed, even low blue light intensity resulted in 25% fewer embryos than in the dark control. Blue light, however, was observed to enhance the differentiation of globular embryos into torpedo-shaped embryos. After 16 days of culturing, 76% of somatic embryos developed into torpedo-shaped embryos under blue light, while only 6 and 18% did in the dark control and red light treatment, respectively. All somatic embryos induced under various light treatments, however, showed significant morphological changes with respect to the somatic embryos grown in the dark. These include the following: leafy cotyledons that were not observed in the dark control, but were observed in more than 80% of somatic embryos in all the light treatments; abnormal somatic embryos with multiple cotyledons under red light treatment and in the dark control (more than 7% in red light and in the dark, while less than 5% in other treatments); orange-pigmented radicles under red light (71% in red light and 0% in other light), while branched radicles were produced under white and blue light (67 and 49% in white light and blue light, respectively, and 0% in other light); and elongated hypocotyls under blue light (88% in blue light, while less than 10% in other light). Similar morphological changes, such as enhanced development of leaves, cotyledons and roots, under light treatments were reported by Ammirato and Steward (1971).

D'Onofrio et al. (1998) investigated the effects of blue light (450 ± 60 nm), red light (670 ± 50 nm), far-red light (> 700 nm), white light, and various combinations of these light qualities on somatic embryogenesis of quince (*Sidonia* sp.) leaves. They reported positive and negative effects of red light and blue light, respectively, on the differentiation of somatic embryos, with more than 0.4 embryos per leaf observed under red light, and fewer than 0.1 embryos per leaf observed in the dark or under blue light. They further correlated the rate of somatic embryo differentiation with photoequilibrium. Photoequilibrium, which is the fraction of physiologically active phytochrome to the total phytochrome, was calculated based on the theory suggested by Mancinelli (1995). The results showed that the ratio of the leaves with embryos was increased exponentially from 0% to approximately 30% as photoequilibrium increased from 0 to 1. Thus, phytocrome activation for somatic embryo induction was suggested. In addition, the blue light treatment resulted in less than half the number of embryo-producing leaves than those exposed to red light plus far-red light even though both treatments had the same photoequilibrium value of 0.43. Since the inhibition occurred at a low photoequilibrium, it implied that less phytochrome was activated. Thus, an interactive mechanism involving phytochrome and a blue-absorbing photoreceptor that caused negative effects on somatic embryo induction was suggested. Similar promotive and inhibitory effects due to the amount of activated phytochrome by red light and far-red light were reported for cruel plant (*Araujia sericifera* L.) somatic embryos by Torne et al. (2001).

Bach and Krol (2001) reported the effects of various light qualities on Hyacinth (*Hyacinthus orientalis* L.) somatic embryogenesis, focusing on callus proliferation and development of somatic embryos. Greater callus proliferation, expressed as "medium rate or strong reaction of proliferation", was obtained under both red light (647–770 nm, 20 μmol m^{-2}s^{-1}) and dark similarly. By contrast, strong inhibition was observed, expressed as "medium or low rate of, or no proliferation", under blue light (450–492 nm, 60 μmol m^{-2}s^{-1}) and especially white light (390–770 nm, 60 μmol m^{-2}s^{-1}). At the same time, however, greater numbers of developed somatic embryos were observed under blue light. Moreover, when 5.0 μM BAP (6-benzylaminopurine) and 0.5 μM NAA (α-naphtalene acetic acid) was added to the culture medium, the greatest number of somatic embryos, 6–10 embryos per one callus clump, was obtained, compared to only 1–2 embryos per one callus clump was obtained in other treatments. A change in chlorophyll content during both cell proliferation and somatic embryo development was observed under blue and white lights. Indeed, the total amounts of chlorophyll under blue (20.62 mg per 100 g embryo) and white light (18.90 mg per 100 g embryo) treatments exceeded by 3 and 40 times those under red light (6.12 mg per 100 g embryo) and darkness (0.48 mg per 100 g embryo), respectively, when 5.0 μM BAP and 0.5 μM NAA was added to the culture medium.

Latkowska et al. (2000) investigated the effects on somatic embryogenesis of three different genotypes of Norway spruce of red light (670 $\pm$ 50 nm) and blue light (450 $\pm$ 60 nm) supplied at 30 μmol m^{-2} s^{-1} for 18 h per day. The cell growth of one genotype was inhibited under red light (38% of control) and especially under blue light (10% of control). Such effects, however, were moderated (85 and 65% of control under red light and blue light, respectively) in the case of a second genotype, and were not observed at all with the third genotype. The results indicated that the effects of light quality vary significantly depending on the species or cultivars. Kvaalen and Appelgren (1999) reported higher sensitivity to various light qualities of somatic embryos and seedlings derived from somatic embryos of Norway spruce (*Picea abies* L.) compared with that for seedlings derived from natural seeds. Germination was promoted (98%) and inhibited (50%) when somatic embryos were exposed under red light (670 $\pm$ 50 nm) and blue light (450 $\pm$ 80 nm), respectively. By contrast, no effect on germination was observed when natural seeds were exposed under various light qualities.

Addressing the previously mentioned three major issues that make it challenging to analyze the specific effects of light quality and intensity on somatic embryogenesis, Takanori and Cuello (2005) determined and optimized the effects of radiation quality and intensity on the induction and development of somatic embryos from carrot (*D. carota*) embryogenic calli using light-emitting diodes (LEDs), which emit precise narrow-waveband radiation. The specific objectives of their study were as follows: (1) to determine the effects of red light and blue light up to 20 μmol m^{-2} s^{-1} emitted from LEDs on the induction of somatic embryos from carrot embryogenic calli and on the resulting distribution of the embryos among the globular, heart-shaped, torpedo-shaped and cotyledonary stages; and (2) to determine the effects of red light and blue light up to 20 μmol m^{-2} s^{-1} on the development of somatic embryos from carrot embryogenic calli by calculating the developmental coefficients of the somatic embryos.

Their results after 14 days of exposure pertaining to somatic embryo induction showed that (1) red radiation at 10 μmol m^{-2} s^{-1} significantly increased the density of total somatic embryos induced from carrot embryogenic calli, (2) lower and higher intensities of red radiation (1–5 and 20 μmol m^{-2} s^{-1}) did not significantly influence the density of induced total somatic embryos, and (3) increasing the intensity of blue radiation (up to 20 μmol m^{-2} s^{-1}) appeared to have reduced the density of induced total somatic embryos. In regard to somatic embryo development, the results showed that (1) red radiation (up to 20 μmol m^{-2} s^{-1}) had virtually no effect on the development of the carrot somatic embryos, and (2) blue radiation (10 or 20 μmol m^{-2} s^{-1}) exerted positive effects on the development of the carrot somatic embryos, especially in the globular and heart-shaped stages.

The foregoing underscores that critical environmental factors, including DO concentration, cell density, and light quality and intensity significantly influence both the production (or induction) and the development (or differentiation) of somatic embryogenesis. Thus, designing for the practical applications of somatic embryogenesis, i.e., artificial seed technology and automated plant mass production using bioreactors, necessitates careful design of their environmental conditions.

References

Ammirato PV, Steward FC (1971) Some effects of environment of the development of embryos from cultured free cells. Bot Gaz 132(2):149–158

Archambault J, Williams RD, Lavoile L, Pepin MF, Chavarie C (1994) Production of somatic embryos in a helical ribbon impeller bioreactor. Biotech Bioeng 44:930–943

Bach A, Krol A (2001) Effect of light quality on somatic embryogenesis in *Hyacinthus orientalis* L. "Delft's Blue". Biol Bull Poznan 38(1):103–107

Bellincampi D , Morpurgo G (1987) Conditioning factor affecting growth in plant cells in culture. Plant Sci 51:83–91

Bellincampi D, Morpurgo G (1989) Evidence for the presence of a second conditioning factor in plant cell cultures. Plant Sci 65:125–130

Bomal C, Tremblay FM (1999) Effect of desiccation to low moisture content on germination synchronization of root emergence, and plantlet regeneration of black spruce somatic embryos. Plant Cell Tissue Organ Cult 56:193–200
D'Onofrio C, Morini S, Bellocchi (1998) Effect of light quality on somatic embryogenesis of quince leaves. Plant Cell Tissue Organ Cult 53:91–98
Feria M, Jimenez E, Barbon R, Capote A, Chavez M, Quiala E (2003) Effect of dissolved oxygen concentration on differentiation of somatic embryo of *Coffea arabica* cv. catimor 9722. Plant Cell Tissue Organ Cult 72:1–6
Fridborg G, Eriksson T (1975) Effects of activated charcoal on growth and morphogenesis in cell cultures. Physiol Plant 34:306–308
Godbole S, Sood A, Sharma M, Nagar PK, Ahuja PS (2004) Starch deposition and amylase accumulation during somatic embryogenesis in bamboo (*Dendrocalamus hamiltonii*). J Plant Phisiol 161(2):245–248
Halperin W (1966) Alternative morphogenetic events in cell suspensions. Am J Bot 53(5):443–453
Halperin W (1967) Population density effects on embryogenesis in carrot-cell cultures. Exp Cell Res 48:170–173
Hari V (1980) Effect of cell density changes and conditioned media on carrot cell embryogenesis. Z Pflanzenphysiol 96:227–231
Higashi K, Daita M, Kobayashi T, Sasaki K, Harada H, Kamada H (1998) Inhibitory conditioning for carrot somatic embryogenesis in high-cell-density cultures. Plant Cell Rep 18:2–6
Hofmann N, Nelson RL, Korban SS (2004) Influence of media components and pH on somatic embryo induction in three genotypes of soybean. Plant Cell Tissue Organ Cult 77(2):157–163
Hohe A, Winkelmann T, Schwenkel HG (1999) The effect of oxygen partial pressure in bioreactors on cell proliferation and subsequent differentiation of somatic embryos of *Cyclamen persicum*. Plant Cell Tissue Organ Cult 59:39–45
Jay V, Genestier S, Courduroux JC (1992) Bioreactor studies on the effect of dissolved oxygen concentrations on growth and differentiation of carrot (*Daucus carota* L.) cell cultures. Plant Cell Rep 11:605–608
Jimenez VM (2001) Regulation of in vitro somatic embryogenesis with emphasis on the role of endogenous hormones. R Bras Fisiol Veg 13(2):196–223
Kessekk RHJ, Carr AH (1972) The effect of dissolved oxygen concentration on growth and differentiation of carrot (*Daucus carota*) tissue. J Exp Bot 23(77):996–1007
Konradova H, Lipavska H, Albrechtova J, Vreugdenhil D (2002) Sucrose metabolism during somatic and zygotic embryogeneses in Norway spruce: content of soluble saccharides and localization of key enzyme activities. J Plant Physiol 159:387–396
Kvaalen H, Appelgren M (1999) Light quality influences germination, root growth and hypocotyl elongation in somatic embryos but not in seedlings of Norway spruce. In Vitro Cell Dev Biol Plant 35:437–441
Latkowska MJ, Kvaalen H, Appelgren M (2000) Genotype dependent blue and red light inhibition of the proliferation of the embryogenic tissue of Norway spruce. In Vitro Cell Dev Biol Plant 36:57–60
Mancinelli AL (1994) The physiology of phytochrome action. In: Kendrick RE, Kronenberg HM (eds) Photomorphogenesis in plants, 2nd edn. Kluwer, Netherlands pp 211–269
Meskaoui AE, Tremblay FM (1999) Effects of sealed and vented gaseous microenvironments on the maturation of somatic embryos of black spruce with a special emphasis on ethylene. Plant Cell Tissue Organ Cult 56(3):201–209

Micheler CH, Lineberger RD (1987) Effects of light on somatic embryo development and abscisic levels in carrot suspension cultures. Plant Cell Tissue Organ Cult 11:189–207
Onishi N, Sakamoto Y, Hirosawa T (1994) Synthetic seeds as an application of mass production of somatic embryos. Plant Cell Tissue Organ Cult 39:137–145
Osuga K, Komamine A (1994) Synchronization of somatic embryogenesis from carrot cells at high frequency as a basis for the mass production of embryos. Plant Cell Tissue Organ Cult 39:125–135
Osuga K, Kamada H, Komamine A (1993) Cell density is an important factor for synchronization of the late stage of somatic embryogenesis at high frequency. Plant Tissue Cult Lett 10(2):180–183
Osuga K, Kamada H, Komamine A (1997) Frequency improvement of somatic embryogenesis at high embryo density by partial replacement of medium in carrot suspension cultures. J Ferment Bioeng 84(3):275–278
Rmarosandratana AV, Staden JV (2004) Effects of auxins and 2,3,5-Triiodobenzoic acid on somatic embryo initiation from Norway spruce zygotic embryos. Plant Cell Tissue Organ Cult 79(1):105–107
Sharma P, Pandey S, Bhattacharya A, Nagar PK, Ahuja PS (2004) ABA associated biochemical changes during somatic embryo development in *Cammelia sinensis* (L.) O. Kuntze. J Plant Phisiol 161(11):1269–1276
Shimazu T, Kurata K (1999) Relationship between production of carrot somatic embryos and dissolved oxygen concentration in liquid culture. Plant Cell Tissue Organ Cult 57:29–38
Sung ZR, Okimoto R (1981) Embryonic proteins in somatic embryos of carrot. Proc Natl Acad Sci 78(6):3683–3687
Takahara K, Takeuchi M, Fujita M, Azuma J, Kamada H, Sato F (2004) Isolation of putative glycoprotein gene from early somatic embryo of carrot and its possible involvement in somatic embryo development. Plant Cell Phisiol 45(11):1165–1668
Takanori T, Cuello J (2005) Regulating radiation quality and intensity using narrow-band LEDs for optimization of somatic embryogenesis. In: Proceedings of the 2005 Annual Meeting of the American Society of Agricultural Engineers.
Torne JM, Moysset L, Santos M, Simon E (2001) Effects of light quality on somatic embryogenesis in *Araujia sericifera*. Physiologia Plantarrun 111:405–411
Vries SC, Booji H, Janssens R, Vogels R, Saris L, LoSchiavo F, Terzi M, Kammen A (1988) Carrot somatic embryogenesis depends on the phytohormone-controlled presence of correctly glycosylated extracellular proteins. Genes Dev 2:462–476

Plant Cell Monogr (2)
A. Mujib · J. Šamaj: Somatic Embryogenesis
DOI 10.1007/7089_024/Published online: 2 December 2005

Importance of Cytoskeleton and Cell Wall in Somatic Embryogenesis

Jozef Šamaj[1,2] (✉) · Milan Bobák[3] · Alžbeta Blehová[3] · Anna Pret'ová[2]

[1]Institute of Cellular and Molecular Botany, University of Bonn, Kirschallee 1, 53115 Bonn, Germany
jozef.samaj@uni-bonn.de

[2]Institute of Plant Genetics and Biotechnology, Slovak Academy of Sciences, Akademická 2, 950 07 Nitra, Slovakia

[3]Department of Plant Physiology, Comenius University, Mlynska dolina B-2, 842 15 Slovakia

Abstract Both the cytoskeleton composed of microtubules and actin microfilaments as well as cell wall components such as arabinogalactan proteins and pectins play crucial roles during somatic and zygotic embryogenesis in plants. These components control proper cell division and expansion during early embryogenesis and later during embryo differentiation. Here we discuss structural, physiological and functional aspects connected to the role of the cytoskeleton and the cell wall during embryogenesis in selected model species including maize, carrot, *Drosera*, *Arabidopsis* and sunflower. Additionally, signalling properties of cell wall components and the cytoskeleton relevant for somatic embryogenesis are also discussed.

1 Introduction

Plant polarity and morphogenesis is controlled via coordinated functions of the cytoskeleton and the cell wall. It was proposed that these two structural entities are interlinked and form a supracellular structure called the cytoskeleton–plasma membrane–cell wall continuum (for a recent review see Baluška et al. 2003).

Somatic embryogenesis requires strict spatio-temporal control over cell division and elongation (Tautorus et al. 1992; Šamaj et al. 1997; Feher et al. 2003). The polarity within the embryo is established through the precisely controlled cell division pattern of embryogenic cells and elongation of supporting suspensor-like and callus cells. Both the cytoskeleton and cell walls appear to play an essential regulatory role during initial and also later steps of embryo development in vitro. Additionally, this process is also controlled by cell wall molecules having signalling properties, such as arabinogalactan proteins (AGPs) and pectins.

2
Structural Markers of Somatic Embryogenesis

Embryogenic cells differ from dedifferentiated callus and differentiated cells (e.g. suspensor cells) in several structural aspects, including their characteristic compartmentation and ultrastructure of organelles and the cell wall. Interestingly, groups of early proembryogenic cells are covered by a network of fibrillar material forming an outer continuous layer. This layer is called the extracellular matrix surface network (ECMSN) and has been found in many dicotyledonous, monocotyledonous and gymnosperm plant species, including *Cofea* (Sondahl et al. 1979), *Cichorium* (Dubois et al. 1992), *Drosera* (Šamaj et al. 1995; Bobák et al. 1995, 1999, 2003), *Zea* (Šamaj et al. 1995), *Pinus* (Jásik et al. 1995), *Papaver* (Ovecka et al. 1997), *Linum* (Dedicova et al. 2000) and *Fagopyrum* (Rumyantseva et al. 2003). Additionally, extracellular layers similar to the ECMSN have also been found to cover meristematic cells during organogenesis both in situ and/or in vitro (reviewed by Šamaj et al. 1997). An ECMSN was observed preferentially in early embryogeneic stages including globular embryos and gradually disappeared when protodermis was formed in torpedo-stage embryos (Dubois et al. 1992; Šamaj et al. 1995). Digestion with enzymes and stabilization with safranine indicated the proteinaceous and/or proteoglycan nature of the ECMSN (Dubois et al. 1991; Šamaj et al. 1995). Actually, some components of the ECMSN have already been identified, such as lectin binding *N*-acetylgalactosamine (Dubois et al. 1991), AGPs (Šamaj et al. 1999a, b; Chapman et al. 2000a) and pectins (Chapman et al. 2000b, and in this volume).

3
Arabinogalactan Proteins and Somatic Embryogenesis

AGPs are ubiquitous plant-specific molecules belonging to the family of highly glycosylated hydroxyprolin-rich glycoproteins. These proteoglycans are supposed to be involved in vegetative, reproductive and cellular growth, as well as in apoptosis (Showalter 2001). In addition to classical and nonclassical AGPs a new subset of AGPs containing an adhesive fascilin-like domain was characterized in *Arabidospis* recently (Johnson et al. 2003). Actually, the first functional studies revealed that AGPs are essential for cell adhesion and expansion, and for female gametogenesis in plants (Shi et al. 2003; Acosta-Garcia and Vielle-Calzada 2004). It is well known that AGPs are developmentally regulated in reproductive organs, and during seed and vegetative development (Šamaj et al. 1998, 1999c; Showalter 2001; van Hengel et al. 2002; Sutherland et al. 2004). Importantly, considerable evidence suggests that these proteoglycans play an essential role in somatic embryogenesis. For example, AGPs released or added to the culture medium are able to induce somatic emryogenesis in diverse plant species (Kreuger and van Holst 1995; Egertsdotter and von Arnold 1995; van Hengel et al. 2001; Borderies et al. 2004). Specific AGP epitopes such as JIM8 released by nonembryogenic cells in suspension culture were found to be important for induction of embryogenic cells and development of somatic embryos (McCabe et al. 1997). AGPs associated with somatic embryogenesis were visualized using β-glucosyl Yariv reagent, a synthetic dye, which specifically binds AGPs, as well as with monoclonal antibodies against diverse AGP epitopes. It was revealed that several AGP epitopes such as JIM4, JIM13, JIM16 and LM2 are developmentally regulated during somatic embryogenesis in carrot, maize and *Cichorium* (Stacey et al. 1990; Šamaj et al. 1999a, 2002a; Chapman et al. 2000a). These epitopes can serve as specific structural markers for embryogenic cells in diverse plant species, and they are likely involved in patterning of globular embryos and their transition to the torpedo stage (Šamaj et al. 1995, 1999a; Chapman et al. 2000a). On the other hand, some other AGP epitopes are not so specific and they are present in both embryogenic and nonembryogenic cells as is the case for Gal4 and JIM15 epitopes in embryogenic maize cultures (Fig. 1A, B). Nevertheless, they might show distinct preferences for certain cell types as is the case for Gal4 which labels embryogenic cells weakly in a spotlike manner while differentiated cells are labelled more strongly at the plasma mebrane and different intracellular compartments (Fig. 1A). In more detailed studies, using correlative epifluorescence and scanning electron microscopy techniques, it was revealed that the ECMSN in maize contain AGPs recognized by JIM4 antibody (Šamaj et al. 1999a, b). On the other hand, young columnar epidermal cells of maize primary root secreted another set of AGPs recognized by MAC207 antibody which was accumulated in the outermost cell wall layer called the outer pellicle (Šamaj et al. 1999b, 2002a). This specific layer is strongly stained with β-glucosyl Yariv reagent, confirming its AGP nature (Bacic et al. 1986). Interestingly, AGPs on the surface of embryogenic cells were enriched in cell–cell contacts and they were localized to fibrillar and filamentous structures (Šamaj et al. 1999b, 2002a). Both the ECMSN in somatic embryogenesis and the pellicle in roots can be considered as protective and water-holding layers. The role of AGPs during somatic embryogeneis was further strengthened by results showing that precipitation of AGPs with β-glucosyl Yariv reagent abolished embryogenic potential (Thompson and Knox 1998; Chapman et al. 2000a).

Some AGPs are associated with programmed cell death and treatment with Yariv reagent was reported to induce programmed cell death in suspension cultures (reviewed by Showalter 2001). Apoptosis also occurs in suspensor, suspensor-like and callus cells during somatic embryogenesis. Here we show that three AGP epitopes, namely JIM8, JIM13 and MAC207, are abundant especially in apoptotic cells in maize embryogenic cultures (Fig. 1C–E).

In suspension cultures, AGPs and endochitinases secreted to the culture medium are required for somatic embryogenesis. It was shown that chiti-

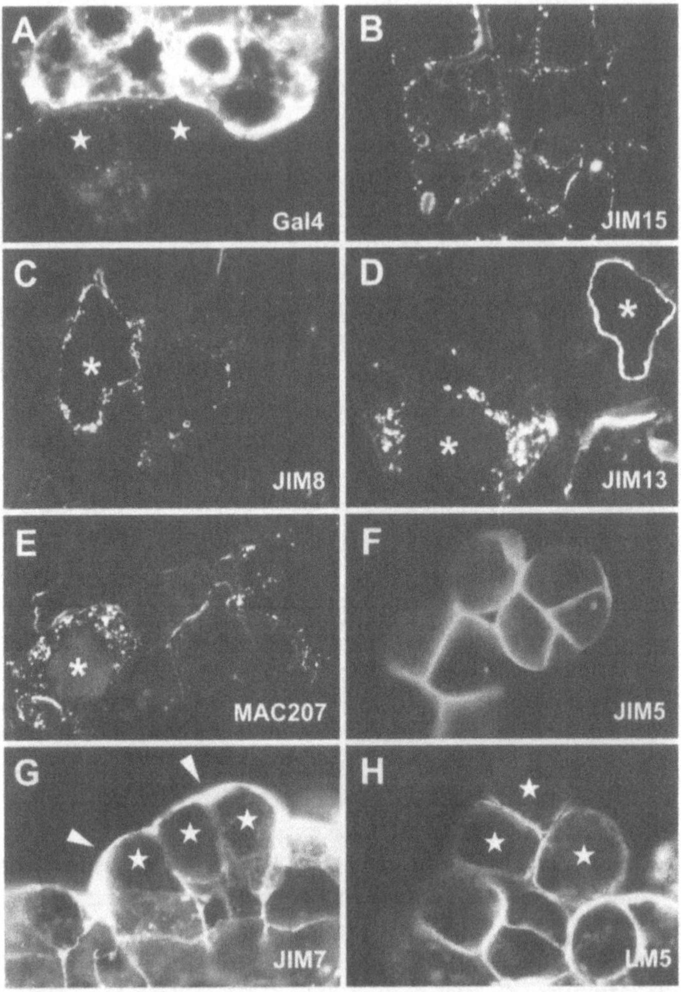

Fig. 1 Immunofluorescence labelling of arabinogalactan proteins (*AGPs*) (**A**–**E**) and pectins (**F**–**H**) in embryogenic maize cultures. **A** Strong presence of AGP epitope Gal4 in differentiated cells and weak presence in embryogenic ones (indicated by *stars*) in the form of fluorescent spots as revealed by immunofluorescence labelling. **B** Universal immunolabelling of all cells within embryogenic maize cultures with JIM15 antibody recognizing spotlike and patchy structures associated with plasma membrane and vacuoles. **C**–**E** Strong preferential labelling of apoptotic cells (indicated by *asterisks*) by JIM8 (**C**), JIM13 (**D**) and MAC207 (**E**) antibodies. Note that the labelling is associated with plasma membrane and intracellular spots and patches indicating AGP degradation. **F** Preferential immunolabelling of cell–cell contacts in pre-embryogenic clumps with JIM5 antibody recognizing low esterified pectins. **G** Immunolabelling of the extracellular matrix surface network (*arrowheads*) and cell–cell contacts in pre-embryogenic units (cells are indicated by *stars*) by JIM7 antibody recognizing highly esterified pectins. **H** Specific immunolabelling of cell–cell contacts in pre-embryogenic clumps (cells are indicated by *stars*) with LM5 antibody recognizing $(1 \rightarrow 4)$-β-D-galactan of pectin rhamnogalacturonan I

nase can cleave a subset of AGPs and that these chitinase-modified AGPs containing *N*-acetylglucosamine and glucosamine are released into the culture medium and are involved, as signal molecules, in the control of somatic embryogenesis in carrot (van Hengel et al. 2001).

4
Pectins and Somatic Embryogenesis

Pectins represent an abundant class of structural cell wall molecules with signalling properties. Similarly to AGPs, for a long time pectins were supposed to have a function in cell–cell adhesion. In maize pre-embryogenic units (clumps of embryogenic cells), highly esterified pectins recognized by JIM7 antibody localize both to the outer ECMSNs and to cell–cell adhesion sites (Fig. 1G), while JIM5 recognizing low-esterified pectins and LM5 specific for $(1 \rightarrow 4)$-β-D-galactan of pectin rhamnogalacturonan I are not present in the ECMSN, but preferentially or solely in cell–cell adhesion sites (Fig. 1F, H). Opposite results with JIM5 and JIM7 antibodies were reported for chicory somatic embryos (Chapman et al. 2000b), indicating that pectin localization within the ECMSN is differently regulated in monocotyledonous versus dicotyledonous plant species.

5
Callose

Callose deposition around embryo-competent cells of *Camellia japonica* was proposed to serve as an early structural marker for these cells (Pedroso and Pais 1995). In sunflower protoplasts, callose deposition was faster in agarose than in liquid medium; however, this concerned only 30–40% of the protoplasts and was not related to embryogenic competence of protoplasts acquired exclusively in agarose culture (Caumont et al. 1997).

6
Glycine-Rich Proteins

In carrot, the gene encoding cell wall associated glycine-rich protein was isolated and reported as a specific marker for embryogenic cells (Sato et al. 1995). In addition, another glycine-rich protein gene *Atgrp*-5 was found to be associated with globular and torpedo stages during somatic embryogenesis in *Arabidopsis* (Magioli et al. 2001). These data suggest that in addition to AGPs glycine-rich proteins might also be involved in early embryogenic development in vitro.

7
Cytoskeleton and Somatic Embryogenesis

Two basic components of the cytoskeleton, namely microtubules and actin microfilaments, participate in embryo polarization and development. Additionally, they control the positioning of cell divisions and regulate cell expansion. Early polarization is also apparent in the egg cell and during zygotic embryogenesis. The egg cell is polarized with a vacuolated micropylar pole and a cytoplasm-rich chalasal pole containing a nucleus (Russell 1993). Unfertilized egg cells of *Plumbago zeylanica* are characterized by longitudinally aligned microtubules in the micropylar pole, while they are enriched and randomly organized around the nucleus at the chalasal pole. This chalasal pole is also enriched with a longitudinally aligned mesh of actin bundles (Huang et al. 1993). Dense micotubular and actin cytoskeleton around the nucleus might be involved in the stabilization of the nuclear position (Russell 1993). Further, it was shown that both the microtubular cytoskeleton and especially the actin cytoskeleton are rearranged (forming a corona structure between the egg and the central cell) during fertilization in order to assist gametic fusion and are also reorganized during early embryogenesis (Huang et al. 1993). Most of our knowledge about the cytoskeleton during zygote polarization and early embryo development comes from the brown algae *Fucus* and *Pelvetia*. Early during zygote development, both cell wall components and actin filaments are required for alignment of the polar axis. A cortical actin patch is located on the entry side of the sperm cell where a tip-growing rhizoid will appear later on (Belanger and Quatrano 200Q). Movement of the actin patch into the most shaded part of the zygote is sensitive to the light gradient along which the polarity is aligned. Thus, actin together with polarized secretion of cell wall molecules play an essential role during early polarization of a fucoid zygote and also in tip-growth of the rhizoid later on (Belanger and Quatrano 2000). On the other hand, microtubules seem to be involved in the rotation and positioning of the nucleus before the first division occurs in the zygote (Kropf 1997).

In comparison with fucoid early embryogenesis little is known about the cytoskeleton during zygotic and somatic embryogenesis in higher plants. Several *Arabidospis* mutants of the PILZ group show very small embryos consisting of only one (*porzino*) or a few large cells (*champignon, pfifferling, hallimasch*) having enlarged nuclei and showing severe microtubule and cytokinesis defects. Spindles are generally absent from mitotic nuclei and interphase cells have no cortical microtubules in these mutants, suggesting that products of these four mutant genes might be involved in regulation of microtubular organization required for proper mitosis and/or cytokinesis (Mayer et al. 1999). More recently, all these genes together with another gene *KIESEL* showing a weakened embryo-lethal phenotype were cloned, and were revealed to encode tubulin-folding cofactors and related G-protein Arl2 involved in the formation of tubulin heterodimers (Steinborn et al. 2002; Tzafrir et al. 2002). Additionally, *CHAMPIGNON* was found to be identical to *TITAN1*. Mutant embryos of the PILZ group lack microtubules or they are disorganized in *KIESEL* mutant, while actin seems to be present, although it is also disorganized from fine meshwork into patchy structures (Steinborn et al. 2002).

Cytokinesis represents the final stage of cell division and it is dependent on the cytoskeletal structure called phragmoplast which is involved in cell plate formation. Here we show that in embryogenic maize cells phragmoplasts are enriched both with microtubules and with actin. Interestingly, microtubules were present along the whole length of the phragmoplast, while actin filaments were most abundant at its growing edges (Fig. 2A, B). This indicates that actin is involved in proper positioning of cell plates, which is important for embryogenic development.

A developmental switch occurs during transition of pre-embryogenic units to polarized transition units possessing both embryogenic and suspensor-like cells during maize somatic embryogenesis. This switch is dependent on deprivation of exogenous auxin and spectacular redistributions of both microtubular and actin cytoskeletons (Šamaj et al. 2003). Loosely attached pre-embryogenic cells (Šamaj et al. 1995) are characterized by an abundant endoplasmic cytoskeleton arranged in the form of perinuclear radiating microtubules and actin filaments (Fig. 2C, D; Šamaj et al. 1995). On the other hand, cytoskeletal rearrangements leading to the more abundant cortical cytoskeleton, composed of both cortical microtubules and actin filaments, seem to be essential for further cell adhesions, polarization and development of somatic embryos (Fig. 2E, F; Šamaj et al. 2003). At this stage, auxin is likely synthesized in preglobular and globular embryos (Friml et al. 2003) and would be satisfactory for embryo development on its own. In contrast, high-exogenous auxin which is not required anymore for cell activation, clearly prevents polarization and elongation, which are necessary for further progression of embryogenesis. One of the most conspicuous cytoskeletal features during embryo initiation are endoplasmic microtubules radiating from nuclear surfaces towards the cell cortex in embryogenic cells induced by exogenous auxin (Šamaj et al. 2003). This phenomenon seems to be more general, because similar microtubular arrangements were found in postmitotic cells of intact roots treated with exogenous auxin (Baluška et al. 1996) or with the microtubule-stabilizing drug taxol (Baluška et al. 1997). Moreover, radiating endoplasmic microtubules are typical for noncellularized endosperm, some isolated cells, such as microspores, and also for cells under environmental or biotic stress (Dickinson and Sheldon 1984; Brown et al. 1994; Caumont et al. 1997; Sivaguru et al. 1999; Timmers et al. 1999; Gervais et al. 2000).

The actin cytoskeleton is known to be involved in auxin transport, signalling and in the regulation of cell polarity (Staiger 2001; Šamaj et al. 2002b). Recently, mutual interactions were found between the actin cytoskeleton,

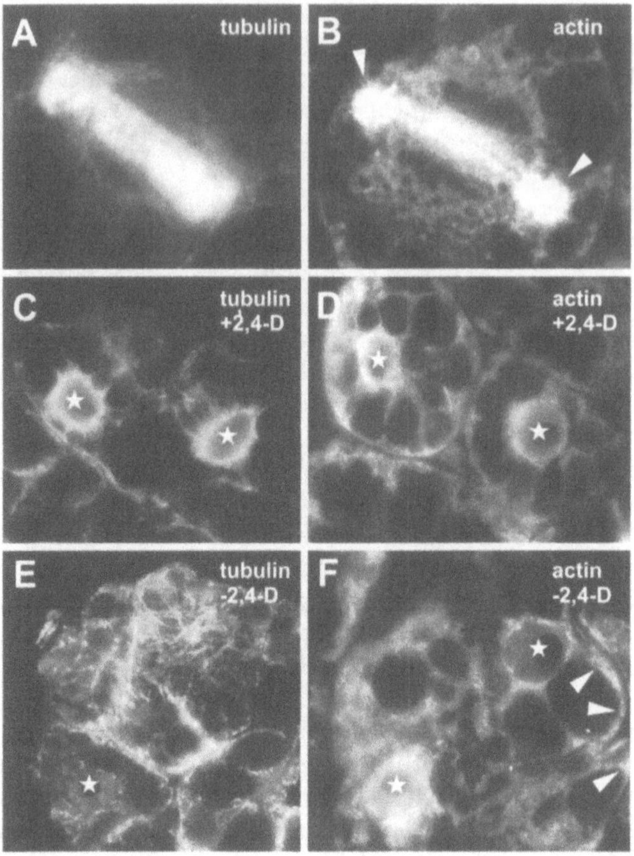

Fig. 2 Immunlocalization of microtubules (**A, C, E**) and the actin cytoskeleton (**B, D, E**) in maize embryogenic cultures. **A** Microtubules are present along the whole phragmoplast length in dividing embryogenic cells. **B** Actin is most abundant at growing edges of the phragmoplast (indicated by *arrowheads*) in dividing embryogenic cells. **C** Prominent endoplasmic microtubules radiating from nuclear surfaces (nuceli are indicated by *stars*) towards cell peripheries in proembryogenic cell cultures supplemented with exogenous 2,4-dichlorophenoxyacetic acid (*2,4-D*). **D** Prominent endoplasmic actin filaments connecting centrally positioned nuclei (indicated by *stars*) with cell peripheries in proembryogenic cell cultures supplemented with exogenous 2,4-D. Note the loose cell–cell contacts between proembryogenic cells. **E** Cortical microtubules organized as parallel bundles and networks in cells of transition embryogenic units upon 2,4-D depletion from the culture medium. Note that endoplamic microtubules radiating from nuclei (the nucleus is indicated by *star*) are depleted in these cells having tight cell–cell contacts. **F** Actin is enriched in the cell cortex (indicated by *arrowheads*) in cells of transition units upon 2,4-D depletion from the culture medium. Nuclei are indicated by *stars*

auxin transport and the establishment of polarity in *Fucus* embryos (Sun et al. 2004). An intact actin cytoskeleton seems to be essential for somatic embryogenesis because depolymerization of actin filaments via latrunculin B clearly inhibited embryo formation and development in diverse plant species (Baluška et al. 2001; Smertenko et al. 2003; Briere et al. 2004). In *Abies* embryogenic cultures latrunculin B prevented elongation of suspensor cells and led to the formation of dwarf embryos (Baluška et al. 2001). Interestingly, microtubules are progressively disrupted while prominent thick actin cables appear during programmed cell death of suspensor cells in *Picea* embryogenic culture, suggesting a role of actin in this process (Smertenko et al. 2003). On the other hand, embryogenic *Picea* cells have prominent microtubule networks and a fine network of actin, which is consistent with the situation in maize proembryogenic cells (Šamaj et al. 2003; Smertenko et al. 2003). Again, these particular cytoskeletal arrangements are well correlated with cell–cell contacts along the embryonal axis with tightly packed small embryogenic cells and with highly elongated suspensor cells having very loose contacts, no cortical microtubules and actin filaments but instead only a few thick actin bundles. This assumption is further supported by results demonstrating that overstabilization of the actin cytoskeleton with jasplakinolide led to very tight cell–cell contacts in intact roots (Baluška et al. 2004). *Medicago* protoplasts induced to form somatic embryos by electrical stimulation have a disordered network of fine microtubules in comparison with thick parallel bundles in nonembryogenic protoplasts (Dijak and Simmonds 1988). Embryo-competent protoplasts in *Medicago* and sunflower divide asymmetrically, resulting in compact embryogenic cell colonies which are composed of small, tightly packed cells. In sunflower, an agarose matrix is required to acquire embryogenic competence and this matrix stabilizes the microtubule cytoskeleton (Caumont et al. 1997). While no difference was observed in cortical arrays of microtubules in nonembryogeneic and embryogenic protoplasts, only agarose-embedded protoplasts showed prominent endoplasmic microtubules forming a basket around nuclei and radiating from the nuclear surface towards the cell periphery (Caumont et al. 1997). This is consistent with our observations on embryogenic maize cells (Šamaj et al. 2003). On the other hand, a narrow preprophase band of microtubules was present only in symmetrically dividing nonembryogenic protoplasts cultured in liquid medium (Caumont et al. 1997). Moreover, embryo-competent sunflower protoplasts embedded in agarose regenerate a new cell wall together with an interconnected network of endogenous and cortical actin filaments within the first few days in the culture. Interestingly, this actin regeneration and organization is disrupted by RGD peptides, putative inhibitors of cell–cell contacts, which also reduce embryoid formation (Briere et al. 2004).

It is well known that microtubular and actin cytoskeletons interact with each other and that they are mutually reorganized upon diverse stimuli (Tominaga et al. 1997; Collings and Allen 2000; Šamaj et al. 2000b). Recently it was shown that microtubules and actin undergo parallel reorientations in response to auxin deprivation in maize embryogenic cultures or after treatment by auxin or light exposure in rice coleptiles (Šamaj et al. 2003; Holweg et al. 2004).

8 Cytoskeleton–Plasma Membrane-Cell Wall Continuum and Its Putative Role in Somatic Embryogenesis

Microtubular drugs such as colchicine and trifluralin as well as cold treatment reversibly disrupted the structural integrity of the ECMSN, suggesting a link between the cytoskeleton and the cell wall during somatic embryogenesis (Bobák et al. 1999). The cytoskeleton–plasma membrane–cell wall continuum was suggested to play an important role in plant cell morphogenesis (Baluška et al. 2003). In spite of initial immunolocalization studies in plants with heterologous antibodies against animal adhesive plasma membrane spanning proteins such as integrins, cadherins/vinculins and catenins (Gens et al. 1996; Katembe et al. 1997; Endlé et al. 1998; Baluška et al. 1999), homologous proteins were not found in plant genome databases. Plants likely developed their own strategies and use other types of plasma membrane proteins for adhesion of plasma membrane to the cell wall. The most-favoured among these putative candidates are wall-associated kinases (WAKs), cellulose and callose synthase complexes, and plant formins (Kohorn 2000; Baluška et al. 2003). Other adhesive linker molecules associated with the apoplastic (cell wall) side of the plasma membrane are AGPs (Šamaj et al. 2000) and pectins. On the other cytoplasmic side, phospholipase D links microtubules to the plasma membrane, and myosin VIII may associate with callose synthase (Baluška et al. 2003; Dhonukshe et al. 2003). WAKs interact with pectins and eventually also with AGPs within the cell wall and have a cytoplasmic kinase domain potentially involved in signalling (Gens et al. 2000; Kohorn 2001). On the other hand, a subset of plant formins have an unusual extracellular domain similar to extensins, a membrane-spanning domain, and all plant formins have conserved FH1 and FH2 domains which interact with the actin cytoskeleton. Nevertheless, some structural modules might be conserved among animal proteins such as vitronectin and fibronectin, both of which use the RGD peptide (composed of arginine–glycine–aspartic acid) domain to bind integrin, and plant adhesive proteins because synthetic RGD peptides interfere with cell growth, protoplast adhesion and membrane-wall adhesion in plants and fungi (Schindler et al. 1989; Henry et al. 1996; Canut et al. 1998). RGD motifs involved in interaction between the plasma membrane and the cell wall were reported to be important for plant–pathogen interactions and plant defence (Mellersh and Heath 2001). Moreover, it was shown that treatment with RGD peptides also prevents embryogenic development from sunflower protoplasts (Barthou et al. 1999). Since animal integrins interact with the actin cytoskeleton it is interesting to note that treatment with RGD disrupts the actin cytoskeleton in embryo-competent protoplasts (Briere et al. 2004). All these data clearly indicate that the cytoskeleton–plasma membrane–cell wall continuum and associated adhesive domains between the plasma membrane and the cell wall as well as between neighbouring cells play a crucial role in the control of somatic embryogenesis.

9
Conclusions and Future Prospects

Embryogenic cells show specific structural features related to the composition of their cell walls and cytoskeletal arrangements. For example, the ECMSN, which represents a thin outer cell wall layer, can be considered as a specific structural marker for embryogenic cells in diverse plant species. This layer is composed of both AGPs and pectins. Cell wall molecules such as glycine-rich proteins, but especially AGPs, can serve not only as specific molecular markers for cells having embryogenic competence, but they also play an important role in the intracellular and intercellular signalling, and participate in apoptotic events during embryogenic development. Cytoskeletal elements including both microtubules and actin microfilaments respond via their dynamic rearrangements to developmental signals and switches triggering somatic embryogenesis, such as stress and exogenous auxin. Moreover, cytoskeletal and cell wall changes seem to be interrelated and coordinated during formation of somatic proembryos and also during progression of embryogenic development. Nevertheless, our information about molecules linking the cytoskeleton across the plasma membrane to the cell wall remains very elusive. Much has to be done in order to characterize cytoskeleton–cell wall adhesion domains in plants.

Acknowledgements This work was supported by a grant from the Slovak Grant Agency APVT (grant no. APVT-51-002302), Bratislava, Slovakia.

References

Acosta-Garcia G, Vielle-Calzada JP (2004) A classical arabinogalactan protein is essential for the initiation of female gametogenesis in Arabidospis. Plant Cell 16:2614–2628

Bacic A, Moody SF, Clarke AE (1986) Structural analysis of secreted root slime from maize (*Zea mays* L.). Plant Physiol 80:771–777

Baluška F, Barlow PW, Volkmann D (1996) Complete disintegration of the microtubular cytoskeleton precedes its auxin-mediated reconstructions in post-mitotic maize root cells. Plant Cell Physiol 37:1013–1021

Baluška F, Šamaj J, Volkmann D, Barlow PW (1997) Impact of taxol-mediated stabilization of microtubules on nuclear morphology, ploidy levels and cell growth in maize roots. Biol Cell 89:221–231

Baluška F, Šamaj J, Volkmann D (1999) Proteins reacting with cadherin and catenin antibodies are present in maize showing tissue-, domain-, and development-specific associations with endoplasmic-reticulum membranes and actin microfilaments in root cells. Protoplasma 206:174–187

Baluška F, Jasik J, Edelmann HG, Salajova T, Volkmann D (2001) Latrunculin B-induced plant dwarfism: plant cell elongation is F-actin-dependent. Dev Biol 231:113–124

Baluška F, Šamaj J, Wojtaszek P, Volkmann D, Menzel D (2003) Cytoskeleton-plasma membrane-cell wall continuum in plants. Emerging links revisited. Plant Physiol 133:482–491

Baluška F, Šamaj J, Hlavačka A, Kendrick-Jones J, Volkmann D (2004) Myosin VIII and F-actin enriched plasmodesmata in maize root inner cortex cells accomplish fluid-phase endocytosis via an actomyosin-dependent process. J Exp Bot 55:463–473

Barthou H, Petitprez M, Briere C, Souvre A, Alibert G (1999) RGD-mediated membrane-matrix adhesion triggers agarose-induced embryooid formation in sunflower protoplasts. Protoplasma 206:143–151

Belanger KD, Quatrano RS (2000) Polarity: the role of localized secretion. Curr Opin Plant Biol 3:67–72

Bobák M, Blehová A, Krištín J, Ovečka M, Šamaj J (1995) Direct plant regeneration from leaf explants of *Drosera rotundifolia* cultured in vitro. Plant Cell Tissue Org Cult 43:43–49

Bobák M, Hlavačka A, Ovečka M, Šamaj J (1999) Effect of trifluralin and colchicine on the extracellular matrix surface networks during early stages of direct somatic embryogenesis of Drosera rotundifolia L. J Plant Physiol 155:387–392

Bobák M, Šamaj J, Hlinková E, Hlavačka A, Ovecka M (2003) Extracellular matrix in early stages of direct somatic embryogenesis in leaves of *Drosera spathulata*. Biol Plant 47:161–166

Borderies G, le Bechec M, Rossignol M, Lafitte C, Le Deunff E, Beckert M, Dumas C, Matthys-Rochon E (2004) Characterization of proteins secreted during maize microspore culture: arabinogalactan proteins (AGPs) stimulate embryo development. Eur J Cell Biol 83:205–212

Briere C, Barthou H, Jauneau A, Alibert G, Petitprez M (2004) The actin cytoskeleton is involved in signalling protoplast embryogenesis induced by agarose embedding. Physiol Plant 122:115–122

Brown RC, Lemmon BE, Olsen OA (1994) Endosperm development in barley: microtubule involvement in the morphogenetic pathway. Plant Cell 6:1241–1252

Canut H, Carrasco A, Galaud JP, Cassan C, Bouyssou H, Vita N, Ferrara P, Pont-Lezica R (1998) High-affinity RGD-binding sites at the plasma membrane of *Arabidopsis thaliana* links the cell wall. Plant J 16:63–71

Caumont C, Petitprez M, Woynaroski S, Barthou H, Briere C, Kallerhoff J, Borin C, Souvre A, Alibert G (1997) Agarose embedding affects cell wall regeneration and microtubule organization in sunflower hypocotyls protoplasts. Physiol Plant 99:129–134

Chapman A, Blervacq AS, Vasseur J, Hilbert JL (2000a) Arabinogalactan-proteins in Cichorium somatic embryogenesis: effect of beta-glucosyl Yaiv reagent and epitope localization during embryo development. Planta 211:305–314

Chapman A, Blervacq AS, Hendriks T, Slomianny C, Vasseur J, Hilbert JL (2000b) Cell wall differentiation during early somatic embryogenesis in plants. II. Ultrastructural study and pectin immunolocalization on chicory embryos. Can J Bot 78:824–831

Collings DA, Allen NS (2000) Cortical actin interacts with the plasma membrane and microtubules. In: Staiger CJ, Baluška F, Volkmann D, Barlow PW (eds) Actin: a dynamic framework for multiple plant cell functions. Kluwer, Dordrecht, pp 145–164

Dedicová B, Hricová A, Šamaj J, Obert B, Bobák M, Pretová A (2000) Shoots and embryo-like structures regenerated from cultured flax (*Linum usitatissimum* L.) hypocotyls segments. J Plant Physiol 157:327–334
Dhonukshe P, Laxalt AM, Goedhart J, Gadella TW, Munnik T (2003) Phospholipase D activation correlates with microtubule reorganization in living plant cells. Plant Cell 15:2666–2679
Dickinson HG, Sheldon JM (1984) A radial system of microtubules extending between the nucleus envelope and the plasma membrane during early male haplophase in flowering plants. Planta 161:86–90
Dijak M, Simmonds DH (1988) Microtubule organization during early direct embryogenesis from mesophyll protoplasts of *Medicago sativa* L. Plant Sci 58:183–191
Dubois T, Guedira M, Dubois J, Vasseur J (1991) Direct somatic embryogenesis in leaves of *Cichorium*. A histological and SEM study of early stages. Protoplasma 162:120–127
Dubois T, Dubois J, Guedira M, Diop A, Vasseur (1992) SEM characterization of an extracellular matrix around somatic proembryos in roots of *Cichorium*. Ann Bot 70:119–124
Egertsdotter U, von Arnold S (1995) Importance of arabinogalactan proteins for the development of somatic embryos of Norway spruce (*Picea abies*). Physiol Plant 93:334–345
Endlé MC, Stoppin V, Lambert AM, Schmit AC (1998) The growing cell plate of higher plants is a site of both actin assembly and vinculin-like antigen recruitment. Eur J Cell Biol 77:10–18
Feher A, Pasternak TP, Dudits D (2003) Transition of somatic plant cells to an embryogenic state. Plant Cell Tissue Org Cult 74:201–228
Friml J, Vieten A, Sauer M, Weijers D, Schwarz H, Hamann T, Offringa R, Jurgens G (2003) Efflux-dependent auxin gradients establish the apical-basal axis of *Arabidopsis*. Nature 426:147–153
Gens JS, Reuzeau C, Doolittle KW, McNally JG, Pickard BG (1996) Covisualization by computational optical-sectioning microscopy of integrin and associated proteins at the cell membrane of living onion protoplasts. Protoplasma 194:215–230
Gens JS, Fujiki M, Pickard BG (2000) Arabinogalactan protein and wall-associated kinase in a plasmalemmal reticulum with specialized vertices. Protoplasma 212:115–134
Gervais C, Newcomb W, Simmonds DH (2000) Rearrangement of the actin filament and microtubule cytoskeleton during induction of microspore embryogenesis in *Brassica napus* L. cv. Topas. Protoplasma 213:194–202
Huang BQ, Pierson ES, Russell SD, Tiezzi A, Cresti M (1993) Cytoskeletal organization and modification during pollen tube arrival, gamete delivery and fertilization in Plumbago zeylanica. Zygote 1:143–154
Henry CA, Jordan JR, Kropf DL (1996) Localized membrane-wall adhesions in Pelvetia zygotes. Protoplasma 190:39–52
Holweg C, Süsslin C, Nick P (2004) Capturing in vivo dynamics of the actin cytoskeleton stimulated by auxin or light. Plant Cell Physiol 45:855–863
Jásik J, Salajová T, Salaj J (1995) Developmental anatomy and ultrastructure of early somatic embryos in european black pine (*Pinus nigra* Arn.). Protoplasma 185: 205–211
Johnson KL, Jones BJ, Bacic A, Schultz CJ (2003) The fasciclin-like arabinogalactan proteins of *Arabidopsis*. A multigene family of putative cell adhesion molecules. Plant Physiol 133:1911–1925
Katembe WJ, Swatzell LJ, Makroff CA, Kiss JZ (1997) Immunolocalization of integrin-like proteins in *Arabidopsis* and *Chara*. Physiol Plant 99:7–14
Kohorn BD (2000) Plasma membrane-cell wall contacts. Plant Physiol 124:31–38

Kohorn BD (2001) Waks: cell wall associated kinases. Curr Opin Cell Biol 13:529–533

Kreuger M, van Holst GJ (1995) Arabinogalactan-proteins in somatic embryogenesis of *Daucus carota* L. Planta 197:135–141

Kropf D (1997) Induction of polarity in fucoid zygotes. Plant Cell 9:1011–1020

Magioli C, Barroco RM, Rocha CA, de Santiago-Fernandes LD, Mansur E, Engler G, Margis-Pinheiro M, Sachetto-Martins G (2001) Somatic embryo formation in *Arabidopsis* and eggplant is associated with expression of a glycine-rich protein gene *Atgrp-5*. Plant Sci 161:559–567

Mayer U, Herzog U, Berger F, Inze D, Jurgens G (1999) Mutations in the pilz group genes disrupt the microtubule cytoskeleton and uncouple cell cycle progression from cell division in *Arabidopsis* embryo and endosperm. Eur J Cell Biol 78:100–108

McCabe PF, Valentine TA, Forsberg LS, Pennell RI (1997) Soluble signals from cells identified at the cell wall establish a developmental pathway in carrot. Plant Cell 9:2225–2241

Mellersh D, Heath M (2001) Plasma membrane-cell wall adhesion is required for expression of plant defence responses during fungal penetration. Plant Cell 13:413–424

Ovecka M, Bobák M, Blehová A, Kristin J (1997) *Papaver somniferum* regeneration by somatic embryogenesis and shoot organogenesis. Biol Plant 40:321–328

Pedroso MC, Pais MS (1995) Factors controlling somatic embryogenesis—cell wall changes as an *in vivo* marker of embryogenic competence. Plant Cell Tissue Org Cult 43:147–154

Rumyantseva NI, Šamaj J, Ensikat HJ, Sal'nikov VV, Kostyukova YA, Baluška F, Volkmann D (2003) Changes in the extracellular matrix surface network during cyclic reproduction of proembryonic cell complexes in the *Fagopyrum tataricum* (L.) gaertn callus. Dokl Biol Sci 391:375–378

Russell SD (1993) The egg cell: Development and role in fertilization and early embryogenesis. Plant Cell 5:1349–1359

Šamaj J, Bobák M, Blehová A, Krištín J, Auxtová-Šamajová O (1995) Developmental SEM observations of an extracellular matrix in embryogenic calli of *Drosera rotundifolia* and *Zea mays*. Protoplasma 186:45–49

Šamaj J, Bobák M, Ovecka M, Blehová A, Pretová A (1997) Structural features of plant morphogenesis in vitro. Veda, Bratislava, pp 1–122

Šamaj J, Baluška F, Volkmann D (1998) Cell specific expression of two arabinogalactan-protein epitopes recognized by monoclonal antibodies JIM8 and JIM13 in maize roots. Protoplasma 204:1–12

Šamaj J, Baluška F, Bobák M, Volkmann D (1999a) Extracellular matrix surface network of embryogenic units of friable maize callus contains arabinogalactan-proteins recognized by monoclonal antibody JIM4. Plant Cell Rep 18:369–374

Šamaj J, Ensikat HJ, Baluška F, Knox JP, Barthlott W, Volkmann D (1999b) Immunogold-localization of plant surface arabinogalactan-proteins using glycerol liquid substitution and scanning electron microscopy. J Microsc 193:150–157

Šamaj J, Braun M, Baluška F, Ensikat HJ, Tsumuraya Y, Volkmann D (1999c) Specific localization of arabinogalactan-protein epitopes at the surface of maize root hairs. Plant Cell Physiol 40:874–883

Šamaj J, Šamajová O, Peters M, Baluška F, Lichtscheidl I, Knox JP, Volkmann D (2000a) Immunolocalization of LM2 arabinogalactan-protein epitope associated with endomembranes of plant cells. Protoplasma 212:186–196

Šamaj J, Peters M, Volkmann D, Baluška F (2000b) Effects of myosin ATPase inhibitor 2,3-butanedione monoxime on distributions of myosins, F-actin, microtubules, and cortical endoplasmic reticulum in maize root apices. Plant Cell Physiol 41:571–582

Šamaj J, Ensikat HJ, Barthlott W, Volkmann D (2002a) Immunogold-silver scanning electron microscopy using glycerol liquid substitution. In: Hacker GW, Gu J (eds) Gold and silver staining: techniques in molecular morphology. CRC, Boca Raton, pp 223–234

Šamaj J, Ovecka M, Hlavačka A, Lecourieux F, Meskiene I, Lichtscheidl I, Lenart P, Salaj J, Volkmann D, Bögre L, Baluška F, Hirt H (2002b) Involvement of the mitogen-activated protein kinase SIMK in regulation of root hair tip-growth. EMBO J 21:3296–3306

Šamaj J, Baluška F, Pretova A, Volkmann D (2003) Auxin deprivation induces a developmental switch in maize somatic embryogenesis involving redistribution of microtubules and actin filaments from endoplasmic to cortical cytoskeletal arrays. Plant Cell Rep 21:940–945

Sato S, Toya T, Kawahara R, Whittier RF, Fukuda H, Komamine A (1995) Isolation of a carrot gene expressed specifically during early-stage somatic embryogenesis. Plant Mol Biol 28:39–46

Schindler M, Meiners S, Cheresh D (1989) RGD-dependent linkage between plant cell wall and plasma membrane: consequences for growth. J Cell Biol 108:1955–1965

Shi H, Kim YS, Guo Y, Stevenson B, Zhu JK (2003) The *Arabidopsis* SOS5 locus encodes a putative cell surface adhesion protein and is required for normal cell expansion. Plant Cell 15:19–32

Showalter AM (2001) Arabinogalactan-proteins: structure, expression and function. Cell Mol Life Sci 58:1399–1417

Sivaguru M, Yamamoto Y, Matsumoto H (1999) Differential impacts of aluminium on microtubule organization depends on growth phase in suspension-cultured tobacco cells. Physiol Plant 107:110–119

Smertenko AP, Bozhkov PV, Filonova LH, von Arnold S, Hussey PJ (2003) Re-orientation of the cytoskeleton during developmental programmed cell death in *Picea abies* embryos. Plant J 33:813–824

Sondahl MR, Salisbury JL, Sharp WR (1979) SEM characterization of embryogenic tissue and globular embryos during high-frequency somatic embryogenesis in coffee callus cells. Z Pflanzenphysiol 94:185–188

Stacey NJ, Roberts K, Knox JP (1990) Patterns of expression of the JIM4 arabinogalactan-protein epitope in cell cultures and during somatic embryogenesis in *Daucus carota* L. Planta 180:285–292

Staiger CJ (2001) Signaling to the actin cytoskeleton in plants. Annu Rev Plant Physiol Plant Mol Biol 51:257–288

Steinborn K, Maulbetsch C, Priester B, Trautmann S, Pacher T, Geiges B, Küttner F, Lepiniec L, Stierhof Y-D, Schwarz H, Jürgens G, Mayer U (2002) The *Arabidopsis* PILZ group genes encode tubulin-folding cofactor orthologs required for cell division but not cell growth. Gen Dev 16:959–971

Sun H, Basu S, Brady SR, Luciano RL, Muday GK (2004) Interactions between auxin transport and the actin cytoskeleton in developmental polarity of *Fucus distichus* embryos in response to light and gravity. Plant Physiol 135:266–278

Sutherland PW, Hallett IC, MacRae E, Fischer M, Redgwell RJ (2004) Cytochemistry and immunolocalisation of polysaccharides and proteoglycan in the endosperm of green *Arabica* coffee beans. Protoplasma 223:203–211

Tautorus TE, Wang H, Foke LC, Dunstan DI (1992) Microtubule pattern and the occurrence of pre-prophase bands in embryogenic cultures of black spruce (*Picea mariana* Mill.) and non-embryogenic cultures of jack pine (*Pinus banksiana* Lamb.). Plant Cell Rep 11:419–423

Thompson HJM, Knox JP (1998) Stage-specific responses of embryogenic carrot cell suspension cultures to arabinogalactan protein-binding beta-glucosyl Yariv reagent. Planta 205:32–38

Timmers ACJ, Auriac MC, Truchet G (1999) Refined analysis of early symbiotic steps of the *Rhizobium-Medicago* interaction in relationship with microtubular cytoskeleton rearrangements. Development 126:3617–3628

Tominaga M, Morita K, Sonobe S, Yokota E, Shimmen T (1997) Microtubules regulate the organization of actin filaments at the cortical region in root hair cells of *Hydrocharis*. Protoplasma 199:83–92

Tzafrir I, McElver JA, Liu C-M, Yang LJ, Wu JQ, Martinez A, Patton DA, Meinke DW (2002) Diversity of TITAN functions in *Arabidopsis* seed development. Plant Physiol 128:38–51

Van Hengel AJ, Tadesse Z, Immerzeel P, Schols H, van Kammen A, de Vries SC (2001) N-Acetylglucosamine and glucosamine-containing arabinogalactan proteins control somatic embryogenesis. Plant Physiol 125:1880–1890

Van Hengel AJ, van Kammen A, de Vries SC (2002) A relationship between seed development, arabinogalactan-proteins (AGPs) and AGP mediated promotion of somatic embryogenesis. Physiol Plant 144:637–644

Plant Cell Monogr (2)
A. Mujib · J. Šamaj: Somatic Embryogenesis
DOI 10.1007/7089_027/Published online: 2 December 2005

Comparison of Molecular Mechanisms of Somatic and Zygotic Embryogenesis

Miho Ikeda[1] (✉) · Hiroshi Kamada[2]

[1]Gene Research Center, Institute of Biological Sciences, University of Tsukuba, Tsukuba, Ibaraki, 305-8572, Japan
mi-ikeda@aist.go.jp

[2]Graduate School of Life and Environmental Sciences, University of Tsukuba, Tsukuba, Ibaraki, 305-8572, Japan

Abstract Somatic embryogenesis has been used as a model system to understand the mechanisms regulating plant embryogenesis. The morphological and physiological characteristics of somatic embryos are similar to those of zygotic embryos. However, what are the patterns of gene expression during somatic embryogenesis? Here, we review molecular events involved in embryogenesis. Four important transcription factors were isolated from a defective-embryo mutant (*LEC1*, *LEC2*, *FUS3* and *ABI3*), and three factors were isolated using deferential screening (*SERK*, *AGL15* and *BBM*); all are expressed during both somatic and zygotic embryo development. These genes may be important in regulating phytohormone synthesis and phytohormone response during somatic and zygotic embryogenesis. Regulation of embryo-specific *LEA* gene expression is similar in both somatic and zygotic embryos. Recent research involves examination of new mutants that form embryonic structures.

1 Introduction

In 1958, the first somatic embryogenesis was performed using carrot tissue cultures (Reinert 1958; Steward et al. 1958). Somatic embryogenesis is unusual, because differentiated somatic cells dedifferentiate to form new embryos, which develop into plantlets. This is good evidence that plant somatic cells have differentiation totipotency. Moreover, somatic embryogenesis produces new, perfect plantlets that have both shoots and roots. Therefore, somatic embryogenesis tissue-culture systems are very useful for plant regeneration and transformation. Since the first somatic embryogenesis in carrot, somatic embryo induction has been attempted for many plant species. Various conditions that may affect somatic embryo induction have been examined (e.g., phytohormones, osmotic stress, temperature and nitrogen sources; reviewed in Kamada 1980, 1996). Tissue-culture systems for somatic embryo induction have been established for many plant species.

Somatic embryo development closely resembles that of zygotic embryos, both morphologically and physiologically; therefore, somatic embryogene-

Table 1 Characteristics of somatic embryogenesis (*SE*)-related genes

Gene name	Protein	Target motif[a]	Expression[b] ZE	Expression[b] SE	SE formation[b]	Phytohormone regulating expression of each gene	Regulation of the phytohormone by each gene	Reference
ABI3 (*ABA INSENSITIVE 3*)	Transcription factor (B2, B3 domain)	G-box (ABRE) RY motif	+	+	–	Auxin (up) (in root)	ABA (sensitivity up)	Giraudat et al. 1992; Shiota et al. 1998; Suzuki et al. 2001, 2003; Brady 2003; Ikeda-Iwai et al. 2003, 2004
FUS3 (*FUSCA3*)	Transcription factor (B3 domain)	RY motif	+	+	–	Auxin (up)	ABA (up) GA (synthesis down)	Gazzarrini et al. 2004; Curaba et al. 2004; Ikeda-Iwai et al. 2003, 2004
LEC1 (*LEAFY COTYLEDON 1*)	Transcription factor (HAP3)	CCAAT box	+	+	Over-expression (without hormone)	ND	ND	Lotan et al. 1998; Zhang et al. 2002; Ikeda-Iwai et al. 2003; Yazawa et al. 2003
LEC2 (*LEAFY COTYLEDON 2*)	Transcription factor (B3 domain)	RY motif	+	ND	Over-expression (without hormone)	ND	GA (down)	Stone et al. 2001; Curaba et al. 2004; Kroj et al. 2003

– the factor does not encode the transcription factor, *ZE* zygotic embryogenesis, *ABA* abscisic acid, *ABRE* ABA-response element, *GA* gibberellic acid, *ND* not determined, *2,4-D* 2,4-dichlorophenoxyacetic acid

[a]Target motif is the promoter *cis*-element that is related to the encoded transcription factors

[b]Somatic embryo formation is observed under each culture condition.

Table 1 (continued)

Gene name	Protein	Target motif[a]	Expression[b] ZE	Expression[b] SE	SE formation[b]	Phytohormone regulating expression of each gene	Regulation of the phytohormone by each gene	Reference
SERK (*SOMATIC EMBRYO RECEPTOR KINASE*)	Receptor kinase (leucine-rich repeat)	–	+	+	Over-expression (with 2,4-D)	ND	ND	Schmidt et al. 1997; Hecht et al. 2001; Shah et al. 2001; Baudino et al. 2001; Thomas et al. 2003; Nolan et al. 2003
AGL15	Transcription factor (MADS box)	GC[A/T]$_8$GG	+	+	Over-expression (with 2,4-D)	Auxin (up) (in seedling)	GA (down)	Heck et al. 1995; Perry et al. 1999; Harding et al. 2003; Zhu and Perry 2005
WUSCHEL	Transcription factor (homeo domain)	ND	ND	ND	Over-expression (without hormone)	ND	ND	Zuo et al. 2002

– the factor does not encode the transcription factor, *ZE* zygotic embryogenesis, *ABA* abscisic acid, *ABRE* ABA-response element, *GA* gibberellic acid, *ND* not determined, *2,4-D* 2,4-dichlorophenoxyacetic acid

[a]Target motif is the promoter *cis*-element that is related to the encoded transcription factors

[b]Somatic embryo formation is observed under each culture condition.

Table 1 (continued)

Gene name	Protein	Target motif[a]	Expression[b] ZE	SE	SE formation[b]	Phytohormone regulating expression of each gene	Regulation of the phytohormone by each gene	Reference
PRIMORDIA TIMING	Glutamate carboxy-peptidase	–	ND	ND	Mutant (with 2,4-D)	ND	ND	Mordhorst et al. 1998; Nogue et al. 2000; von Recklinghausen et al. 2000; Helliwell et al. 2001
PICKLE	CHD3-chromatin-remodeling factor	–	ND	–	Mutant (without hormone)	ND	GA (sensitivity up)	Ogas et al. 1997, 1999; Rider et al. 2003; Henderson et al. 2004
BBM (BABY BOOM)	Transcription factor (AP2/ERF)	ND	+	+	Over-expression (without hormone)	ND	ND	Boutilier et al. 2002

– the factor does not encode the transcription factor, *ZE* zygotic embryogenesis, *ABA* abscisic acid, *ABRE* ABA-response element, *GA* gibberellic acid, *ND* not determined, *2,4-D* 2,4-dichlorophenoxyacetic acid
[a]Target motif is the promoter *cis*-element that is related to the encoded transcription factors
[b]Somatic embryo formation is observed under each culture condition.

sis is used extensively as an experimental system to examine physiological, biochemical and morphological events during embryogenesis (Zimmerman 1993). Increasingly, somatic embryogenesis tissue-culture systems are used as model systems to examine the mechanisms regulating gene expression and other molecular events during zygotic embryogenesis.

Here, we first summarize what is known about gene expression in somatic embryos compared with gene expression in zygotic embryos. We then describe the relationship between gene expression and phytohormones in somatic embryogenesis. Lastly, we discuss some mutants that form somatic embryos. Table 1 shows characteristics of the genes described in this review.

2 Gene Expression During Somatic Embryogenesis

The genes expressed during somatic embryogenesis are identified using two different techniques. First, genes or proteins involved in somatic embryogenesis are isolated using comparisons of gene or protein expression patterns in embryonic and non-embryonic culture. Second, genes involved in zygotic embryogenesis, which are identified using defective-embryo mutants, are examined for expression during somatic embryogenesis.

2.1 Regulation of *LEA* Gene Expression

2.1.1 *LEA* Gene Expression in Embryonic Culture

Since the late 1980s, many researchers have attempted to isolate genes and proteins involved in somatic embryogenesis (Franz et al. 1989; Kiyosue et al. 1992, 1993a). In most cases, general differential screening methods have been used to identify genes and proteins. Most of the genes identified in these experiments encode late-embryogenesis abundant (LEA) proteins. LEA proteins are very hydrophilic and are expressed abundantly late in zygotic embryogenesis in many plant species. *LEA* gene expression in zygotic embryos is regulated by abscisic acid (ABA). *DC8, DC59, ECP31, ECP40* and *ECP63* isolated from carrot embryonic culture encode the LEA protein, and expression of this gene is found both in embryonic cultures and in immature seeds of carrot. Expression of these genes only occurs in embryonic tissues, and is not observed in vegetative tissue. Additionally, expression is induced by treatment with ABA in somatic and zygotic embryos (Zimmerman 1993). Similarly, the *Arabidopsis* homologs of carrot *ECP31* and *ECP63* (*AtECP31* and *AtECP63*) are expressed in somatic embryos and immature seeds, but not in vegetative tissue, and their expression is induced by ABA in somatic embryos (Yang et al. 1996, 1997; Ikeda-Iwai et al. 2002). Thus, the embryo-specific *LEA* genes that are expressed during zygotic embryogenesis are also expressed in somatic embryonic cultures. Moreover, *LEA* gene expression is induced by ABA in both types of embryos.

2.1.2
ABA and *ABI3* Control *LEA* Gene Expression

Expression of the *LEA* genes during zygotic embryogenesis is regulated by various factors. ABA is the most important phytohormone controlling *LEA* gene expression, and *ABA-INSENSITIVE3/VIVIPAROUS1* (*ABI3/VP1*) is one of the important transcriptional factors regulating *LEA* gene expression in zygotic embryos (Giraudat et al. 1992; Suzuki et al. 2003).*Arabidopsis abi3* and maize *vp1* are seed-specific ABA-insensitive mutants. Seeds of these mutants undergo viviparous germination; they fail to exhibit dormancy, desiccation tolerance, and accumulation of seed storage proteins. In these mutants, expression levels of some *LEA* genes are low; thus, *ABI3/VP1* may be an important factor in the control of *LEA* gene expression (Parcy et al. 1994). *ABI3/VP1* contains three conserved domains (B1, B2 and B3), of which B2 and B3 may be important for seed-specific ABA signal transduction. Analyses of the mechanisms regulating seed-specific ABA-inducible gene (*Em* and *Osem*) expression suggest that *ABI3/VP1* is involved in expression of these genes via the ABA-response element (ABRE) (Marcotte et al. 1989; Hattori et al. 1995). In this regulatory system, ABI3/VP1 cannot bind directly with ABRE. ABI3/VP1 may form complexes with bZIP proteins, which then bind with ABRE (Fig. 1; Gultinan et al. 1990; Nakagawa et al. 1996; Nantel and Quatrano 1996; Lopez-Molina et al. 2002; Lara et al. 2003).

Ikeda-Iwai et al. (2002, 2003) found *ABI3* gene expression in somatic embryos and embryonic cultures in *Arabidopsis*, while Shiota et al. (1998) reported expression of *C-ABI3* (carrot homolog of *ABI3*) during both zygotic

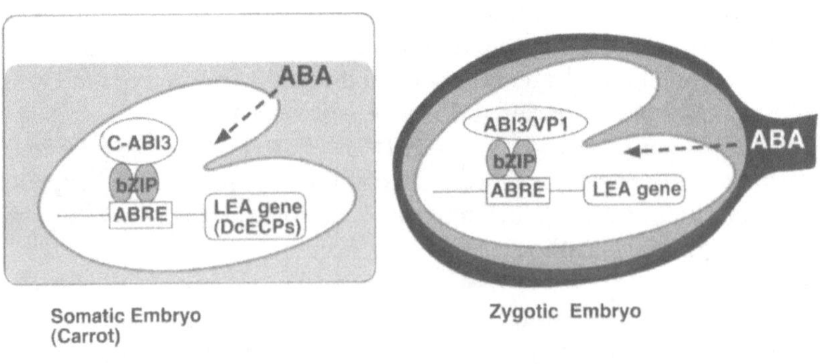

Fig. 1 The regulation of *LEA* gene expression by abscisic acid (*ABA*) and *ABA-INSENSITIVE3* (*ABI3*)/carrot homolog of *ABI3* (*C-ABI3*)

and somatic embryogenesis in carrot. Moreover, expression of *ECP31* and *ECP63* is induced in *C-ABI3*-overexpressed leaves treated with ABA (Shiota et al. 1998). This indicates that *C-ABI3* and ABA are involved in the regulation of *ECP31* and *ECP63* gene expression in carrot.

Promoter analyses show that ABRE is also important for regulation of *ECP31* and *ECP63* expression by *C-ABI3* and ABA during carrot somatic embryogenesis (Ko et al. 2001a, b). Indeed, C-ABI3 does not bind directly with ABRE (Ko and Shiota, unpublished data). Ko and Kamada (2002) isolated two bZIP proteins (clone 22 and DcBZ43) from the carrot embryonic cell library that bind to the *ECP31* promoter *cis*-element. It is possible that these bZIP proteins and C-ABI3 form a complex, and the complex combines with ABRE on the *ECP31* promoter, inducing expression of *ECP31* during somatic embryogenesis in carrot. The regulation system of *LEA* gene expression may be similar for zygotic and somatic embryogenesis (Fig. 1).

2.2 Expression of Transcriptional Factor Genes Isolated from Zygotic Defective-Embryo Mutants

2.2.1 *LEC1* Gene Expression

LEAFY COTYLEDON1 (*LEC1*) is a seed-specific transcriptional factor. Embryos of *lec1* mutants have abnormal morphology, with trichomes on the cotyledons, and fail to exhibit desiccation tolerance and accumulation of seed storage proteins (Vicient et al. 2000; Brocard-Gifford et al. 2003). Expression of the *LEC1* gene occurs in developing seeds, and the ectopic expression of *LEC1* in transgenic plants induces the formation of somatic embryo-like structures (Lotan et al. 1998). This suggests that *LEC1* has an important function in plant embryogenesis. The *LEC1* gene encodes a HAP3 subunit of the CCAAT binding transcription factor (Lotan et al. 1998; Lee et al. 2003) and may be involved in the gene expression control system related to the CCAAT promoter *cis*-element.

Expression of *LEC1* and *LEC1* homologs is observed during somatic embryogenesis in *Arabidopsis*, maize and carrot (Ikeda-Iwai et al. 2002; Zhang et al. 2002; Yazawa et al. 2003). In situ hybridization analysis revealed that the expression patterns of *ZmLEC1* and *C-LEC1* are similar in zygotic and somatic embryos (Zhang et al. 2002; Yazawa et al. 2003). This indicates that *LEC1* may also play an important role in somatic embryogenesis.

2.2.2 *FUS3* and *LEC2* Gene Expression

Embryos of the *fusca3* (*fus3*) mutant show increased accumulation of anthocyanin and decreased accumulation of seed storage proteins compared with the wild type (Luerssen et al. 1998). Introduction of the *AtML1::FUS3* gene into *Arabidopsis* induced expression of *FUS3* in the L1 layer of the shoot apical meristem (SAM), resulting in the production of cotyledon-like organs in the SAM of transgenic plant (Gazzarrini et al. 2004). Embryos of *lec2* mutants produce trichomes on the cotyledons and have abnormal suspensor morphology. Ectopic expression of the *LEAFY COTYLEDON2* (*LEC2*) gene induces the formation of somatic embryo-like and other organ-like structures, and often confers embryonic characteristics to the seedling (Stone et al. 2001). *FUS3* and *LEC2* genes encode the B3 domain-containing protein and that domain is conserved in *ABI3*-type transcription factors. FUS3 and LEC2 proteins bind directly with the RY motif and regulate expression of some embryonic genes (Kroj et al. 2003; Monke et al. 2004).

Although *FUS3* expression is known to occur in somatic embryos of *Arabidopsis* (Ikeda-Iwai et al. 2002, 2003), the actual functions of *FUS3* during somatic embryogenesis have not been elucidated. Expression of *LEC2* during somatic embryogenesis has not been examined.

2.2.3
Regulation of Gene Expression in Somatic Embryos

Analyses of *Arabidopsis* defective-embryo mutants (*lec1*, *fus3*, *lec2* and *abi3*) have shown that *LEC1*, *ABI3*, *FUS3* and *LEC2* regulate the expression of many genes during embryogenesis and seed germination (Fig. 2; Parcy et al. 1997; Wobus and Weber 1999; Ezcurra et al. 2000; Vicient et al. 2000; Nambara et al. 2000; Kroj et al. 2003; Monke et al. 2004; Tsuchiya et al. 2004). These regulation mechanisms are related to the production of phytohormones (e.g., ABA, GA, auxin, ethylene), carbohydrate metabolism, photoreactions, and cell di-

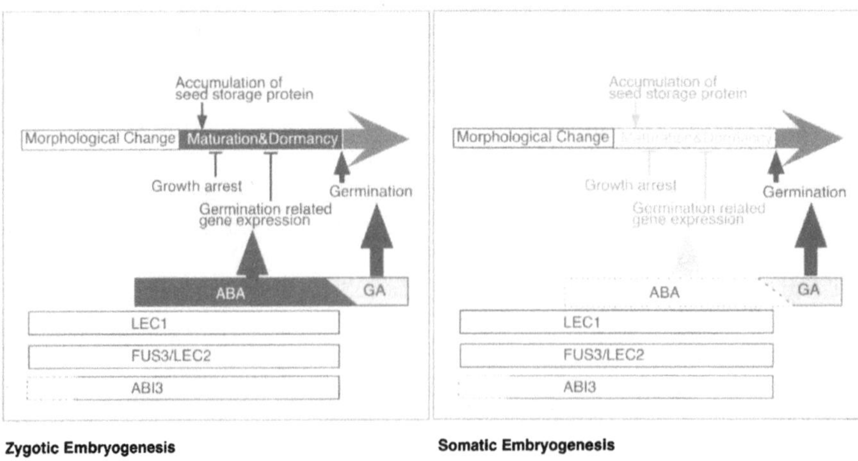

Fig. 2 Seed development regulated by embyogenesis-related factors

vision (Raz et al. 2001; Brocard-Gifford et al. 2003; Gazzarrini et al. 2004; Curaba et al. 2004). The regulatory systems for expression of each gene are specialized and complicated. For somatic embryogenesis, however, little is known regarding the regulatory mechanisms of gene expression. Expression of *LEC1, ABI3* and *FUS3* is similar in somatic and zygotic embryogenesis (Shiota et al. 1998; Ikeda-Iwai et al. 2002, 2003; Zhang et al. 2002; Yazawa et al. 2003); therefore, common regulatory mechanisms may function during both somatic and zygotic embryogenesis.

On the other hand, the mechanisms controlling the expression of *LEC1, ABI3, FUS3* and *LEC2* have been examined using *Arabidopsis* mutants and transgenic plants. But, the regulatory mechanisms are still unclear. However, a carrot embryonic tissue-culture system has been used to analyze the 5′ upstream region of *C-ABI3*, and carrot embryonic element 1 (CEE1) was identified as a promoter *cis*-element that regulates gene expression in carrot somatic embryos and in *Arabidopsis* zygotic embryos (Ikeda, unpublished data). CEE1-like elements are found on the promoter region of *Arabidopsis ABI3* and rice *OsVP1*, named AEE1-1 and OsEE1. AEE1-1 and OsEE1 can bind with CEE1-binding factors, which are found in the embryonic cell nucleus of carrot, and AEE1-1 regulates gene expression in *Arabidopsis* zygotic embryos from a very early stage of embryogenesis (Ikeda, unpublished data).

2.3 Factors Isolated from Embryonic Tissue

The screening of the new embryogenesis-related genes has been made easier by advances in molecular-genetic research technology. New factors have been isolated using differential display and microarray analysis methods (Thibaud-Nissen et al. 2003; Takahata et al. 2004). Future elucidation of the developmental system involved in somatic embryogenesis is expected through the isolation and examination of these new factors. Next, we describe three representative factors (*SOMATIC EMBRYOGENESIS RECEPTOR KINASE*; *SERK*, *AGAMOUS-like 15*; *AGL15* and *BABY BOOM*; *BBM*) isolated from plant embryonic tissues.

2.3.1 *SOMATIC EMBRYOGENESIS RECEPTOR KINASE*

SERK (*DcSERK*) was isolated from carrot embryonic tissue culture. Expression of *DcSERK* is found in somatic and zygotic embryos, but not in any other plant tissues. In addition, the expression is observed at a very early stage in the developing somatic embryo, ie., from the single-cell stage to the globular stage (Schmidt et al. 1997). This suggests that the *SERK* gene is suitable as a marker gene for embryonic-competent cells in somatic embryogenesis. *SERK* encodes a leucine-rich repeat containing receptor-like kinase proteins

(Schmidt et al. 1997). *DcSERK* homologous genes were isolated from *Arabidopsis* (*AtSERK1*), maize (*ZmSERK1, ZmSERK2*) and *Meicago truncatula* (*MtSERK1*) (Hecht et al. 2001; Shah et al. 2001; Baudino et al. 2001; Nolan et al. 2003). Although the expression of *SERK* homologs is detected during somatic embryogenesis in *Arabidopsis* (*pt1*mutant), maize, *M. truncatula*, sunflower and Poaceae, most of the homologous genes expression are not embryo-specific (Somleva et al. 2000; Hecht et al. 2001; Shah et al. 2001; Baudino et al. 2001; Nolan et al. 2003; Thomas et al. 2004). Ectopic expression of the *AtSERK1* gene under the control of the CaMV35S promoter did not result in an altered plant phenotype. However, when *AtSERK1* overexpressed seedlings are germinated in medium containing 2,4-dichlorophenoxyacetic acid (2,4-D), the embryonic structure is formed at 3–4 times the rate in the wild type (Hecht et al. 2001). Thus, *SREK* may be involved in the early stages of plant somatic embryogenesis, but its actual function is still unknown.

2.3.2
AGAMOUS-Like 15

AGL15 was isolated as a MADS box gene expressed in tissues of *Arabidopsis* and *Brassica napus* derived by double fertilization (i.e., zygotic embryo, endosperm and suspensor; Heck et al. 1995). Although expression of *AGL15* is observed in the vegetative tissue, the expression is especially strong in embryo-related tissues (Heck et al. 1995; Fernandez et al. 2000). The AGL15 protein is detected in apomictic embryogenesis in dandelion, microspore embryogenesis in *B. napus*, and somatic embryogenesis in alfalfa. Thus, the AGL15 protein is widely found in various embryonic tissues of various plant species (Perry et al. 1999). Ectopic expression of the full-length *AGL15* under the control of the CaMV35S promoter promotes somatic embryo formation from SAMs of germinated seedlings in culture, at low frequency (Harding et al. 2003). *AGL15* encodes the MADS box family transcription factor and appears to control the expression of many genes during somatic embryogenesis. One of the genes regulated by *AGL15* encodes *AtGA2ox6*(Sauer and Friml, this volume). The results of the promoter analysis of *AGL15* indicate that the expression of *AGL15* is regulated by 2,4-D and *AGL15* itself (Zhu and Perry 2005).

2.3.3
BABY BOOM

The *BBM* gene was isolated from the somatic embryo-inducible condition in the pollen-derived somatic embryogenesis tissue-culture system of *B. napus*. This gene encodes an *AP2/ERF* family transcriptional factor. Its expression is observed during pollen-derived somatic and zygotic embryogenesis. Ectopic expression of *BBM* or *Arabidopsis BBM* (*AtBBM*) in transgenic plants induces the formation of somatic embryo-like structures from the edges of cotyledons and leaves. However, ectopic *BBM* expression induces additional pleiotropic phenotypes, including neoplastic growth, phytohormone-free regeneration of explants, and abnormal leaf and flower morphology. Thus, *BBM* is thought to promote cell proliferation and morphogenesis during embryogenesis (Boutilier et al. 2002). In addition to the functional analysis of the *BBM* gene, the chromatin structure in somatic embryogenesis was examined using *35S:BBM*. Expression of the HD2-type histone deacetylases (*HD2A* and *HD2B*) occurs in the somatic embryo-like structures of *BBM* (Zhou et al. 2004).

3
Gene Expression and Phytohormones (ABA and GA)

ABA and GA are important phytohormones that regulate seed dormancy and germination. In this section, we describe the relationship between these two phytohormones and the expression of genes related to somatic embryogenesis.

3.1
ABA Regulates the Acquisition of Desiccation Tolerance and Dormancy

ABA is synthesized at a late stage of embryogenesis and controls the acquisition of desiccation tolerance and seed dormancy. ABA controls the expression of many genes that are expressed during the late stage of embryogenesis (including *LEA*). And *ABI3, LEC1, FUS3* and *LEC2* are related to ABA signaling in embryogenesis.

In zygotic embryogenesis, embryo development is arrested during seed dormancy (Fig. 2). In contrast, arrested development and seed dormancy are not observed in somatic embryogenesis; rather, the somatic embryos germinate directly. This may be caused by a deficiency in ABA synthesis during somatic embryogenesis. In somatic embryos of carrot and *Arabidopsis*, the expression of some ABA-inducible genes (e.g., *ECP31* and *ECP63*) is low, but is increased by treatment with exogenous ABA (Kiyosue et al. 1993b; Ikeda-Iwai et al. 2002). The data suggest that there are insufficient quantities of ABA to induce expression of some *LEA* genes in embryonic cultures of carrot and *Arabidopsis*. ABA may be supplied from the mother plant or other tissues during zygotic embryogenesis; this tissue does not exist in the somatic embryogenesis tissue-culture system (Fig. 1). Moreover, carrot somatic embryos treated with exogenous ABA exhibit desiccation tolerance and dormancy. Somatic embryos that were desiccated after treatment with ABA can survive for several years at – 25 °C, and can germinate when returned to culture medium at room temperature (Shiota et al. 1999). Thus, somatic embryos can acquire seedlike desiccation tolerance via treatment with ABA. Therefore, the responses of somatic and zygotic embryos to ABA may be similar.

3.2 GA Regulates the Transition from Embryogenesis to Germination

GA is the phytohormone antagonistic to ABA. Quantities of GA increase during germination, and GA regulates the transition from embryogenesis to germination.

3.2.1 GA Response and *PKL*

The *pickle* (*pkl*) mutant forms a somatic embryo-like structure from the root of the seedling, and the major seed storage proteins accumulate in the pickle root (Ogas et al. 1997; Rider et al. 2004). These characteristics indicate that the *pkl* mutant cells may have failed in the transition from embryo to seedling. The *PKL* gene encodes CHD3, a type of chromatin-remodeling factor. Treatment with uniconazole (a GA-synthesis inhibitor) increases the frequency of pickle root formation in *pkl*, suggesting that *PKL* functions in GA synthesis or signaling (Ogas et al. 1997). In *pkl* mutants, reactivity to GA is decreased, and the quantity of bioactive GA is increased compared with that in the wild-type. Therefore, *PKL* is involved in the GA response during germination (Henderson et al. 2004).

Expression of the *LEC1*, *LEC2* and *FUS3* genes occurs in the somatic embryo-like structure of the *pkl* root. This indicates that the *PKL* gene may be involved in the regulation of these genes via chromatin remodeling. (Ogas et al. 1997; Rider et al. 2003). In contrast, expression of *ABI3* and *WUSCHEL* (*WUS*) is not affected by the mutation. A new factor for which expression was increased in *pkl* was isolated and named *AtWLIM2*. *AtWLIM2* encodes a transcriptional factor with an LIM domain and is strongly expressed in *Arabidopsis* siliques. Thus, *AtWLIM2* may regulate some gene expression during plant embryogenesis, and the *PKL*-related chromatin-remodeling system may regulate expression of *AtWLIM2* (Rider et al. 2003).

3.2.2 Relationships between GA Synthesis and *FUS3*, *LEC2* and *AGL15*

In *lec2* and *fus3* mutants, the quantity of endogenous bioactive GA is increased. This may be caused by increased expression of the *AtGA3ox2* gene (encodes the key enzyme for bioactive GA synthesis; Curaba et al. 2004). In addition, ectopic expression of *FUS3* represses expression of *AtGA3ox2* (Gazzarrini et al. 2004). *AtGA3ox2* promoter analysis indicates that the expression of *AtGA3ox2* was directly regulated by binding of LEC2 and FUS3 proteins with the RY motif on the promoter of *AtGA3ox2* (Curaba et al. 2004). *FUS3* and *LEC2* negatively regulate bioactive GA synthesis.

AGL15 directly binds to the promoter region of *AtGA2ox6*, and positively regulates *AtGA2ox6* expression. *AtGA2ox6* encodes the enzyme that converts bioactive GA into inactive GA. Although *AGL15* overexpression transformants form somatic embryo-like structures when germinated in medium containing 2,4-D, the frequency of somatic embryo formation is increased in the *AGL15*, *AtGA2ox6* double-overexpresser. In *35S:AGL15*, *atga2ox6*, the somatic embryo formation rate is decreased (Wang et al. 2004). Therefore, conversion of bioactive GA into inactive GA is enhanced by *AGL15*. In addition, the quantity of bioactive GA is strongly related to somatic embryo formation in *Arabidopsis*.

4 Arabidopsis Mutants that Form Somatic Embryos

4.1 *WUSCHEL* and *CLAVATA*

Somatic embryos form on the *WUS* overexpression mutant under phytohormone-free conditions. *WUS* functions to maintain stem cells in the shoot meristems and works in cooperation with *CLAVATA* (*CLV*), which controls cell differentiation in shoot meristems (Zuo et al. 2002). In addition, *clv* (*clv1* and *clv3*) mutants form somatic embryo-like structures at low frequencies when germinated in liquid medium containing 2,4-D (Mordhorst et al. 1998). Although *CLV* and *WUS* function in somatic embryo formation, they may not be involved in the acquisition of embryonic competence. However, *WUS* and *CLV* may regulate cell differentiation in the SAM. In the SAM of *clv* mutants, cell populations are high; these additional non-committed cells may form somatic embryos. Although *WUS* suppresses the expression of *LEC1* during somatic embryogenesis, *WUS* may not directly control expression of *LEC1*; *LEC1* expression changes as somatic embryo development progresses (Zuo et al. 2002).

4.2 *PRIMORDIA TIMING*

The *primordia timing* (*pt*) mutant (*hpt*, *cop2* and *amp1*) forms somatic embryo-like structures when it is germinated in liquid culture medium containing 2,4-D (Mordhorst et al. 1998; von Recklinghausen et al. 2000). In zygotic embryos of *pt*, SAM cell populations are increased, and the number of cotyledons is often altered. *pt clv* double mutants (both mutants possess enlarged SAMs) show additive effects on the size of the SAM and an even higher frequency of seedling-producing embryonic cell lines. This indicates that *PT* controls SAM size, and increased populations of non-committed SAM cells may facilitate somatic embryogenesis (Mordhorst et al. 1998). On the other hands, the quantity of cytokinin is increased in *pt* (*amp1*) mutants (Nogue et al. 2000), and cytokinin-induced gene expression is observed in these mutants. *AMP1* encodes the glutamate carboxypeptidase-like gene (Helliwell et al. 2001).

5 Conclusions

Somatic embryogenesis begins with dedifferentiation and redifferentiation of somatic cells, whereas zygotic embryogenesis begins with double fertilization. Though the starting points of these two types of embryogenesis differ, the molecular events that occur during somatic and zygotic embryogenesis are similar from a very early stage of embryo development. Four important transcription factors (*LEC1*, *LEC2*, *FUS3* and *ABI3*) express and regulate both types of embryo development, and the regulatory mechanisms of gene expression by these transcription factors may be similar.

Now, many questions regarding the regulatory mechanisms of embryo development still remain. Somatic embryogenesis is one of the best model systems with which to examine the details of plant embryogenesis. We expect that new findings, such as the identification of new factors controlling embryogenesis-related gene expression will be made in the future, as a result of the careful examination of embryo-defective mutants in combination with somatic embryogenesis.

References

Baudino S, Hansen S, Brettschneider R, Hecht VEG, Dresselhaus T, Lorz H, Dumas C, Rogowsky PM (2001) Molecular characterization of two novel maize LRR receptor-like kinases, which belong to the *SERK* gene family. Planta 213:1–10

Boutilier K, Offringa R, Sharma VK, Kieft H, Ouellet T, Zhang L, Hattori J, Liu C-M, van Lammeren AAM, Miki BLA, Custers JBM, van Lookeren Campage MM (2002) Ectopic expression of BABY BOOM triggers a conversion from vegetative to embryonic growth. Plant Cell 14:1737–1749

Brady SM, Sarkar SF, Bonetta D, McCourt P (2003) The *ABSCISIC ACID INSENSITIVE3* (*ABI3*) gene is modulated by farnesylation and is involved in auxin signaling and lateral root development in *Arabidopsis*. Plant J 34:67–75

Brocard-Gifford IM, Lynch TJ, Finkelstein RR (2003) Regulatory networks in seeds integrating developmental, abscisic acid, sugar, and light signaling. Plant Physiol 131:78–92

Curaba J, Moritz T, Blervaque R, Parcy F, Raz V, Herzog M, Vachon G (2004) *AtGA3ox2*, a key gene responsible for bioactive gibberellin biosynthesis, is regulated during embryogenesis by *LEAFY COTYLEDON2* and *FUSCA3* in *Arabidopsis*. Plant Physiol 136:3660–3669

Ezcurra I, Wycliffe P, Nehlin L, Ellerstrom M, Rask L (2000) Transactivation of the *Brassica napus* napin promoter by ABI3 requires interaction of the conserved B2 and B3 domains of ABI3 with different *cis*-elements: B2 mediates activation through an ABRE, whereas B3 interacts with an RY/G-box. Plant J 24:57–66

Fernandez DE, Heck GR, Perry SE, Patterson SE, Bleecker AB, Fang S-C (2000) The embryo MADS domain factor AGL15 acts post-embryonically: inhibition of perianth senescence and abscission via constitutive expression. Plant Cell 12:183–197

Franz G, Hatzopoulos P, Jones TJ, Krauss M, Sung ZR (1989) Molecular and genetic analysis of an embryonic gene, DC8, from *Daucus carota* L. Mol Gen Genet 218:143–151

Gazzarrini S, Tsuchiya Y, Lumba S, Okamoto M, McCourt P (2004) The transcription factor *FUSCA3* controls developmental timing in *Arabidopsis* through the hormones gibberellin and abscisic acid. Dev Cell 7:373–385

Giraudat J, Hauge BM, Valon C, Smalle J, Parcy F, Goodman HM (1992) Isolation of the *Arabidopsis ABI3* gene by positional cloning. Plant Cell 4:1251–1261

Guiltinan MJ, Marcotte WR, Quatrano RS (1990) A plant leucine zipper protein that recognizes an abscisic acid response element. Science 12:267–271

Harding EW, Tang W, Nichols KW, Fernandez DE, Perry SE (2003) Expression and maintenance of embryogenic potential is enhanced through constitutive expression of *AGAMOUS-like 15*. Plant Physiol 133:653–663

Hattori T, Terada T, Hamasuna S (1995) Regulation of the *Osem* gene by abscisic acid and the transcriptional activator VP1: analysis of *cis*-acting promoter elements required for regulation by abscisic acid and VP1. Plant J 7:913–925

Hecht V, Vielle-Calzada J-P, Hartog MV, Schmidt EDL, Boutilier K, Grossniklaus U, de Vries SC (2001) The *Arabidopsis SOMATIC EMBRYOGENESIS RECEPTOR KINASE1* gene is expressed in developing ovules and embryos and enhances embryogenic competence in culture. Plant Physiol 127:803–816

Heck GR, Perry SE, Nichols KW, Fernandez DE (1995) AGL15, a MADS domain protein expressed in developing embryos. Plant Cell 7:1271–1282

Helliwell CA, Chin-Atkins AN, Wilson IW, Chapple R, Dennis ES, Chaudhury A (2001) The *Arabidopsis AMP1* gene encodes a putative glutamate carboxypeptidase. Plant Cell 13:2115–2125

Henderson JT, Li H-C, Rider SD, Mordhorst AP, Romero-Severson J, Cheng J-C, Robey J, Sung ZR, de Vries SC (2004) *PICKLE* acts throughout the plant to repress expression of embryonic traits and may play a role in gibberellin-dependent responses. Plant Physiol 134:995–1005

Ikeda-Iwai M, Satoh S, Kamada H (2002) Establishment of a reproducible tissue-culture system for the induction of *Arabidopsis* somatic embryos. J Exp Bot 53:1575–1580

Ikeda-Iwai M, Umehara M, Satoh S, Kamada H (2003) Stress-induced somatic embryogenesis in vegetative tissues of *Arabidopsis thaliana*. Plant J 34:107–114

Kamada H (1980) Control of in vitro asexual embryogenesis in higher plants. Chem Reg Plants 15:62–78

Kamada H (1996) Physiological and molecular biological studies on somatic embryogenesis. Chem Reg Plants 31:1–11

Kiyosue T, Yamaguti-Shinozaki K, Shinozaki K, Higashi K, Satoh S, Kamada H, Harada H (1992) Isolation and characterization of a cDNA that encodes ECP31, an embryogenic-cell protein from carrot. Plant Mol Biol 19:239–249

Kiyosue T, Yamaguti-Shinozaki K, Shinozaki K, Kamada H, Harada H (1993a) cDNA cloning of ECP40, an embryogenic-cell protein in carrot, and its expression during somatic and zygotic embryogenesis. Plant Mol Biol 21:1053–1068

Kiyosue T, Satoh S, Kamada H, Harada H (1993b) Somatic embryogenesis in higher plants. J Plant Res 3:75–82

Ko S, Kamada H (2002) Isolation of carrot basic leucine zipper transcription factor using yeast one-hybrid screening. Plant Mol Biol Rep 20:1–8

Ko S, Thitamadee S, Yang H, Eun C-H, Sage-ono K, Higashi K, Satoh S, Kamada H (2001a) Comparison and characterization of *cis*-regulatory region in some embryo-specific and ABA-responsive carrot genes, *DcECPs*. Plant Biotech 18:45–54

Ko S, Eun C-H, Satoh S, Kamada H (2001b) Analysis of *cis*-regulatory elements in carrot embryo-specific and ABA-responsive gene, *DcECP31*. Plant Biotech 18:55–60

Kroj T, Savino G, Valon C, Giraudat J, Parcy F (2003) Regulation of storage protein gene expression in *Arabidopsis*. Development 130:6065–6073

Lara P, Onate-Sanchez L, Abraham Z, Ferrandiz C, Diaz I, Carbonero P, Vicente-Carbajosa J (2003) Synergistic activation of seed storage protein gene expression in *Arabidopsis* by ABI3 and two bZIPs related to OPAQUE2. J Biol Chem 278:21003–21011

Lee H, Fisher RL, Goldberg RB, Harada JJ (2003) *Arabidopsis* LEAFY COTYLEDON1 represents a functionally specialized subunit of the CCAAT binding transcription factor. Proc Natl Acad Sci USA 18:2152–2156

Lopez-Molina L, Mongrand S, McLachlin DT, Chait BT, Chua N-H (2002) ABI5 acts downstream of ABI3 to execute an ABA-dependent growth arrest during germination. Plant J 32:317–328

Lotan T, Ohto M, Yee KM, West MAL, Lo R, Kwong RW, Yamagishi K, Fisher RL, Goldberg RB, Harada JJ (1998) *Arabidopsis* LEAFY COTYLEDON1 is sufficient to induce embryo development in vegetative cells. Cell 93:1195–1205

Luerssen H, Kirik V, Herrmann P, Misera S (1998) *FUSCA3* encodes a protein with a conserved VP1/ABI3-like B3 domain which is of functional importance for the regulation of seed maturation in *Arabidopsis thaliana*. Plant J 15:755–764

Marcotte WR, Russell SH, Quatrano RS (1989) Abscisic acid-responsive sequences from the Em gene of wheat. Plant Cell 1:969–976

Mönke G, Altschmied L, Tewes A, Reidt W, Mock H-P, Bäumlein H, Conrad U (2004) Seed-specific transcription factors ABI3 and FUS3: molecular interaction with DNA. Planta 219:158–166

Mordhorst AP, Voerman KJ, Hartog MV, Meijer EA, van Went J, Koornneef M, de Vries SC (1998) Somatic embryogenesis in *Arabidopsis thaliana* is facilitated by mutations in genes repressing meristematic cell divisions. Genetics 149:549–563

Nakagawa H, Ohmiya K, Hattori T (1996) A rice bZIP protein, designated OSBZ8, is rapidly induced by abscisic acid. Plant J 9:217–227

Nambara E, Hayama R, Tsuchiya Y, Nishimura M, Kawaide H, Kamiya Y, Naito S (2000) The role of *ABI3* and *FUS3* loci in *Arabidopsis thaliana* on phase transition from late embryo development to germination. Dev Biol 220:412–423

Nantel A, Quatrano R (1996) Characterization of three rice basic/leucine zipper factors, including two inhibitors of EmBP-1 DNA binding activity. J Biol Chem 271:31296–31305

Nogue N, Hocart H, Letham DS, Dennis ES, Chaudhury AM (2000) Cytokinin synthesis is higher in the *Arabidopsis amp1* mutant. Plant Growth Reg 32:267–273

Nolan KE, Irwanto RR, Rose RJ (2003) Auxin up-regulates *MtSERK1* expression in both *Medicago truncatula* root-forming and embryogenic cultures. Plant Physiol 133:218–230

Ogas J, Cheng J-C, Sung ZR, Somerville C (1997) Cellular differentiation regulated by gibberellin in the *Arabidopsis thaliana pickle* mutant. Science 277:91–94

Ogas J, Kaufmann S, Henderson J, Somerville C (1999) PICKLE is a CHD3 chromatin-remodeling factor that regulates the transition from embryonic to vegetative development in *Arabidopsis*. Proc Natl Acad Sci USA 23:13839–13844

Parcy F, Valon C, Raynal M, Gaubier-Comella P, Delseny M, Giraudat J (1994) Regulation of gene expression programs during *Arabidopsis* seed development: Roles of the *ABI3* locus and of endogenous abscisic acid. Plant Cell 6:1567–1582

Parcy F, Valon C, Kohara A, Misera S, Giraudat J (1997) The *ABSCISIC ACID-INSENSITIVE3, FUSCA3,* and *LEAFY COTYLEDON1* loci act in concert to control multiple aspects of *Arabidopsis* seed development. Plant Cell 9:1265–1277

Perry SE, Lehti MD, Fernandez DE (1999) The MADS-domain protein AGAMOUS-like 15 accumulates in embryonic tissues with diverse origins. Plant Physiol 120:121–129

Raz V, Bergervost JHW, Koornneef M (2001) Sequential steps for development arrest in *Arabidopsis* seeds. Development 128:243–252

Reinert J (1958) Untersuchungen über die Morphogenese an Gewebekulturen. Ber Dtsch Bot Ges 71:15

Rider SD, Henderson JT, Jerome RE, Edenberg HJ, Romero-Severson J, Ogas J (2003) Coordinated repression of regulators of embryonic identity by *PICKLE* during germination in *Arabidopsis*. Plant J 35:33–43

Rider SD, Hemm MR, Hostetler HA, Li H-C, Chapple C, Ogas J (2004) Metabolic profiling of the *Arabidopsis pkl* mutant reveals selective derepression of embryonic traits. Planta 219:489–499

Schmidt EDL, Guzzo F, Toonen MA, de Vries SC (1997) A leucine-rich repeat containing receptor-like kinase marks somatic plant cells competent to form embryos. Development 124:2049–2062

Shah K, Gadella TWJ, van Erp H, Hecht V, de Vries SC (2001) Subcellular localization and oligomerization of the *Arabidopsis thaliana* somatic embryogenesis receptor kinase 1 protein. J Mol Biol 309:641–655

Shiota H, Satoh R, Watabe K, Harada H, Kamada H (1998) *C-ABI3*, the carrot homologue of the *Arabidopsis ABI3*, is expressed during both zygotic and somatic embryogenesis and functions in the regulation of embryo-specific ABA-inducible genes. Plant Cell Physiol 39:1184–1193

Shiota H, Tachikawa K, Watabe K, Kamada H (1999) Successful long-term preservation of abscisic-acid-treated and desiccated carrot somatic embryos. Plant Cell Rep 18:749–753

Somleva MN, Schmidt EDL, de Vries SC (2000) Embryogenic cells in *Dactylis glomerata* L. (Poaceae) explants identified by cell tracking and by SERK expression. Plant Cell Rep 19:718–726

Steward FC, Mapes MO, Mears K (1958) Growth and organized development of cultured cells. II. Organization in cultures grown from freely suspended cells. Am J Bot 45:705–708

Stone SL, Kwong LW, Yee KM, Pelletier J, Lepiniec L, Fischer RL, Goldberg RB, Harada JJ (2001) *LEAFY COTYLEDON2* encodes a B3 domain transcription factor that induces embryo development. Proc Natl Acad Sci USA 98:11806–11811

Suzuki M, Kao C-Y, Cocciolone S, McCarty DR (2001) Maize VP1 complements *Arabidopsis abi3* and confers a novel ABA/auxin interaction in roots. Plant J 28:409–418

Suzuki M, Ketterling MG, Li Q-B, McCarty DR (2003) Viviparus1 alters global gene expression patterns through regulation of abscisic acid signaling. Plant Physiol 132:1664–1677

Takahata K, Takeuchi M, Fujita M, Azuma J, Kamada H, Sato F (2004) Isolation of putative glycoprotein gene from early somatic embryo of carrot and its possible involvement in somatic embryo development. Plant Cell Physiol 45:1658–1668

Thibaud-Nissen F, Shealy RT, Khanna A, Vodkin LO (2003) Clustering of microarray data reveals transcript patterns associated with somatic embryogenesis in soybean. Plant Physiol 132:118–136

Thomas C, Meyer D, Himber C, Steinmetz A (2004) Spatial expression of a sunflower *SERK* gene during induction of somatic embryogenesis and shoot organogenesis. Plant Physiol Biochem 42:35–42

Tsuchiya Y, Nambara E, Naito S, McCourt P (2004) The *FUS3* transcription factor functions through the epidermal regulator *TTG1* during embryogenesis in *Arabidopsis*. Plant J 37:73–81

Wang H, Caruso LV, Downie AB, Perry SE (2004) The embryo MADS domain protein AGAMOUS-like 15 directly regulates expression of a gene encoding an enzyme involved in gibberellin metabolism. Plant Cell 16:1206–1219

Wobus U, Weber H (1999) Seed maturation: genetic programmes and control signals. Curr. Opin. Plant Biol 2:33–38

Vicient CM, Bies-Etheve N, Delseny M (2000) Changes in gene expression in the *leafy cotyledon1* (*lec1*) and *fusca3* (*fus3*) mutants of *Arabidopsis thaliana* L. J Exp Bot 51:995–1003

von Recklinghausen IR, Iwanowska A, Kieft H, Mordhorst AP, Schel JHN, van Lammeren AAM (2000) Structure and development of somatic embryos formed in *Arabidopsis thaliana pt* mutant callus cultures derived from seedlings. Protoplasma 211:217–224

Yang H, Saitou T, Komeda Y, Harada H, Kamada H (1996) Late embryogenesis abundant protein in *Arabidopsis thaliana* homologous to carrot ECP31. Physiol Plant 98:661–666

Yang H, Saitou T, Komeda Y, Harada H, Kamada H (1997) *Arabidopsis thaliana* ECP63 encoding a LEA protein is located in chromosome 4. Gene 184:83–88

Yazawa K, Takahata K, Kamada H (2003) Isolation of the gene that encodes carrot leafy cotyledon 1 and expression analysis during somatic and zygotic embryogenesis. Plant Physiol Biochem 42:215–223

Zimmerman JL (1993) Somatic embryogenesis: A model for early development in higher plants. Plant Cell 5:1411–1423

Zhang S, Wong L, Meng L, Lemaux PG (2002) Similarity of expression patterns of *knotted1* and *ZmLEC1* during somatic and zygotic embryogenesis in maize (*Zea mays* L.). Planta 215:191–194

Zhou C, Labbe H, Sridha S, Wang L, Tian L, Latoszek-Green M, Yang Z, Brown D, Miki B, Wu K (2004) Expression and function of HD2-type histone deacetylases in *Arabidopsis* development. Plant J 38:715–724

Zhu C, Perry SE (2005) Control of expression and autoregulation of *AGL15*, a member of the MADS-box family. Plant J 41:583–594

Zuo J, Niu Q-W, Frugis G, Chua N-H (2002) The *WUSCHEL* gene promotes vegetative-to-embryonic transition in *Arabidopsis*. Plant J 30:349–359

Plant Cell Monogr (2)
A. Mujib · J. Šamaj: Somatic Embryogenesis
DOI 10.1007/7089_039/Published online: 20 October 2005

Genome-Wide Expression Analysis of Genes Involved in Somatic Embryogenesis

W. Tang (✉) · R. J. Newton

Department of Biology, East Carolina University, Howell Science Complex, Greenville, NC 27858-4353, USA
tangw@mail.ecu.edu

Abstract Genome-wide expression analysis is an important tool for identifying and analysing genes involved in various biological processes, including cell division, growth and development, signal transduction, transcript regulation, and responses to environmental cues. In this review, we discuss and compare the merits and limitations of the different genome-wide expression analysis technologies, including (1) complementary DNA (cDNA) microarrays, (2) oligonucleotide microarrays, (3) serial analysis of gene expression, (4) massively parallel signature sequencing, and (5) cDNA-amplified fragment-length polymorphism. Particular attention will be given to the genome-wide expression analysis of genes involved in somatic embryogenesis.

1 Introduction

Genome-wide expression analysis is an important tool for analysing genes involved in cellular, molecular, and developmental biological processes in microorganisms, plants, and animals (Hegde et al. 2000; Schena et al. 1995). Somatic embryogenesis is an asexual form of plant propagation in nature that mimics many of the events of sexual reproduction. The control of somatic embryo development involves the temporal expression of different sets of genes that allow the dividing cell to progress through the different stages of somatic embryogenesis. DNA microarrays provide a convenient tool for genome-wide expression analysis (Hegde et al. 2000; Schena et al. 1995). Studies using DNA microarrays to follow the patterns of genes allowed the identification of thousands or hundreds of genes that are involved in specific developmental processes. Although DNA microarrays are rapidly becoming the standard tool for genome-wide expression analysis, their application is still limited to a restricted number of experimental systems where the complete genome sequence or a large complementary DNA (cDNA) collection is available (Breyne and Zabeau 2001; Hegde et al. 2000; Schena et al. 1995). Several alternative technologies for expression profiling based on DNA sequencing or cDNA fragment analysis have been developed and successfully used in other biological systems, including plant species. DNA fragment analysis based methods, such as cDNA-amplified fragment-length polymorphism (AFLP), provide

Table 1 Comparison of methods used for genome-wide gene expression analysis

	cDNA microarray	Oligonucleotide microarray	SAGE	MPSS	cDNA-AFLP
Sensitivity	Moderate	Moderate	Moderate/ high	Moderate/ high	High
Specificity	Low	Low	High	High	High
Expression-level measurement	Relative	Relative	Absolute	Absolute	Relative
Possibility to integrate data	Yes	Yes	Yes	Yes	No
Necessity of molecular resources	Yes	Yes	Yes	Yes	No
Labour intensity	Low	Low	High	High	High
Cost	High	High	High	High	Low

a more appropriate tool for genome-wide expression analysis. Moreover, cDNA-AFLP exhibits properties that complement DNA microarrays and can be a useful tool for gene discovery (Breyne and Zabeau 2001). In this study, we overview the different genome-wide expression analysis technologies, including (1) cDNA microarrays, (2) oligonucleotide microarrays, (3) serial analysis of gene expression, (4) massively parallel signature sequencing (MPSS), and (5) cDNA-AFLP (Table 1). Particular attention will be given to the genome-wide expression analysis of genes involved in somatic embryogenesis.

2 Somatic Embryogenesis

Somatic embryogenesis is an important prerequisite for the use of many biotechnological tools for genetic improvement, as well as for clonal propagation (Schenk and Hildebrandt 1972; Yeung and Meinke 1993). Somatic embryogenesis may be induced by the manipulation of tissues and cells in vitro. Some of the most important factors for a successful plant regeneration are the culture medium and environmental incubation conditions. In angiosperms, the zygote divides transversally into two cells. The apical cell is small and dense with an intense activity of DNA synthesis (Yeung and Meinke 1993). This cell gives rise to the embryo head that will be the new plant. The basal cell is a large and highly vacuolated one that will form the suspensor complex, which plays an important role during the early stages of the young embryo (Yeung and Sussex 1979). Somatic embryos generally follow the same pattern and are initiated from a somatic cell. Somatic embryos are formed from single cells cultivated in liquid or solid medium. Embryos can be distinguished from adventitious shoots, because they are bipolar, having both a shoot and root pole, and they do not have any vascular connections with the underlying parental tissue (Haccius 1978). Somatic embryo production is steadily being increased as essential factors become better understood (Williams and Maheswaran 1986). The ability to recover plants from single cells has made possible the genetic improvement. The most important advantages of somatic embryogenesis used in plant biology, including the ability to handle large numbers of individual cells in very small spaces and genetic variability, can be created deliberately in cultured cells by using genetic-engineering techniques (Yeung and Meinke 1993).

3
Late Embryogenesis Abundant Proteins

Late embryogenesis abundant (LEA) proteins are developmentally induced during the different stages of embryogenesis and are environmentally induced in embryos by desiccation or culture with abscisic acid (ABA) or high osmoticum (Hughes and Galau 1991). LEA proteins comprise a large group of probable desiccation protectants that are induced by similar stresses in vegetative tissues of different plant species (Skriver and Mundy 1991). In cotton (*Gossypium hirsutum*), 18 *Lea* and *LeaA* messenger RNAs (mRNAs) were cloned and identified to be environmentally induced by water stress; two of them, *Lea5* (cDNA D73) and *Leal4* (cDNA D95) are highly induced in mature leaves of water-stressed plants (Galau et al. 1986). In *Craterostigma plantagineum*, the desiccation-induced cDNA pcC27-45 were identified to encode proteins that are very hydrophilic (Baker et al. 1988; Piatkowski et al. 1990). *Lea* genes encode proteins with significant hydropathic character. Their hydropathic profiles are unremarkable; the amino-terminal half is somewhat hydropathic, possibly with a membrane-spanning region, and the carboxy-terminal half is somewhat hydrophilic (Galau et al. 1993). The proteins encoded by cotton *Leal4* and *Craterostigma* pcC27-45 thus define an additional family of water-stress-related proteins (Baker et al. 1988), the group 4 LEA proteins. An ACGT-containing element has been shown to be involved in the ABA induction of a wheat *Lea* gene (Guiltinan et al. 1990). *Leal4-A* contains sequences at nucleotides – 58 and – 14 from the transcription start that are similar to this element and similar sequences that are in many cotton Leu genes (Galau et al. 1992). *LeaZ4-A* encodes a 16.4-kD protein that is exactly collinear, with 66% identity, with that encoded by the *Craterostigma* cDNA pcC27-45, which is induced in leaves and roots during desiccation and in ABA-treated and NaCl-treated callus (Piatkowski et al. 1990). These proteins are slightly hydropathic throughout.

4
cDNA Microarray

Microarray expression analysis has become one of the most widely used functional genomics tools (Schaffer et al. 2000). Efficient application of this technique requires the development of robust and reproducible protocols, including PCR amplification of target cDNA clones, microarray printing, probe labelling, and hybridization cDNA microarrays (Hegde et al. 2000; Schena et al. 1995). cDNA microarrays have been developed that allow mRNA expression to be assessed on a global scale, allowing the parallel assessment of gene expression for hundreds or thousands of genes in a single experiment (Baldwin et al. 1999). The commonest use of these is for the determination of patterns of differential gene expression, comparing differences in mRNA expression levels between identical cells subjected to different stimuli or between different cellular phenotypes or developmental stages (Laub et al. 2000).

Microarray expression analysis is the most widely used method for profiling mRNA expression (Laub et al. 2000). cDNA segments representing the collection of genes are amplified by PCR and mechanically spotted at high density on glass microscope slides using robotic systems, creating a microarray containing thousands of elements (Hegde et al. 2000). Microarrays containing tens of thousands of cDNA clones can be easily constructed. The kinetics of hybridization allows relative expression levels to be determined based on the ratio with which each probe hybridizes to an individual array element. Hybridization is assayed using a confocal laser scanner to measure fluorescence intensities, allowing simultaneous determination of the relative expression levels of all the genes represented in the array (Hegde et al. 2000; Schena et al. 1995). The process of expression analysis can be broadly divided into three stages: (1) array fabrication; (2) probe preparation and hybridization; (3) data collection, normalization, and analysis (Hegde et al. 2000; Schena et al. 1995).

The cDNA microarrays (Schena et al. 1995) have proven powerful and are now widely used for genome-wide expression analysis in a wide range of organisms, including plants (Baldwin et al. 1999; Richmond and Somerville 2000; Schaffer et al. 2000). cDNA microarrays allow up to tens of thousands of genes to be analysed simultaneously. Microarrays comprising complete gene sets are available for a number of organisms, such as yeast (Wodicka et al. 1997), a number of bacteria (Laub et al. 2000; Selinger et al. 2000), and *Caenorhabditis elegans* (Jiang et al. 2001), for which the entire genome sequence has been determined. For example, it was reported that gene expression during the cell cycle in bacteria is strictly regulated at the level of transcription and that the expression profiles of cell cycle modulated genes are coincident with the functional activity of the genes (Laub et al. 2000). For a few other well-studied animal and plant species, the current generation of microarrays is limited to a subset of the genes, namely those for which a cDNA clone or an expressed sequence tag (EST) sequence is available. Hegde et al. (2000) developed protocols that had been standardized and that had been used regularly in many laboratories for microarray analysis. The procedures described have been tested and refined over the past year and have been optimized using hybridization of RNA derived from cell lines to give reproducible and consistent results. It should be noted that a number of alternative protocols have been published (Eisen and Brown 1999), but the system developed by Hegde et al. (2000) has a number of advantages over these. In particular, the combination of printing, labelling, and hybridization conditions that have allowed a significant reduction in the quantity of starting total RNA required for analysis.

5
Oligonucleotide Microarrays

Oligonucleotide microarray based hybridization analysis is a promising new technology which potentially allows rapid and cost-effective screens for all possible mutations and sequence variations in genomic DNA (Roberts et al. 2000; Saiki et al. 1989). Identifying and cataloguing these variations is a critical part of approaches that seek to identify the genetic basis for resistance to disease. These sequence variations will serve as genetic markers in studies of diseases and traits with complex inheritance patterns (Golub et al. 1999; Roberts et al. 2000). Large-scale sequence analysis is needed for population-based genetic risk assessment and diagnostic tests once mutations have been identified, because traditional technologies cannot easily meet the demands for rapid and cost-effective large-scale comparative sequence and mutational analysis (Hacia 1999). To perform thousands of separate hybridization reactions to evaluate each sample makes an oligonucleotide microarray more amenable to a large-scale clinical diagnostic laboratory than a common research laboratory setting (Lockhart et al. 1996). The current scientific literature largely centres on arrays manufactured using photolithographic-based methodologies developed by Affymetrix (Fodor et al. 1991; Hacia 1999). However, technologies such as mass spectroscopy based hybridization detection, could have an important role in coming years.

Oligonucleotide array based detection of known genomic DNA sequence variations was first reported in 1989 (Saiki et al. 1989). Probes complementary to six *HLA-DQA* alleles as well as nine mutations in *HBB* (encoding β-globin) were spotted onto nylon filters and incubated with biotin-labelled PCR products (Yershov et al. 1996). Advanced oligonucleotide array manufacturing processes have opened the way to evaluating more complex systems (Yershov et al. 1996). Arrays of 1480 oligonucleotide probes synthesized in situ by photolithographic-based processes were designed to detect 37 known mutations in the coding region of *CFTR*, as well as all possible single-nucleotide substitutions (Yershov et al. 1996). In a blinded study, ten genomic DNA samples were successfully genotyped by characterizing fluorescent hybridization signals from test and wild-type reference samples at mutation-specific probes relative to those from wild-type samples. In a separate study, arrays of six oligonucleotide probes, generated by spotting oligonucleotides onto activated surfaces, were used to detect three different mutations in *HBB* (Yershov et al. 1996).

In *Arabidopsis*, defence and wounding responses have been analysed using cDNA microarrays (Schenk et al. 2000), whereas oligonucleotide arrays were used to study circadian-rhythm-modulated gene expression (Harmer et al. 2000). The analysis of the processes underlying fruit ripening in strawberries (Aharoni et al. 2000) was the first application of microarrays in a non-model plant species. The most important advantage of microarray-based technology is that gene expression profiles from either different samples or samples obtained using different treatments can be compared with each other and analysed together (Golub et al. 1999). Another striking example is presented in the landmark paper that describes the construction of a compendium of yeast expression profiles, combining data from both a number of mutant strains and treatments with different chemical compounds (Hughes et al. 2000). The power of microarrays was clearly illustrated by the characterization of a number of novel yeast genes solely on the basis of the gene expression profiles of the mutant strains. Similarly, the crosstalk and interaction among multiple mitogen-activated protein kinase pathways could be revealed by integrating gene expression profiles obtained under different experimental conditions (Roberts et al. 2000).

6 Serial Analysis of Gene Expression

Serial analysis of gene expression (SAGE) is a technique designed to take advantage of high-throughput sequencing technology to obtain a quantitative profile of cellular gene expression (Fig. 1). The SAGE technique measures not the expression level of a gene, but quantifies a tag that is a nucleotide sequence of a defined length adjacent to the 3′-most restriction site for a particular restriction enzyme and represents the transcription product of a gene (Velculescu et al. 1995). The SAGE technique is based on counting sequence tags of 14–15 bases from cDNA libraries (Velculescu et al. 1995; Zhang et al. 1997). This technology has been widely used to monitor gene expression in human cell cultures and tissue samples (Lash et al. 2000; Velculescu et al. 2000), but not in other organisms. In plants, this method has been applied only sporadically (Matsumura et al. 1999). The principle advantage of SAGE is that it gives an absolute measure of gene expression instead of measuring

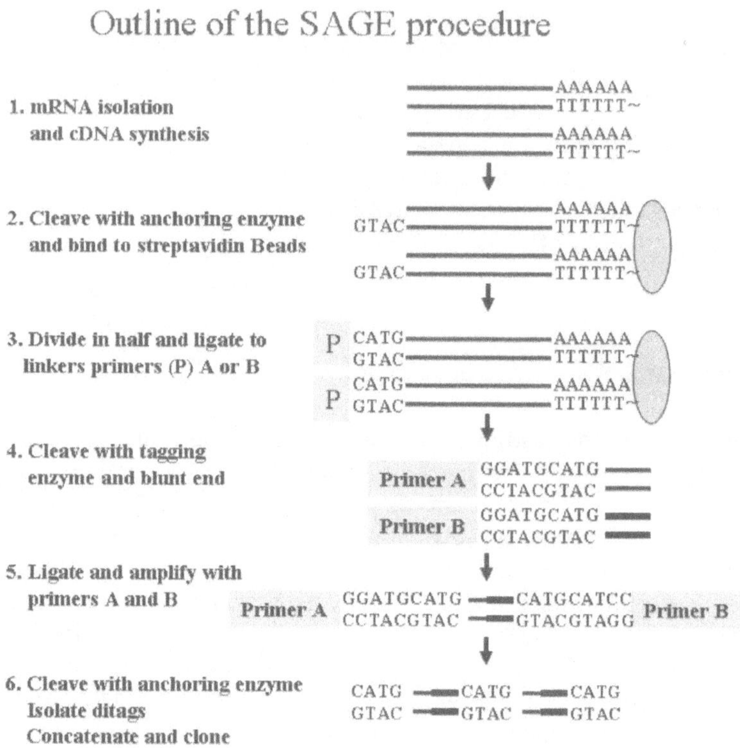

Fig. 1 Schematic of the serial analysis of gene expression (*SAGE*)

relative expression levels. Indeed, by counting the number of tags from each cDNA, one obtains an accurate measure of the number of transcripts present in the mRNA sample. As in the case of microarrays, independent data sets can be compiled in a single database, allowing the comparative analysis of data from different experiments (Lash et al. 2000; Velculescu et al. 2000). The public database SAGEmap already contains a comprehensive quantity of SAGE data from different cDNA libraries (Lash et al. 2000). Newly obtained data can be merged with the records already present in the database, enabling a more significant analysis of gene expression profiles.

SAGE required high amounts of input RNA, restricting its utility to large tissue samples. Recent improvements, however, now allow the use of 500–5000-fold less starting material and permit work with minute quantities of tissue containing only a few hundred or thousand cells (Datson et al. 1999; Matsumura et al. 1999). Although *Nla*III remains the most widely used restriction enzyme, enzyme substitutions are possible. The data product of the SAGE technique is a list of tags, with their corresponding count values, and thus is a digital representation of cellular gene expression. The principal limitation of SAGE is the need to sequence large numbers of tags in order to monitor the scarcely expressed genes. Another drawback of SAGE is that the tags obtained are very short and hence not always unambiguous. Gene identification on the basis of short sequence tags relies on the availability of large databases of well-characterized ESTs. So there are two problems to be tackled when dealing with SAGE data in the form of tags and counts. The first deals with ensuring that the tags and their counts are a valid representation of transcripts and their levels of expression, and the second with making valid tag-to-gene assignments.

7
Massively Parallel Signature Sequencing

The recently developed MPSS technology holds the promise of a major improvement over SAGE (Brenner et al. 2000). MPSS is a parallel sequencing method that can generate hundreds of thousands of short sequence signatures in a single analysis, thus overcoming the principal shortcoming of SAGE (Brenner et al. 2000). Because the method generates longer, 16–20-base signatures, it should also be more accurate. Technically, however, the method is rather complex and not yet readily available to the broad scientific community (Brenner et al. 2000). The genomic sequence of *A. thaliana* has been completed in recent years (*Arabidopsis* Genome Initiative 2000). Experimental analyses and comprehensive descriptions of plant transcriptomes continue in parallel (Haas et al. 2003; Yamada et al. 2003). No plant transcriptome has been extensively characterized experimentally with both quantitative and qualitative expression data. Computational approaches to genome annotation can miss or incorrectly predict many genes, and validation of genome annotations with experimental data is essential (Andrews et al. 2000; Guigo et al. 2000).

As genomic sequencing becomes faster and more economical, it is critically important that methods are developed to detect and quantify every gene and alternatively spliced transcript within a genome (Adams et al. 1995). Large-scale sequencing of short mRNA-derived tags can establish the qualitative and quantitative characteristics of a complex transcriptome (Meyers et al. 2004). Meyers et al. (2004) sequenced 12 304 362 tags from five diverse libraries of *A. thaliana* using MPSS. A total of 48 572 distinct signatures, each representing a different transcript, were expressed at significant levels (Meyers et al. 2004). These signatures were compared with the annotation of the *A. thaliana* genomic sequence; in the five libraries, this comparison yielded between 17 353 and 18 361 genes with sense expression, and between 5487 and 8729 genes with antisense expression (Meyers et al. 2004). An additional 6691 MPSS signatures mapped to unannotated regions of the genome. Expression was demonstrated for 1168 genes for which expression data were previously unknown (Meyers et al. 2004). Alternative polyadenylation was observed for more than 25% of *A. thaliana* genes transcribed in these libraries. The MPSS expression data suggest that the *A. thaliana* transcriptome is complex and contains many as-yet uncharacterized variants of normal coding transcripts (Meyers et al. 2004).

8
cDNA-Amplified Fragment-Length Polymorphism

The differential display technique developed by Liang and Pardee (1992) has been widely used to screen for genes that are differentially expressed. After the first publication of the differential display technique (Liang and Pardee 1992), several improved PCR-based methods, using restriction enzymes to generate cDNA specific tags, were described (Bachem et al. 1996; Kawamoto et al. 1999; Shimkets et al. 1999; Sutcliffe et al. 2000). The most widely used method, cDNA-AFLP, has been applied with success to the systematic analysis of genes involved in particular biological processes (Breyne and Zabeau 2001; Durrant et al. 2000). The cDNA-AFLP is based on the principle that a complex starting mixture of cDNAs is fractionated into smaller subsets, after which cDNA tags are PCR-amplified and separated on high-resolution gels (Breyne and Zabeau 2001; Durrant et al. 2000). The observed differences in the intensity of the bands provide a good measure of the relative differences in the levels of gene expression (Breyne and Zabeau 2001; Durrant et al. 2000). In a study of fungal pathogen response in tobacco cells, the screening of approximately 30 000 transcript tags identified a total of 273 modulated gene tags (Durrant et al. 2000). These differential display methods have proven useful for discovering differentially expressed genes, but not for quantitative genome-wide transcription analysis (Breyne and Zabeau 2001).

cDNA-AFLP analysis has been used to reveal early gene expression associated with the commitment and differentiation of a plant tracheary element by Milioni et al. (2002). The exogenous growth factors, auxin and cytokinin, are not required in the first 48 h after isolation of *Zinnia* mesophyll cells; furthermore, as little as 10 min of exposure to the growth factors at 48 h is both necessary and sufficient to commit cells to the tracheary element's differentiation pathway (Milioni et al. 2001). These findings suggest that the first 48 h of culture represents a time in which the cells adapt to liquid culture and acquire the competence to respond to the inductive signals (McCann 1997; Milioni et al. 2001). The precise transdifferentiation process provides a new and improved context in which to discover the earliest genes involved in switching on the developmental programme. In this project, a total of 652 differentially accumulated transcript-derived fragments (TDFs), ranging in length from 50 to 450 bp, were recovered from gels and reamplified, subcloned, and sequenced (Milioni et al. 2002). A total of 349 fragments (53.5%) of the differentially expressed genes showed close matches to database entries with assigned identities. Thirteen groups were classified from these sequences based on functional categories established for *Arabidopsis* (*Arabidopsis* Genome Initiative 2000). The major group is involved in primary and secondary metabolism and energy generation (19.2%), whereas a slightly higher proportion (8%) is cell-wall-related. An additional 9.7% of the TDFs are involved in information processing and constitute genes involved in transcriptional control and signal transduction. In addition, 12.4% of the sequences share significant similarity to unknown or hypothetical genes with no assigned function from various genome projects, which represent new candidate proteins involved in cell fate determination, differentiation, cell wall remodelling, and cell death.

To understand how embryonic cells differentiate into the 40 or so cell types that constitute plants (Hulskamp and Kirik 2000), one approach is to study mutants in which meristematic function has been compromised (Haecker and Laux 2001). Another approach is to study mutants in which a clear developmental phenotype for a particular cell type can be identified, for example, root hairs (Parker et al. 2000), trichomes (Hulskamp and Kirik 2000), or xylem (McCann and Roberts 2000), based on identification of genes that are differentially expressed. Global gene expression technologies may permit the dissection of downstream events through comparisons of mutants in these pathways; however, to date, only a few genes have been identified that are specific to particular cell types (Milioni et al. 2001). Genes involved in vascular cell fates have been identified in cDNA-sequencing projects using material derived from young xylem tissue of loblolly pine (Allona et al. 1998) and poplar (Sterky et al. 1998). Tissue-specific transcript profiles have been obtained using DNA microarray analysis of 3000 ESTs of poplar (Hertzberg et al. 2001). To elucidate genetic programmes that control embryogenesis and regeneration of rice, Ito et al. (2002) conducted genome-wide expression analysis of genes involved in somatic embryogenesis. Functional analyses of genes demonstrated that five *KNOX* family class 1 homeobox genes were involved in somatic embryogenesis (Ito et al. 2002). The *KNOX* family class 1 homeobox genes encode transcription factors and protein kinases. Expression patterns of these genes during early embryogenesis and regeneration were analysed by reverse transcription PCR and in situ hybridization (Ito et al. 2002). It was found that constitutive expression of these genes is sufficient to maintain cells in a meristematic undifferentiated state (Ito et al. 2002).

9 Conclusion

Genome-wide expression analysis allows scientists to identify genes that are involved in somatic embryogenesis in plants. The control of somatic embryogenesis involves the temporal expression of different sets of genes through the different phases of the embryo development. A landmark study using genome-wide expression analysis to follow the patterns of gene expression in rice has allowed the identification of hundreds of genes that are involved in somatic embryogenesis (Ito et al. 2002). Different genome-wide expression analysis technologies, including (1) cDNA microarray, (2) oligonucleotide microarrays, (3) serial analysis of gene expression, (4) MPSS, and (5) cDNA-AFLP, provide opportunities to explore the mechanism of somatic embryogenesis. DNA microarrays provide a convenient tool for genome-wide expression analysis; however, their use is limited to organisms for which the complete genome sequence or a large cDNA collection is available. Alternative technologies for expression profiling based on DNA sequencing or cDNA fragment analysis have been developed and successfully used in different biological systems. For example, cDNA-AFLP exhibits properties that complement DNA microarrays and may provide a more appropriate tool for genome-wide expression analysis, gene discovery, and transcript profiling. Somatic embryogenesis has been induced in some pine species (Tang 2000; Tang et al. 2001). We are using different genome-wide expression analysis technologies to identify genes involved in somatic embryogenesis.

References

Adams MD, Kerlavage AR, Fleischmann RD, Fuldner RA, Bult CJ, Lee NH, Kirkness EF, Weinstock KG, Gocayne JD, White O, Sutton G, Blake JA, Brandon RG, Chiu MW, Clayton RA, Cline RT, Cotton MD, Earlehughes J, Fine LD, Fitzgerald LM, Fitzhugh WM, Fritchman JL, Geoghagen NSM, Glodek A, Gnehm CL, Hanna MC, Hedblom E, Hinkle PS, Kelley JM, Klimek KM, Kelley JC, Liu LI, Marmaros SM, Merrick JM, Morenopalanques RF, Mcdonald LA, Nguyen DT, Pellegrino SM, Phillips CA, Ryder SE, Scott JL, Saudek DM, Shirley R, Small KV, Spriggs TA, Utterback TR, Weldman JF, Li Y, Barthlow R, Bednarik DP, Cao LA, Cepeda MA, Coleman TA, Collins EJ, Dimke D, Feng P, Ferrie A, Fischer C, Hastings GA, He WW, Hu JS, Huddleston KA, Greene JM, Gruber J, Hudson P, Kim A, Kozak DL, Kunsch C, Ji HJ, Li HD, Meissner PS, Olsen H, Raymond L, Wei YF, Wing J, Xu C, Yu GL, Ruben SM, Dillon PJ, Fannon MR, Rosen CA, Haseltine WA, Fields C, Fraser CM, Venter JC (1995) Initial assessment of human gene diversity and expression patterns based upon 83 million nucleotides of cDNA sequence. Nature 377:3–174

Aharoni A, Keizer LCP, Bouwmeester HJ, Sun ZK, Alvarez-Huerta M, Verhoeven HA, Blaas J, van Houwelingen AMML, De Vos RCH, van der Voet H, Jansen RC, Guis M, Mol J, Davis RW, Schena M, van Tunen AJ, O'Connell AP (2000) Identification of the SAAT gene involved in strawberry flavor biogenesis by use of DNA microarrays. Plant Cell 12:647–662

Allona I, Quinn M, Shoop E, Swope K, St Cyr S, Carlis J (1998) Analysis of xylem formation in pine by cDNA sequencing. Proc Natl Acad Sci USA 95:9693–9698

Andrews J, Bouffard GG, Cheadle C, Lu JN, Becker KG, Oliver B (2000) Gene discovery using computational and microarray analysis of transcription in the Drosophila melanogaster testis. Genome Res 10:2030–2043

Arabidopsis Genome Initiative (2000) Analysis of the genome sequence of the flowering plant Arabidopsis thaliana. Nature 408:796–815

Bachem CW, van der Hoeven RS, de Bruijn SM, Vreugdenhil D, Zabeau M, Visser RG (1996) Visualization of differential gene expression using a novel method of RNA fingerprinting based on AFLP: analysis of gene expression during potato tuber development. Plant J 9:745–753

Baker J, Steele C, Dure L (1988) Sequence and characterization of 6 Leu proteins and their genes from cotton. Plant Mol Biol 11:277–291

Baldwin D, Crane V, Rice D (1999) A comparison of gel-based, nylon filter and microarray techniques to detect differential RNA expression in plants. Curr Opin Plant Biol 2:96–103

Brenner S, Johnson M, Bridgham J, Golda G, Lloyd DH, Johnson D, Luo SJ, McCurdy S, Foy M, Ewan M, Roth R, George D, Eletr S, Albrecht G, Vermaas E, Williams SR, Moon K, Burcham T, Pallas M, DuBridge RB, Kirchner J, Fearon K, Mao J, Corcoran K (2000) Gene expression analysis by massively parallel signature sequencing (MPSS) on microbead arrays. Nat Biotechnol 18:630–634

Breyne P, Zabeau M (2001) Genome-wide expression analysis of plant cell cycle modulated genes. Curr Opin Plant Biol 4:136–142

Datson NA, van der Perk-de Jong J, van den Berg MP, de Kloet ER, Vreugdenhil E (1999) MicroSAGE: a modified procedure for serial analysis of gene expression in limited amounts of tissue. Nucleic Acids Res 27:1300–1307

Durrant WE, Rowland O, Piedras P, Hammond-Kosack KE, Jones JD (2000) cDNA-AFLP reveals a striking overlap in race-specific resistance and wound response gene expression profiles. Plant Cell 12:963–977

Eisen MB, Brown PO (1999) DNA arrays for analysis of gene expression. Methods Enzymol 303:179–205

Fodor SPA, Read JL, Pirrung MC, Stryer L, Lu AT, Solas D (1991) Light-directed spatially addressable parallel chemical synthesis. Science 251:767–773

Galau GA, Hughes DW, Dure L (1986) Abscisic acid induction of cloned cotton late embryogenesis-abundant (Leu) mRNAs. Plant Mol Biol 7:155–170

Galau GA, Wang HYC, Hughes DW (1992) Cotton Leu4 (D19) and LeaA2 (D132) Group 1 Leu genes encoding water stress-related proteins containing a 20-amino acid motif. Plant Physiol 99:783–788

Galau GA, Wang HYC, Hughes DW (1993) Cotton fea5 and fea74 encode typical late embryogenesis-abundant proteins. Plant Physiol 101:695–696

Golub TR, Slonim DK, Tamayo P, Huard C, Gaasenbeek M, Mesirov JP, Coller H, Loh ML, Downing JR, Caligiuri MA, Bloomfield CD, Lander ES (1999) Molecular classification of cancer: class discovery and class prediction by gene expression monitoring. Science 286:531–537

Guigo R, Agarwal P, Abril JF, Burset M, Fickett JW (2000) An assessment of gene prediction accuracy in large DNA sequences. Genome Res 10:1631–1642

Guiltinan MJ, Marcotte WR Jr, Quatrano RS (1990) A plant leucine zipper protein that recognizes an abscisic acid response element. Science 250:267–271

Haas BJ, Delcher AL, Mount SM, Wortman JR, Smith RK, Hannick LI, Maiti R, Ronning CM, Rusch DB, Town CD, Salzberg SL, White O (2003) Improving the Arabidopsis genome annotation using maximal transcript alignment assemblies. Nucleic Acids Res 31:5654–5666

Haccius B (1978) Question of unicellular origin of non-zygotic embryos in callus cultures. Phytomorphology 28:74–81

Hacia JG (1999) Resequencing and mutational analysis using oligonucleotide microarrays. Nat Genet 21(S):42–47

Haecker A, Laux T (2001) Cell–cell signaling in the shoot meristem. Curr Opin Plant Biol 4:441–446

Harmer SL, Hogenesch JB, Straume M, Chang HS, Han B, Zhu T, Wang X, Kreps JA, Kay SA (2000) Orchestrated transcription of key pathways in Arabidopsis by the circadian clock. Science 290:2110–2113

Hegde P, Qi R, Abernathy K, Gay C, Dharap S, Gaspard R, Hughes JE, Snesrud E, Lee N, Quackenbush J (2000) A concise guide to cDNA microarray analysis. Biotechniques 29:548–562

Hertzberg M, Aspeborg H, Schrader J, Andersson A, Erlandsson R, Blomqvist K, Bhalerao R, Uhlen M, Teeri TT, Lundeberg J, Sundberg B, Nilsson P, Sandberg G (2001) A transcriptional roadmap to wood formation. Proc Natl Acad Sci USA 98:14732–14737

Hughes DW, Galau GA (1991) Developmental and environmental induction of Leu and LeuA mRNAs and the postabscission program during embryo culture. Plant Cell 3:605–618

Hughes TR, Marton MJ, Jones AR, Roberts CJ, Stoughton R, Armour CD, Bennett HA, Coffey E, Dai HY, He YDD, Kidd MJ, King AM, Meyer MR, Slade D, Lum PY, Stepaniants SB, Shoemaker DD, Gachotte D, Chakraburtty K, Simon J, Bard M, Friend SH (2000) Functional discovery via a compendium of expression profiles. Cell 102:109–126

Ito Y, Hirochika H, Kurata N (2002) Organ specific alternative transcripts of KNOX family class 2 homeobox genes of rice. Gene 288:41–47

Hulskamp M, Kirik V (2000) Trichome differentiation and morphogenesis in Arabidopsis. Adv Bot Res 31:237–260

Jiang M, Ryu J, Kiraly M, Duke K, Reinke V, Kim SK (2001) Genome-wide analysis of developmental and sex-regulated gene expression profiles in Caenorhabditis elegans. Proc Natl Acad Sci USA 98:218–223

Kawamoto S, Ohnishi T, Kita H, Chisaka O, Okubo K (1999) Expression profiling by iAFLP: a PCR-based method for genome-wide gene expression profiling. Genome Res 9:1305–1312

Lash AE, Tolstoshev CM, Wagner L, Schuler GD, Strausberg RL, Riggins GJ, Altschul SF (2000) SAGEmap: a public gene expression resource. Genome Res 10:1051–1060

Laub MT, McAdams HH, Feldblyum T, Fraser CM, Shapiro L (2000) Global analysis of the genetic network controlling a bacterial cell cycle. Science 290:2144–2148

Liang P, Pardee AB (1992) Differential display of eukaryotic messenger RNA by means of the polymerase chain reaction. Science 257:967–971

Lockhart DJ, Dong HL, Byrne MC, Follettie MT, Gallo MV, Chee MS, Mittmann M, Wang CW, Kobayashi M, Horton H, Brown EL (1996) Expression monitoring by hybridization to high-density oligonucleotide arrays. Nat Biotechnol 14:1675–1680

Matsumura H, Nirasawa S, Terauchi R (1999) Technical advance: transcript profiling in rice (Oryza sativa L.) seedlings using serial analysis of gene expression (SAGE). Plant J 20:719–726

McCann MC (1997) Tracheary element formation: Building up to a dead end. Trends Plant Sci 2:333–338

McCann MC, Roberts K (2000) Xylogenesis: the birth of a corpse. Curr Opin Plant Biol 3:517–522

Meyers BC, Vu TH, Tej SS, Ghazal H, Matvienko M, Agrawal V, Ning J, Haudenschild CD (2004) Analysis of the transcriptional complexity of Arabidopsis thaliana by massively parallel signature sequencing. Nat Biotechnol 22:1006–1011

Milioni D, Sado PE, Stacey NJ, Domingo C, Roberts K, McCann MC (2001) Differential expression of cell-wall-related genes during the formation of tracheary elements in the Zinnia mesophyll cell system. Plant Mol Biol 47:221–238

Milioni D, Sado PE, Stacey NJ, Roberts K, McCann MC (2002) Early gene expression associated with the commitment and differentiation of a plant tracheary element is revealed by cDNA-Amplified Fragment Length Polymorphism analysis. Plant Cell 14:2813–2824

Parker JS, Cavell AC, Dolan L, Roberts K, Grierson CS (2000) Genetic interactions during root hair morphogenesis in Arabidopsis. Plant Cell 12:1961–1974

Piatkowski D, Schneider K, Salamini F, Bartels D (1990) Characterization of five abscisic acid-responsive cDNA clones isolated from the desiccation-tolerant plant Craterostigma plantagineum and their relationship to other water-stress genes. Plant Physiol 94:1682–1688

Richmond T, Somerville S (2000) Chasing the dream: plant EST microarrays. Curr Opin Plant Biol 3:108–116

Roberts CJ, Nelson B, Marton MJ, Stoughton R, Meyer MR, Bennett HA, He YDD, Dai HY, Walker WL, Hughes TR, Tyers M, Boone C, Friend SH (2000) Signaling and circuitry of multiple MAPK pathways revealed by a matrix of global gene expression profiles. Science 287:873–880

Saiki RK, Walsh PS, Levenson CH, Erlich HA (1989) Genetic analysis of amplified DNA with immobilized sequence-specific oligonucleotide probes. Proc Natl Acad Sci USA 86:6230–6234

Schaffer R, Landgraf J, Perez-Amador M, Wisman E (2000) Monitoring genome-wide expression in plants. Curr Opin Biotechnol 11:162–167

Schena M, Shalon D, Davis RW, Brown PO (1995) Quantitative monitoring of gene expression patterns with a complementary DNA microarray. Science 270:467–470

Schenk RV, Hildebrandt HC (1972) Medium and techniques for induction and growth of monocotyledoneous and dicotyledoneous plant cell cultures. Can J Bot 50:199–204

Schenk PM, Kazan K, Wilson I, Anderson JP, Richmond T, Somerville SC, Manners JM (2000) Coordinated plant defense responses in Arabidopsis revealed by microarray analysis. Proc Natl Acad Sci USA 97:11655–11660

Selinger DW, Cheung KJ, Mei R, Johansson EM, Richmond CS, Blattner FR, Lockhart DJ, Church GM (2000) RNA expression analysis using a 30 base pair resolution Escherichia coli genome array. Nat Biotechnol 18:1262–1268

Shimkets RA, Lowe DG, Tai JTN, Sehl P, Jin HK, Yang RH, Predki PF, Rothberg BEG, Murtha MT, Roth ME, Shenoy SG, Windemuth A, Simpson JW, Simons JF, Daley MP, Gold SA, McKenna MP, Hillan K, Went GT, Rothberg JM (1999) Gene expression analysis by transcript profiling coupled to a gene database query. Nat Biotechnol 17:798–803

Skriver K, Mundy J (1991) Gene expression in response to abscisic acid and osmotic stress. Plant Cell 2:503–512

Sterky F, Regan S, Karlsson J, Hertzberg M, Rohde A, Holmberg A, Amini B, Bhalerao R, Larsson M, Villarroel R, Van Montagu M, Sandberg G, Olsson O, Teeri TT, Boerjan W, Gustafsson P, Uhlen M, Sundberg B, Lundeberg J (1998) Gene discovery in the wood-forming tissues of poplar: Analysis of 5692 expressed sequence tags. Proc Natl Acad Sci USA 95:13330–13335

Sutcliffe JG, Foye PE, Erlander MG, Hilbush BS, Bodzin LJ, Durham JT, Hasel KW (2000) TOGA: an automated parsing technology for analyzing expression of nearly all genes. Proc Natl Acad Sci USA 97:1976–1981

Tang W (2000) Peroxidase activities of desiccation tolerant loblolly pine somatic embryos. In Vitro Cell Dev Biol Plant 36:488–491

Tang W, Guo Z, Ouyang F (2001) Plant regeneration from embryogenic cultures initiated from mature loblolly pine zygotic embryos. In Vitro Cell Dev Biol Plant 37:558–563

Velculescu VE, Zhang L, Vogelstein B, Kinzler KW (1995) Serial analysis of gene expression. Science 270:484–487

Velculescu VE, Vogelstein B, Kinzler KW (2000) Analysing uncharted transcriptomes with SAGE. Trends Genet 16:423–425

Williams EG, Maheswaran G (1986) Somatic embryogenesis: factors influencing coordinated behaviour of cells as an embryogenic group. Ann Bot 57:443–462

Wodicka L, Dong H, Mittmann M, Ho MH, Lockhart DJ (1997) Genomewide expression monitoring in Saccharomyces cerevisiae. Nat Biotechnol 15:1359–1367

Yamada K, Lim J, Dale JM, Chen HM, Shinn P, Palm CJ, Southwick AM, Wu HC, Kim C, Nguyen M, Pham P, Cheuk R, Karlin-Newmann G, Liu SX, Lam B, Sakano H, Wu T, Yu GX, Miranda M, Quach HL, Tripp M, Chang CH, Lee JM, Toriumi M, Chan MMH, Tang CC, Onodera CS, Deng JM, Akiyama K, Ansari Y, Arakawa T, Banh J, Banno F, Bowser L, Brooks S, Carninci P, Chao QM, Choy N, Enju A, Goldsmith AD, Gurjal M, Hansen NF, Hayashizaki Y, Johnson-Hopson C, Hsuan VW, Iida K, Karnes M, Khan S, Koesema E, Ishida J, Jiang PX, Jones T, Kawai J, Kamiya A, Meyers C, Nakajima M, Narusaka M, Seki M, Sakurai T, Satou M, Tamse R, Vaysberg M, Wallender EK, Wong C, Yamamura Y, Yuan SL, Shinozaki K, Davis RW, Theologis A, Ecker JR (2003) Empirical analysis of transcriptional activity in the Arabidopsis genome. Science 302:842–846

Yershov G, Barsky V, Belgovskiy A, Kirillov E, Kreindlin E, Ivanov I, Parinov S, Guschin D, Drobishev A, Dubiley S, Mirzabekov A (1996) DNA analysis and diagnostics on oligonucleotide microchips. Proc Natl Acad Sci USA 93:4913–4918

Yeung EC, Meinke DW (1993) Embryogenesis in angiosperms: development of the suspensor. Plant Cell 5:1371–1381

Yeung EC, Sussex IM (1979) Embryogeny of Phaseolus coccineus: the suspensor and the growth of the embryo-proper in vitro. Z Pflanzenphysiol 91:423–433

Zhang L, Zhou W, Velculescu VE, Kern SE, Hruban RH, Hamilton SR, Vogelstein B, Kinzler KW (1997) Gene expression proles in normal and cancer cells. Science 23:1268–1272

Plant Cell Monogr (2)
A. Mujib · J. Šamaj: Somatic Embryogenesis
DOI 10.1007/7089_019/Published online: 9 December 2005

Why Somatic Plant Cells Start to form Embryos?

Attila Fehér

Institute of Plant Biology, Biological Research Center, Hungarian Academy of Sciences, P.O. Box 521, 6701 Szeged, Hungary
fehera@brc.hu

Abstract Embryogenesis in plants is not restricted to the fertilized egg cell but can be naturally or artificially induced in many different cell types, including somatic cells. Although genetic components clearly determine the potential of species/genotypes to form somatic embryos, the expression of embryogenic competence at the cellular level is defined by developmental and physiological cues. Competent cells can respond to a variety of conditions by the initiation of embryogenic development. In general, these conditions include alterations in auxin (exogenous and/or endogenous) levels and evoke stress responses. Recent experimental results in the field of developmental and molecular plant biology emphasize the role of chromatin remodelling in the coordination of overall gene expression patterns associated with developmental switches. It can be hypothesized that the initiation of somatic embryogenesis is a general response to a multitude of parallel signals (including auxin and stress factors). This response includes, in addition to cellular and physiological reorganization, the extended remodelling of the chromatin and a release of the embryogenic programme otherwise blocked in vegetative cells by chromatin-mediated gene silencing. In this review I attempt to give a general overview of experimental results supporting the aforementioned hypothesis, leaving the detailed elaboration of special subjects to other chapters.

1
Embryogenesis in Plants—Variations on a Theme

In higher plants, double fertilization generates the embryo and the endosperm simultaneously, the joint development of which leads to a viable seed. Plant zygotic embryogenesis is a process that is deeply hidden in maternal tissues. In addition to the large body of histological data generated in various species, analysis of *Arabidopsis* mutants enlighted the series of events underlying plant embryo development (for a review see Mordhorst et al. 1997). Micromanipulation and in vitro fertilization supplemented by molecular and genomic methods have already revealed additional details and will also contribute to our understanding of plant embryogenesis (Grimanelli et al. 2005; Kranz et al. 1995; Kranz 1999; Sprunck et al. 2005).

However, within higher plants, detours to zygotic embryogenesis became known for a considerable number of species generally referred to as apomixis (more than 400 species belonging to at least 40 different families; Bicknell and Koltunow 2004). During apomixis, the asexual formation of a seed starts from the maternal tissues of the ovule, avoiding the processes of meiosis and fertilization, leading to embryo development (Bicknell and Koltunow 2004). The widely observed phenomenon of apomixis reveals two important aspects of plant embryogenesis: (1) the fertilization trigger can be substituted by endogenous mechanisms (2) in higher plants other cell types in addition to the fertilized egg cell can maintain or regain the capability for embryogenic development. Although apomictic processes are restricted to the cells of the generative apex or the ovule, there is a large variety of somatic plant cells that can also undergo embryogenic development under appropriate conditions. Natural formation of embryos as vegetative propagules can take place, for example, on leaf margins of *Kalanchoë*, *Bryophyllum* (Yarbrough 1932) or *Malaxis* (Taylor 1967) species. There are many more examples for embryogenesis initiated from in vitro cultured somatic (for a comprehensive overview see Thorpe 1995) or gametic (e.g. microspores; for a review see Reynolds 1997) cells.

In all forms of plant embryogenesis (Fig. 1) certain criteria have to be fulfilled before initiation. The species or genotype has to have the genetic potential to form embryos from somatic cells and one or a few cells of the plant/explant have to be competent to receive a signal (endogenous or exogenous) that triggers the pathway of embryogenic development (commitment) leading to embryo formation even in the absence of further signals. For the in vitro forms of somatic embryogenesis, these conditions (potential, competence, induction, commitment) have to be experimentally optimized.

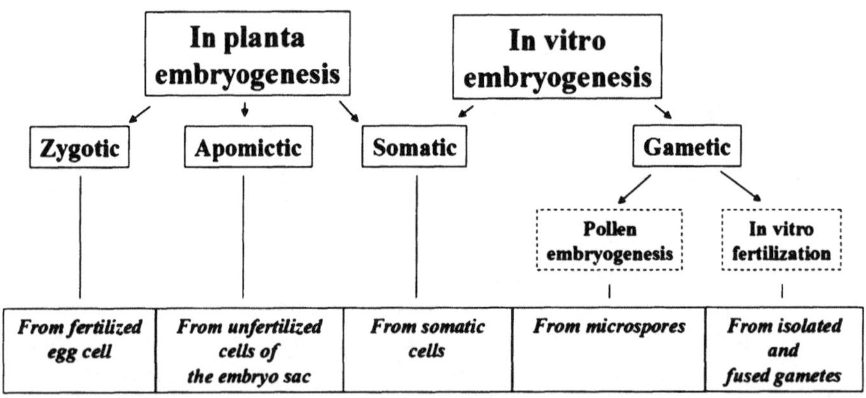

Fig. 1 Various pathways leading to embryo development in higher plants. Embryogenesis in most higher plant species starts with the fertilization of the egg cell that is parallel to the fertilization of the central cell (double fertilization). However, in certain species and in certain conditions, embryogenesis can be initiated in the embryo sac in the absence of fertilization (apomixis). In other species (e.g. in *Kalanchoë* sp.), embryos as vegetative propagules arise on leaf margins (in planta somatic embryogenesis). Embryogenesis can also be artificially induced in somatic or gametic cells in vitro

Although in vitro somatic embryogenesis is practised in many tissue culture laboratories using many species, genotypes and explants, the biological background of the process is still largely unknown. The special conditions required for successful embryo induction are set up experimentally without knowing why a given genotype/explant has embryogenic potential and how and why competence or commitment is achieved or what is the real trigger initiating embryo development.

2 Embryogenic Potential

The potential for somatic embryogenesis is first of all determined at the level of the genotype. It is clearly proved by the successful transfer of the embryogenic capability between embryogenic and recalcitrant genotypes via sexual crossing (Bowley et al. 1993; Kielly and Bowley 1992; Moltrasio et al. 2004). In spite of the continuously increasing group of species where the conditions for somatic embryo induction have been established, there are a number of species that are still recalcitrant to form somatic embryos. Highly embryogenic and recalcitrant genotypes exist even within a given species. It has to be emphasized, however, that in many instances "recalcitrance" could be resolved by optimizing growth conditions of plants or by proper explant selection (Krishna Raj and Vasil 1995). Genetic determinants therefore may only serve to define the conditions when and where embryogenic competence can be expressed (see later). Thus, the embryogenic potential is largely defined by the developmental programme of the plant as well as by environmental cues.

Somatic embryos can develop on all organs of seedlings in certain highly embryogenic genotypes of carrot or alfalfa, indicating a wide expression of embryogenic potential. In most plant species, however, embryogenic competence is restricted to certain tissues of a given genotype. Tissue culture experiences support the view that there exists a kind of gradient in the embryogenic response among the various plant organs. The embryogenic potential is highest in tissues with embryonic origin and decreases towards the hypocotyl, petiole, leaf and root (reviewed by Neumann 2000). But even if embryogenic competence seems to be lost in somatic plant cells, it can potentially be regained. In these "indirect" ways of somatic embryogenesis an intermediate phase of callus formation is required in order to express the embryogenic potential.

Obviously, the embryogenic capability of plant cells continuously decreases during plant ontogenesis, and it is species-dependent. In monocotyledonous plants, including most of the agronomically important cereals, embryogenic competence is mostly restricted to cells with embryogenic or meristematic origin, including immature embryos or seeds, leaf bases (*Gram-*

inae) or tips (*Orchidaceae*), bulb scales (*Liliaceae*), lateral buds, etc. (for a detailed list see Krishna Raj and Vasil 1995). The embryogenic potential of these meristematic cells can be maintained if the explants are cultured in a medium containing 2,4-dichlorophenoxyacetic acid (2,4-D) followed by excessive callus formation. A high frequency of somatic embryogenesis can be achieved after the transfer of these "embryogenic callus" cells to a low-auxin or hormone-free medium.

In contrast to the cells of meristematic tissues, somatic cells of monocotyledonous plants differentiate early and rapidly and this is followed by the loss of their mitotic and morphogenetic capabilities. In this respect it is interesting to note that the regulation of the juvenile-to-adult transition might be different in dicots and monocots (for a review see Chuck and Hake 2005). Although the direct reasons for the early loss of totipotency in monocots are not known, they may be linked to the strict regulation of the synthesis and/or metabolism of endogenous growth regulators such as auxin.

Several attempts have been made to compare embryogenic and closely related recalcitrant genoypes to point out significant differences (for a review see Fehér et al. 2003). In alfalfa (*Medicago sativa* ssp. *varia*), closely related genotypes were selected on the basis of their embryogenic potential (Bögre et al. 1990). Their response to auxin has been compared and characteristic differences could be established. Auxin-responsive genes were induced/repressed at a significantly lower auxin concentration in the embryogenic versus the non-embryogenic genotype (Bögre et al. 1990). Furthermore, auxin inhibited rooting of in vitro grown shoot cuttings also at a much lower concentration (Bögre et al. 1990). Callus growth of the non-embryogenic genotype continued at the same 2,4-D concentration that inhibited cell division in the cells of the embryogenic genotype where this level of 2,4-D triggered somatic embryogenesis. These observations indicated a considerable difference between the auxin sensitivity of the two genotypes. The key role of endogenous hormone metabolism affected by genetic, physiological and environmental cues is well accepted in the induction phase of somatic embryogenesis (Jimenez, this volume).

3
Cellular Competence

Embryogenic competence is expressed finally at the level of single cells. It is very difficult to define, however, what this cellular competence means. According to a widely accepted definition, embryogenic competent cells are those cells which are capable of differentiating into embryos if they receive inducers of differentiation (Halperin 1969). However, embryogenic competence itself needs to be induced in many cases (e.g. during "indirect" somatic embryogenesis, see earlier), and the signals inducing competence and triggering embryogenic development are not easy to separate. Cellular competence is associated with the dedifferentiation of somatic cells that allows them to respond to new developmental signals.

It is well accepted that embryogenic competent cells can be morphologically recognized as small, rounded cells with rich cytoplasm and small vacuoles. In this respect they are very similar to meristematic cells or zygotes and this similarity is further emphasized by their asymmetric division (Fig. 2). Embryogenic competent cells can also be characterized by the central position of the nucleus and by prominent radiating perinuclear microtubules and actin filaments (Šamaj et al. 2003). Additionally, they exhibit a special cell wall composition that is discussed in detail by Šamaj (this volume).

These types of cells either originate from embryonic/meristematic tissues or can be formed from elongated, vacuolized cells under specific conditions, e.g. after treatment with 2,4-D. However, other hormones (abscisic acid, ABA, cytokinin) or stress treatments (Ikeda-Iwai et al. 2003; Kamada et al. 1993; Nishiwaki et al. 2000; Pasternak et al. 2002) can also induce the formation of the embryogenic competent cell type.

Development of embryogenic competent cells can be best documented in systems where single cells were selected (Nomura and Komamine 1985; Osuga et al. 1999) or video-tracked (Toonen et al. 1994) using carrot suspension cells or *Medicago* leaf protoplasts (Bögre et al. 1990; Dudits et al. 1991; Pasternak et al. 2002; Fehér et al. 2005).

Although video cell tracking of individual carrot cells of a heterogeneous cell suspension culture could not clearly assign a morphological type to the initial cells that could form proembryogenic cell clusters, the highest frequency could be observed in the case of small, spherical, densely cytoplasmic cells (Toonen et al. 1994). The same technology was successfully used to demonstrate that the expression of the somatic embryogenesis receptor kinase (*SERK1*) gene is indeed linked to the embryogenic cell fate (Schmidt

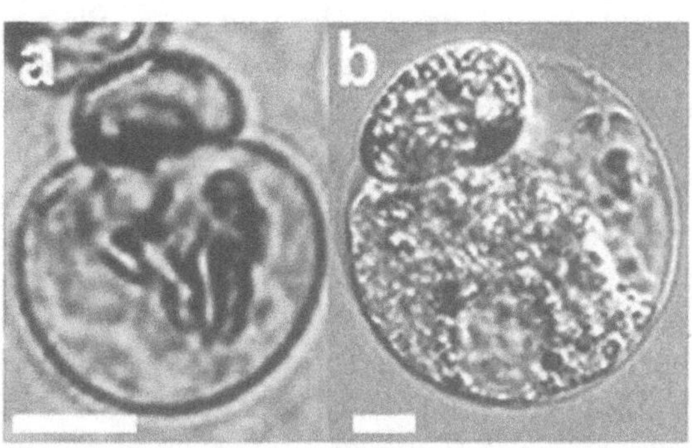

Fig. 2 Morphological similarity of an asymmetrically dividing leaf-protoplast-derived embryogenic alfalfa cell (**a**) and an isolated maize zygote (**b**). The *bar* represents 10 μm

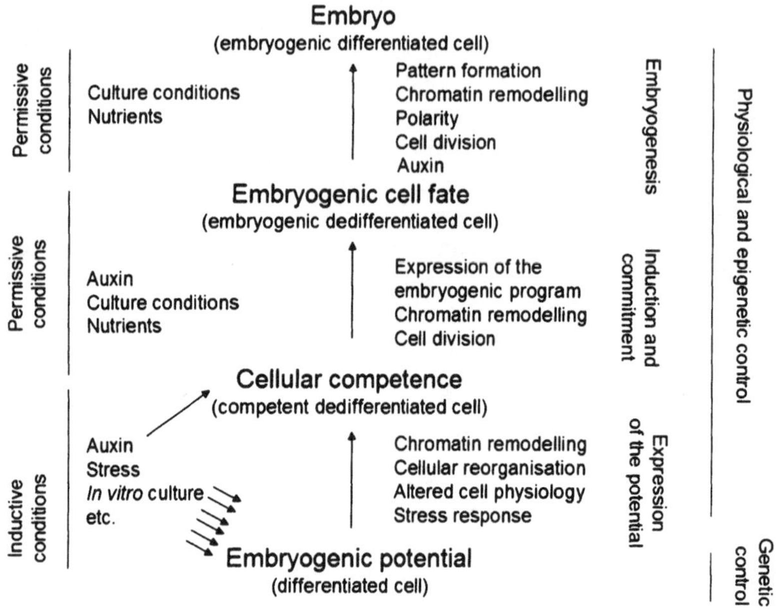

Fig. 3 A hypothetical model of events underlying somatic embryogenesis. A multitude of parallel signals, including auxin (either exogenously supplied or endogenously altered), evoke a wide cellular response including reorganizations at the levels of cell structure, physiology, chromatin and gene expression. As a result, the dedifferentiated cells become competent for embryogenesis. Competent cells will indeed be embryogenic if external and cellular conditions allow the expression of the embryogenic programme that is, in most cases, preceded by or parallel to cell divisions. Further cell divisions together with polarity establishment and pattern formation result in the development of the embryo. The central role of chromatin remodelling can be hypothesised in all phases, including dedifferentiation, embryogenic reprogramming and embryo differentiation. They are all associated with the parallel activation/inactivation of a large number of genes

et al. 1997). Following the division of these small, spherical, dense carrot cells, the JIM8 cell wall epitope was shown to be asymmetrically transferred to the daughter cells from which only those devoid of the epitope remained embryogenic (Toonen et al. 1996).

Another approach was developed by Nomura and Komamine (1985) based on the fractionation of suspension-cultured carrot cells. They could isolate a fraction of small, dense, isodiametric cell type (state 0) that could synchronously develop into somatic embryos under appropriate conditions (Osuga et al. 1999). It was found that the formation of state 1 cells (forming small embryogenic cell clusters) was dependent on auxin, which, however, blocked the further development (Nomura and Komamine, 1985).

Alfalfa leaf protoplasts also represent a rather homogenous and synchronized system that allows detailed investigations both at the single cell and at the cell population level (for a review see Fehér et al. 2005). A further advantage of the system is that the development of the cells is dependent on 2,4-D concentration: 1 μM 2,4-D results in the formation of elongated vacuolated cells, while small, cytoplasmicaly rich, embryogenic cells are formed at a tenfold higher concentration (Dudits et al. 1991; Pasternak et al. 2002). Furthermore, the system can be used to compare genotypes with or without embryogenic potential (Bögre et al. 1990; see also earlier). The comparisons made between embryogenic and non-embryogenic cells revealed that the two types exhibit not only characteristic morphological differences but that their physiology is also altered. Among other differences, the embryogenic competent cells have higher cytoplasmic and vacuolar pH values and an altered auxin metabolism (Pasternak et al. 2002). These protoplast-derived cells were activated earlier as was shown by faster medium acidification and earlier BrdU/thymidine incorporation into their genomic DNA as well as by earlier cell divisions (Bögre et al. 1990; Pasternak et al. 2002). The correlation between the plasma membrane pH gradient, the timing of cell activation and embryogenic cell formation was strengthened by several further observations. For example, buffering of the medium by 2-morpholinoethanesulphonic acid slowed down medium acidification, delayed cell division and prevented embryogenic cell formation in the presence of the embryogenic (10 μM) 2,4-D concentration. On the other hand, gradual medium acidification achieved by L-galactolactone accelerated cell division and promoted embryogenic cell formation under non-embryogenic (1 μM 2,4-D) conditions (Pasternak et al. 2002; Fehér et al. 2005). Oxidative stress (iron, copper, menadione, nitric oxide) was also shown to promote both cell division and embryogenic cell formation under non-embryogenic conditions (Pasternak et al. 2002; Ötvös et al. 2005). Some of these changes could be linked to the timing of endogenous auxin (indole acetic acid, IAA) peaks (Pasternak et al. 2002).

The same system seemed to be useful for the identification of genes differentially expressed in vacuolated, non-embryogenic (1 μM 2,4-D) versus dense, embryogenic competent (10 μM 2,4-D) cells. A PCR-based complementary DNA (cDNA) subtraction approach was used to obtain a cDNA population enriched in sequences preferentially expressed in the embryogenic cell type (Fehér et al., unpublished results). The functional classification of 36 differentially expressed genes revealed that most of the proteins indentified are related to cellular reorganization, including stress responses, intracellular membrane transport and secretion, protein synthesis and nuclear functions. The genes had distinct expression patterns during somatic embryogenesis, indicating their participation in various processes underlying the embryo formation from protoplast-derived cells.

Similar molecular approaches resulting in the identification of genes with similarly diverse functions have also been carried out in other embryogenic systems (for a review see Fehér et al. 2003). Further investigations are needed in order to establish the significance of these genes/proteins in somatic embryogenesis, but their diversity indicates the wide range of cellular changes that are associated with embryogenic cell formation. Further details on differential gene expression during somatic embryogenesis are also given by Suprassanna (this volume).

The best-characterized gene that can be associated with embryogenic competence is the gene coding for the somatic embryogenesis receptor kinase (SERK1) identified first by Schmidt et al. (1997) in carrot. Using the SERK promoter fused to the luciferase gene and video cell tracking, it was shown that SERK-expressing single cells could indeed develop into somatic embryos (Schmidt et al. 1997). Furthermore, the ectopic expression of the *AtSERK* gene could facilitate the formation of somatic embryos (Hecht et al. 2001). SERK expression is therefore widely used as a marker of embryogenic competence (Baudino et al. 2001; Nolan et al. 2003; Somleva et al. 2000; Thomas et al. 2004; Ötvös et al. 2005). It was shown that in planta, AtSERK1 expression was first expressed during megasporogenesis and then in the functional megaspore, in all cells of the embryo sac until fertilization and in the embryo up to the heart stage. After this stage, expression was undetectable in any part of the developing seed. Low expression was, however, detected in adult vascular tissues. AtSERK1 gene expression was also observed in the shoot apical meristem and cotyledons of auxin-grown *Arabidopsis* seedlings used to initiate embryogenic callus cultures (Hecht et al. 2001). In other species (Baudino et al. 2001; Nolan et al. 2003; Somleva et al. 2000; Thomas et al. 2004), *SERK* gene homologues were also identified, but they were found to be even more widely expressed, indicating roles for these genes beyond the regulation of embryogenesis. Therefore, it was suggested that the SERK protein is rather a general morphogenetic than strictly an embryogenic marker (Nolan et al. 2003).

4 Induction of the Developmental Switch

Many tissue culture systems use 2,4-D as an efficient inducer of somatic embryogenesis. If we can answer the question why this synthetic auxin is so efficient in this respect, we may get closer to understanding the processes underlying the induction phase of somatic embryogenesis. The first question to be answered is whether 2,4-D is required for the acquisition of competence or for the initiation of the embryogenic cell fate or both. The question is not easy to answer in the case of cultures which are established in the long-term presence of 2,4-D and where embryos are formed only after the removal of 2,4-D (e.g. in the case of carrot). Does the commitment for embryo development happen before or after 2,4-D removal? Now it is well accepted that cell fate determination takes place in the presence of 2,4-D, which blocks the progression of the development at the same time. That 2,4-D is only a trigger of the cell fate switch is emphasized by experiments with a special *Medicago* cell culture (microcallus suspension culture or MCS) maintained in the presence of another synthetic auxin, namely naphthylacetic acid (NAA) (Dudits et al. 1991; Györgyey et al. 1991,1997). If these cells are transferred to hormone-free medium, they form roots with high frequency. If a large concentration (100 μM) of 2,4-D is applied to the cells for as short a time as a few minutes before the transfer to hormone-free medium, the cells will develop into somatic embryos. However, the first embryos can be observed on the surfaces of the calli only 2–3 weeks following the treatment. On the basis of these experiments, a high efficiency of embryogenesis could also be achieved on carrot hypocotyl surfaces after exposure to 450 μM 2,4-D for 2 h (Kitamiya et al. 2000). Indeed, these observations indicate that 2,4-D is required for the initiation of a programme that can further proceed on its own. Removal of 2,4-D from the induction medium can be important to allow the establishment of cellular polarity, which is one of the first cytological events underlying embryogenic development (Šamaj et al. 2003; for a review see Fehér et al. 2003).

2,4-D is often simply considered as an auxin analogue, but it has distinct and much more diverse effects than natural auxins. For example, 2,4-D has recently been demonstrated to regulate cell elongation and division in a different way from NAA (Campanoni and Nick 2005). That 2,4-D enhances division but simultaneously blocks elongation of cells could also be observed in the case of embryogenic alfalfa leaf protoplasts (Pasternak et al. 2002; Fehér et al. 2005).

As 2,4-D is also used as a herbicide, several attempts have been made to clarify its mode of action. Recent studies have proposed that ethylene is induced in response to auxinic herbicides (Grossmann 2000; Zheng and Hall 2001) and that ethylene in turn triggers ABA biosynthesis (Grossmann and Hansen 2001). The increased expression of the gene coding for 1-aminocyclopropane-1-carboxylic acid synthase which catalyses the rate-limiting step in ethylene biosynthesis as well as the involvement of 9-*cis*-epoxycarotenoid dioxygenase, a key regulator in ABA biosynthesis, has been demonstrated in the action of auxinic herbicides such as 2,4-D (Hansen and Grossmann 2000; Woeste et al. 1999). Further cell damage and death can be attributed to cyanide formation as a co-product of ethylene biosynthesis (Grossmann 1996). A genome-wide analysis of gene expression changes in *Arabidopsis* in response to 1-h treatment with 1 μM 2,4-D (only twice the concentration used to induce somatic embryogenesis in carrot by Kitamiya et al. 2000) has also been reported (Raghavan et al. 2005). In total 148 genes showed increased and 85 genes decreased transcription in response to this treatment. The wide spectrum of 2,4-D action is indicated by the various classes of genes affected, including genes involved in transcription, metabolism, signal transduction, cellular communication, protein turnover, subcellular localization, cellular transport and interaction with the cellular environment in addition to the 25% of the genes indentified that could not be classified.

These findings are in agreement with many observations made in experimental systems where 2,4-D was used to trigger somatic embryogenesis. Additionally, ABA has been reported to induce somatic embryogenesis in seedlings (Nishiwaki et al. 2000). Application of ABA to immature zygotic sunflower embryos resulted in the induction of somatic embryogenesis under sucrose conditions which otherwise allow only caulogenesis to occur (Charriére et al. 1999). Direct experimental evidence of the contribution of endogenous ABA to the induction phase of somatic embryos was provided by Senger et al. (2001). These authors showed that reduced cellular ABA levels in *Nicotiana plumbaginifolia* resulted in disturbed morphogenesis at the preglobular embryoid formation stage, which could be reversed by exogenous ABA application. ABA is considered to be a "stress hormone" in plants. Indeed, it has been widely reported that application of stress conditions can also induce or promote somatic embryo formation (for a review see Fehér et al. 2003). In alfalfa leaf protoplast-derived cells, various oxidative stress-inducing agents were shown to induce embryogenic cell formation under conditions where normally elongated, vacuolated cells develop (Pasternak et al. 2002). H_2O_2 and nitric oxide have also been shown to promote somatic embryogenesis (Kairong et al. 1999; Ötvös et al. 2005).

That oxidative stress and the stress responses are indeed an inherent part of 2,4-D-induced somatic embryogensis is well demonstrated by a microarray study. As a suitable experimental system, soybean cotyledones were placed with their abaxial side down on a medium containing 40 mg l^{-1} (approximately 200 μM) 2,4-D (Thibaud-Nissen et àl. 2003). Embryos appeared only on the adaxial side of explants after 21 days of culture. The gene expression pattern of the separated abaxial and adaxial parts was compared at different time points on a 9280-clone cDNA microarray. Clustering of the microarray data revealed that oxidative burst/detoxification, cell wall modification and cell division related genes significantly increased their expression after 7 days in culture. At 14 days, cell division activity was decreased, but the transcription of stress-responsive genes was enhanced. Proteomic analysis of somatic embryogenesis in *M. truncatula* also resulted in the identification of thioredoxin and 1-Cys-peroxiredoxin among the 16 proteins associated with embryogenic development (Imin et al. 2005).

In addition to induction of ABA and ethylene synthesis, 2,4-D has also been shown to increase endogenous auxin (IAA) levels in plant cells (Michalczuk et al. 1992a, b). The general role of auxin in the initiation of embryogenesis is supported by the findings that an auxin surge has been shown to accompany fertilization in carrot (Ribnicky et al. 2001) and that 2,4-D could induce the development of unfertilized isolated egg cells of wheat in vitro (Kranz et al. 1995). The appropriate endogenous auxin level of explants can be a key requirement for somatic embryogenesis. Even in those systems where no exogenous auxin is required for somatic embryo induction, the importance of the endogenous auxin level can be recognized. For example, ABA could induce embryogenesis in carrot seedlings only if the shoot tips, regions of auxin synthesis, were present (Nishiwaki et al. 2000). Ikeda-Iwai et al. (2003) reported that various stress treatments also promoted subsequent somatic embryo induction in shoot tip and flower bud explants. In alfalfa leaf protoplasts sodium nitroprusside as a NO donor could promote embryogenic cell formation only in the presence of auxin (Ötvös et al. 2005).

5 Determination of Embryogenic Cell Fate

Obviously, the initiation of embryogenic development in a differentiated cell requires a complete cellular reprogramming. Differentiated functions have to be deregulated and, following a transition phase, a new programme leading to embryo development has to be started. Although this reorganization is accompanied by profound morphological and physiological changes, reprogramming of the overall gene expression pattern is of utmost importance. During recent years it has become well accepted that the precise control of chromatin modifications in response to developmental and environmental cues determines the correct spatial and temporal expression of the genes (Li et al. 2002). The higher order of chromatin stabilizes gene expression patterns determining the regions of the genome that are silent or active in a given cell or at a given developmental phase (Wagner 2003). Experimental evidence has highlighted the importance of regulating chromatin structure in embryogenic transition. For example, chromatin-mediated gene silencing has been shown to play key roles in determining embryo and endosperm development in *Arabidopsis*. Mutations in *Arabidopsis* genes coding for similar proteins ("polycomb" group) that have been shown to have chromatin silencing functions during drosophila development have been identified. These mutations resulted in fertilization-independent endosperm (*fie*) or seed (*fis*) formation (Chaudhury et al. 2001; Grossniklaus et al. 2001; Luo et al. 1999; Ohad et al. 1999). Another mutation, *medea*, is defective in the protein involved in the same regulatory pathway (Grossniklaus et al. 1998; Kiyosue et al. 1999). These observations suggest that the embryogenic programme is repressed by chromatin-based gene silencing and becomes released in response to fertilization.

A further *Arabidopsis* mutant, *pickle (pkl)*, has a phenotype characterized by the postembryonic expression of embryo-specific markers and the spontaneous regeneration of somatic embryos in roots (Ogas et al. 1997, 1999). The product of the *pkl* gene was characterized as a chromatin-remodelling factor that represses embryogenesis-related gene expression and regulates the developmental transition from an embryogenic to a vegetative state (Ogas et al.

1999). In addition to chromatin organization, the direct regulation of genes involves specific transcription factors. Until now, several transcription factors (*leafy cotyledon 1 and 2, wuschell, baby boom*) have been identified to be involved in zygotic embryogenesis and to result in ectopic embryo formation if expressed in vegetative tissues (Boutilier et al. 2002; Lotan et al. 1998; Stone et al. 2001; Zuo et al. 2002; Sauer and Friml, this volume). The link between chromatin remodelling and these transcription factors has been demonstrated by the release of the repression of *lec1* expression in *pickle* mutants that can lead to the development of embryos on roots (Ogas et al. 1999). *Pickle* has been shown to repress embryogenic cell fate in all vegetative tissues (Henderson et al. 2004), but it was also demonstrated that the derepression of embryogenic functions in *pickle* mutants is selective (Dean Rider et al. 2003).

On the basis of the aforementioned evidence, one can hypothesize that during the induction of somatic embryogenesis the remodelling of chromatin results in the release of the embryogenic programme otherwise repressed by chromatin-based silencing mechanisms in vegetative plant cells. Polycomb-like chromo-domain-containing proteins have been shown to be expressed during carrot somatic and zygotic embryogenesis (Kiyosue et al. 1998). Furthermore, the expression of *lec1* during somatic embryogenesis has already been demonstrated in carrot and alfalfa (Yazawa et al. 2004; Fehér et al., unpublished results). It is interesting to note that in carrot *c-lec1* transcripts are already present in embryogenic cultures and the gene is strongly expressed 1 day after the removal of 2,4-D from the medium (Yazawa et al. 2004), but in alfalfa where a 1-h 2,4-D shock was followed by several weeks of culturing in hormone-free conditions, *ms-lec1* expression increased only at the time of the differentiation of embryos (3 weeks after induction; Fehér et al., unpublished results). This observation further supports the hypothesis that in the carrot system embryogenic commitment takes place before the removal of 2,4-D.

If we accept the primary role of chromatin remodelling in the initiation of the embryogenic programme, the main question still remains: what is the main signal and how does that signal result in chromatin remodelling and reprogramming of gene expression during somatic embryogenesis? In this respect it is interesting to note that the ectopic expression of the homeotic transcription factor *wuschel* in the root has been shown to induce shoot stem cell identity and leaf development on its own, floral development together with *leafy*, and embryogenesis together with auxin (Gallois et al. 2004). These results indicate that although auxin is required, it is insufficient to initiate embryogenesis in somatic plant cells on its own. A plausible model of the induction of somatic embryogenesis therefore might be be based on (at least) two factors: auxin, which is responsible for an appropriate cellular environment, and other unknown factor(s), including stress, which trigger the embryogenic programme.

6 Conclusions and Future Prospects

While the inducers of somatic embryogenesis are highly variable, the common cellular response has to be rather general. In vitro somatic embryogenesis is associated with artificial conditions, high levels of exogenous growth regulators and many other stress factors. These extreme and stressful conditions may result in a general stress response in cells showing extended chromatin reorganization. The presence of auxin as a growth regulator might also be important in order to provide the cells with the required developmental flexibility, e.g. promoting dedifferentiation. In this view, the general applicability of 2,4-D for the induction of somatic embryogenesis rests on its ability to evoke stress and auxin-responses at the same time (see earlier).

The extended chromatin reorganization caused by the inducing conditions might result in the "accidental" release of the embryogenic programme normally repressed by chromatin-mediated gene silencing mechanisms. Auxin (exogenous and/or endogenous) is also required for the expression of the embryogenic programme by ensuring cell survival, providing the suitable physiological background, inducing cell division and/or providing further necessary pathways. The large number of cellular events that have to be coordinated during the formation of embryogenic cells define together only a narrow window that indeed permits the initiation and progression of embryogenic development. That is why not all cells of an explant subjected to the same treatment are capable of developing into embryos, and why various explants, genotypes and species need different conditions for successful induction. This hypothesis, which should be validated by further experimental data on both zygotic and somatic embryogenesis, is summarized in Fig. 3.

Acknowledgements The research reported by the author was supported by grants BIO-00080/2002 and OTKA T34818. The author is also thankful for the support of the János Bólyai research fellowship.

References

Baudino S, Hansen S, Brettschneider R, Hecht VF, Dresselhaus T, Lorz H, Dumas C, Rogowsky PM (2001) Molecular characterisation of two novel maize LRR receptor-like kinases, which belong to the SERK gene family. Planta 213:1–10

Bicknell RA, Koltunow AM (2004) Understanding apomixis: recent advances and remaining conundrums. Plant Cell 16:228–245

Bögre L, Stefanov I, Ábrahám M, Somogyi I, Dudits D (1990) Differences in the responses to 2,4-dichlorophenoxyacetic acid (2,4-D) treatment between embryogenic and non-embryogenic lines of alfalfa. In: Nijkamp HJJ, van der Plaas LHW, Van Aartrijk J (eds) Progress in plant cellular and molecular biology. Kluwer, Dordrecht, pp 427–436

Boutilier K, Offringa R, Sharma VK, Kieft H, Ouellet T, Zhang L, Hattori J, Liu CM, van Lammeren AA, Miki BL, Custers JB, Lookeren Campagne MM (2002) Ectopic expres-

sion of BABY BOOM triggers a conversion from vegetative to embryonic growth. Plant Cell 14:1737–1749
Bowley SR, Kielly GA, Anandarajah K, McKersie BD, Senaratna T (1993) Field-evaluation following 2 cycles of backcross transfer of somatic embryogenesis to commercial alfalfa. Germplasm. Can J Plant Sci 73:131–137
Campanoni P, Nick P (2005) Auxin-dependent cell division and cell elongation. 1-Naphthaleneacetic acid and 2,4-dichlorophenoxyacetic acid activate different pathways. Plant Physiol 137:939–948
Charriére F, Sotta B, Miginiac É, Hahne G (1999) Induction of adventitious or somatic embryos on in vitro cultured zygotic embryos of Helianthus annuus: variation of endogenous hormone levels. Plant Physiol Biochem 37:751–757
Chaudhury AM, Koltunow A, Payne T, Luo M, Tucker MR, Dennis ES, Peacock WJ (2001) Control of early seed development. Annu Rev Cell Dev Biol 17:677–699
Chuck G, Hake S (2005) Regulation of developmental transitions. Curr Opin Plant Biol 8:67–70
Dean Rider S Jr, Henderson JT, Jerome RE, Edenberg HJ, Romero-Severson J, Ogas J (2003) Coordinate repression of regulators of embryonic identity by PICKLE during germination in Arabidopsis. Plant J 35:33–43
Dudits D, Bögre L, Györgyey J (1991) Molecular and cellular approaches to the analysis of plant embryo development from somatic cells in vitro. J Cell Sci 99:475–484
Fehér A, Pasternak TP, Dudits D (2003) Transition of somatic plant cells to an embryogenic state. Plant Cell Tissue Org Cult 74:201–228
Fehér A, Pasternak TP, Ötvös K, Dudits D (2005) Plant protoplasts: consequences of lost cell walls. In: Murch S, Saxena PK (eds) Journey of a single cell to a plant. Science Publishers Inc., Enfield, NH, USA, pp 59–89
Gallois JL, Nora FR, Mizukami Y, Sablowski R (2004) WUSCHEL induces shoot stem cell activity and developmental plasticity in the root meristem. Genes Dev 18:375–380
Grimanelli D, Perotti E, Ramirez J, Leblanc O (2005) Timing of the maternal-to-zygotic transition during early seed development in maize. Plant Cell 17:1061–1072
Grossmann K (1996) A role for cyanide, derived from ethylene biosynthesis, in the development of stress symptoms. Physiol Plant 97:772–775
Grossmann K (2000) Mode of action of auxinic herbicides: a new ending to a long, drawn out story. Trends Plant Sci 5:506–508
Grossmann K, Hansen H (2001) Ethylene-triggered abscisic acid: a principle in plant growth regulation? Physiol Plant 113:9–14
Grossniklaus U, Vielle-Calzada JP, Hoeppner MA, Gagliano WB (1998) Maternal control of embryogenesis by MEDEA, a polycomb group gene in Arabidopsis. Science 280:446–450
Grossniklaus U, Spillane C, Page DR, Kohler C (2001) Genomic imprinting and seed development: endosperm formation with and without sex. Current Opin Plant Biol 4:21–27
Györgyey J, Gartner A, Nemeth K, Magyar Z, Hirt H, Heberle-Bors E, Dudits D (1991) Alfalfa heat shock genes are differentially expressed during somatic embryogenesis. Plant Mol Biol 16:999–1007
Györgyey J, Nemeth K, Magyar Z, Kelemen Z, Alliotte T, Inze D, Dudits D (1997) Expression of a novel-type small proline-rich protein gene of alfalfa is induced by 2,4-dichlorophenoxiacetic acid in dedifferentiated callus cells. Plant Mol Biol 34:593–602
Halperin W (1969) Morphogenesis in cell cultures. Annu Rev Plant Physiol 20:395–418
Hansen H, Grossmann K (2000) Auxin-induced ethylene triggers abscisic acid biosynthesis and growth inhibition. Plant Physiol 124:1437–1448

Hecht V, Vielle-Calzada JP, Hartog MV, Schmidt ED, Boutilier K, Grossniklaus U, de Vries SC (2001) The Arabidopsis somatic embryogenesis receptor kinase 1 gene is expressed in developing ovules and embryos and enhances embryogenic competence in culture. Plant Physiol 127:803–816

Henderson JT, Li HC, Rider SD, Mordhorst AP, Romero-Severson J, Cheng JC, Robey J, Sung ZR, de Vries SC, Ogas J (2004) PICKLE acts throughout the plant to repress expression of embryonic traits and may play a role in gibberellin-dependent responses. Plant Physiol 134:995–1005

Ikeda-Iwai M, Umehara M, Satoh S, Kamada H (2003) Stress-induced somatic embryogenesis in vegetative tissues of Arabidopsis thaliana. Plant J 34:107–114

Imin N, Nizamidin M, Daniher D, Nolan KE, Rose RJ, Rolfe BG (2005) Proteomic analysis of somatic embryogenesis in Medicago truncatula. Explant cultures grown under 6-benzylaminopurine and 1-naphthaleneacetic acid treatments. Plant Physiol 137:1250–1260

Kairong C, Gengsheng X, Xinmin L, Gengmei X, Yafu W (1999) Effect of hydrogen peroxide on somatic embryogenesis of Lycium barbarum L. Plant Sci 146:9–16

Kamada H, Ishikawa K, Saga H, Harada H (1993) Induction of somatic embryogenesis in carrot by osmotic stress. Plant Tissue Cult Lett 10:38–44

Kielly GA, Bowley SR (1992) Genetic control of somatic embryogenesis in alfalfa. Genome 35:474–477

Kitamiya E, Suzuki S, Sano T, Nagata T (2000) Isolation of two genes that were induced upon the initiation of somatic embryogenesis on carrot hypocotyls by high concentrations of 2,4-D. Plant Cell Rep 19:551–557

Kiyosue T, Shiota H, Higashi K, Kamada H, Shinozaki K (1998) A chromo box gene from carrot (Daucus carota L.): its cDNA structure and expression during somatic and zygotic embryogenesis. Biochim Biophys Acta 1398:42–46

Kiyosue T, Ohad N, Yadegari R, Hannon M, Dinneny J, Wells D, Katz A, Margossian L, Harada JJ, Goldberg RB, Fischer RL (1999) Control of fertilization-independent endosperm development by the MEDEA polycomb gene in Arabidopsis. Proc Natl Acad Sci USA 96:4186–4191

Kranz E (1999) In vitro fertilization with isolated single gametes. Methods Mol Biol 111:259–267

Kranz E, von Wiegen P, Lörz H (1995) Early cytological events after induction of cell division in egg cells and zygote development following in vitro fertilization with angiosperm gametes. Plant J 8:9–23

Krishna Raj S, Vasil IK (1995) Somatic embryogenesis in herbaceous monocots. In: Thorpe TA (ed) In vitro embryogenesis in plants. Kluwer, Dord recht, pp 417–470

Li G, Hall TC, Holmes-Davis R (2002) Plant chromatin: development and gene control. Bioessays 24:234–243

Lotan T, Ohto M, Yee KM, West MAL, Lo R, Kwong RW, Yamagishi K, Fischer RL, Goldberg RB, Harada JJ (1998) Arabidopsis LEAFY COTYLEDON1 is sufficient to induce embryo development in vegetative cells. Cell 93:1195–1205

Luo M, Bilodeau P, Koltunow A, Dennis ES, Peacock WJ, Chaudhury AM (1999) Genes controlling fertilization-independent seed development in Arabidopsis thaliana. Proc Natl Acad Sci USA 96:296–301

Michalczuk L, Cooke TJ, Cohen JD (1992a) Auxin levels at different stages of carrot somatic embryogenesis. Phytochemistry 31:1097–1103

Michalczuk L, Ribnicky DM, Cooke TJ, Cohen JD (1992b) Regulation of indole-3-acetic acid biosynthetic pathways in carrot cell cultures. Plant Physiol 100:1346–1353

Moltrasio R, Robredo CG, Gomez MC, Paleo AHD, Diaz DG, Rios RD, Franzone PM (2004) Alfalfa (Medicago sativa) somatic embryogenesis: genetic control and introduction of favourable alleles into elite Argentinean germplasm. Plant Cell Tissue Organ Cult 77:119–124

Mordhorst AP, Toonen MAJ, de Vries SC (1997) Plant embryogenesis. Crit Rev Plant Sci 16:535–576

Neumann KH (2000) Some studies on somatic embryogenesis: A tool in plant biotechnology. http://geb.uni-giessen.de/geb/volltexte/2000/321/

Nishiwaki M, Fujino K, Koda Y, Masuda K, Kikuta Y (2000) Somatic embryogenesis induced by the simple application of abscisic acid to carrot (Daucus carota L.) seedlings in culture. Planta 211:756–759

Nolan KE, Irwanto RR, Rose RJ (2003) Auxin up-regulates MtSERK1 expression in both Medicago truncatula root-forming and embryogenic cultures. Plant Physiol 133:218–230

Nomura K, Komamine A (1985) Identification and isolation of single cells that produce somatic embryos at a high frequency in a carrot cell suspension culture. Plant Physiol 79:988–991

Ogas J, Cheng JC, Sung ZR, Somerville C (1997) Cellular differentiation regulated by gibberellin in the Arabidopsis thaliana pickle mutant. Science 277:91–94

Ogas J, Kaufmann S, Henderson J, Somerville C (1999) PICKLE is a CHD3 chromatin-remodeling factor that regulates the transition from embryonic to vegetative development in Arabidopsis. Proc Natl Acad Sci USA 96:13839–13844

Ohad N, Yadegari R, Margossian L, Hannon M, Michaeli D, Harada JJ, Goldberg RB, Fischer RL (1999) Mutations in FIE, a WD polycomb group gene, allow endosperm development without fertilization. Plant Cell 11:407–416

Osuga K, Masuda H, Komamine A (1999) Synchronization of somatic embryogenesis at high frequency using carrot suspension cultures: model systems and application in plant development. Methods Cell Sci 21:129–140

Ötvös K, Pasternak TP, Miskolczi P, Domoki M, Dorjgotov D, Szûcs A, Bottka S, Dudits D, Fehér A (2005) Nitric oxide is required for, and promotes auxin-mediated activation of, cell division and embryogenic cell formation but does not influence cell cycle progression in alfalfa cell cultures. Plant J 43:849–860

Pasternak TP, Prinsen E, Ayaydin F, Miskolczi P, Potters G, Asard H, Van Onckelen HA, Dudits D, Fehér A (2002) The role of auxin, pH, and stress in the activation of embryogenic cell division in leaf protoplast-derived cells of alfalfa. Plant Physiol 129:1807–1819

Raghavan C, Ong EK, Dalling MJ, Stevenson TW (2005) Effect of herbicidal application of 2,4-dichlorophenoxyacetic acid in Arabidopsis. Funct Integr Genomics 5:4–17

Reynolds TL (1997) Pollen embryogenesis. Plant Mol Biol 33:1–10

Ribnicky DM, Cohen JD, Hu WS, Cooke TJ (2001) An auxin surge following fertilization in carrots: a mechanism for regulating plant totipotency. Planta 214:505–509

Šamaj J, Baluška F, Pretová A, Volkmann D (2003) Auxin deprivation induces a developmental switch in maize somatic embryogenesis involving redistribution of microtubules and actin filaments from endoplasmic to cortical cytoskeletal arrays. Plant Cell Rep 21:940–945

Schmidt ED, Guzzo F, Toonen MA, de Vries SC (1997) A leucine-rich repeat containing receptor-like kinase marks somatic plant cells competent to form embryos. Development 124:2049–2062

Senger S, Mock HP, Conrad U, Manteuffel R (2001) Immunomodulation of ABA function affects early events in somatic embryo development. Plant Cell Rep 20:112–120

Somleva MN, Schmidt EDL, de Vries SC (2000) Embryogenic cells in Dactylis glomerata L. (Poaceae) explants identified by cell tracking and by SERK expression. Plant Cell Rep 19:718–726

Sprunck S, Baumann U, Edwards K, Langridge P, Dresselhaus T (2005) The transcript composition of egg cells changes significantly following fertilization in wheat (Triticum aestivum L.). Plant J 41:660–672

Stone SL, Kwong LW, Yee KM, Pelletier J, Lepiniec L, Fischer RL, Goldberg RB, Harada JJ (2001) LEAFY COTYLEDON2 encodes a B3 domain transcription factor that induces embryo development. Proc Natl Acad Sci USA 98:11806–11811

Taylor RL (1967) The foliar embryos of Malaxis paludosa. Can J Bot 45:1553–1556

Thibaud-Nissen FO, Shealy RT, Khanna A, Vodkin LO (2003) Clustering of microarray data reveals transcript patterns associated with somatic embryogenesis in soybean. Plant Physiol 132:118–136

Thomas C, Meyer D, Himber C, Steinmetz A (2004) Spatial expression of a sunflower SERK gene during induction of somatic embryogenesis and shoot organogenesis. Plant Physiol Biochem 42:35–42

Thorpe TA (1995) In vitro embryogenesis in plants. Kluwer, Dordrecht

Toonen MAJ, Hendriks T, Schmidt EDL, Verhoeven HA, van Kammen A, de Vries SC (1994) Description of somatic-embryo forming single cells in carrot suspension cultures employing video cell tracking. Planta 194:565–572

Toonen MAJ, Schmidt EDL, Hendriks T, Verhoeven HA, van Kammen A, de Vries SC (1996) Expression of the JIM8 cell wall epitope in carrot somatic embryogenesis. Planta 200:167–173

Wagner D (2003) Chromatin regulation of plant development. Curr Opin Plant Biol 6:20–28

Woeste KE, Ye C, Kieber JJ (1999) Two Arabidopsis mutants that overproduce ethylene are affected in the posttranscriptional regulation of 1-aminocyclopropane-1-carboxylic acid synthase. Plant Physiol 119:521–530

Yarbrough JA (1932) Anatomical and developmental studies of the foliar embryos of Bryophyllum calcynum. Am J Bot 19:443–453

Yazawa K, Takahata K, Kamada H (2004) Isolation of the gene encoding Carrot leafy cotyledon1 and expression analysis during somatic and zygotic embryogenesis. Plant Physiol Biochem 42:215–223

Zheng H, Hall C (2001) Understanding auxinic herbicide resistance in wild mustard: physiology, biochemical, and molecular genetic approaches. Weed Sci 49:276–281

Zuo J, Niu QW, Frugis G, Chua NH (2002) The WUSCHEL gene promotes vegetative-to-embryonic transition in Arabidopsis. Plant J 30:349–359

Plant Cell Monogr (2)
A. Mujib · J. Šamaj: Somatic Embryogenesis
DOI 10.1007/7089_034/Published online: 30 November 2005

Participation of Plant Hormones in Determination and Progression of Somatic Embryogenesis

Víctor M. Jiménez[1] (✉) · Clément Thomas[2]

[1]CIGRAS, Universidad de Costa Rica, 2060 San Pedro, Costa Rica
vjimenez@cariari.ucr.ac.cr

[2]Plant Molecular Biology, CRP-Santé, Bâtiment modulaire,
84 Val Fleuri, 1526 Luxembourg, Luxembourg
clement.thomas@crp-sante.lu

Abstract In vitro culture protocols have been developed for many species, mainly using empirical approaches, to induce somatic embryogenesis from various explant types. However, the underlying biochemical mechanisms governing induction, expression and maturation during somatic embryogenesis are still poorly understood. Among the signals that participate directly in the regulation of the different phases of this process, plant hormones emerged as candidates of choice. In this chapter, studies concerning the role of exogenously added plant growth regulators in somatic embryogenesis are reviewed. In addition, we discuss possible relationships between hormonal contents in starting explants and in cultures derived from them with their embryogenic competence. Moreover, information on evolution of endogenous plant hormone levels during induction and progression of somatic embryogenesis is presented. Finally, an overview of interactions between exogenous plant growth regulators and endogenous hormones in embryogenic systems is also included.

Abbreviations

2,4-D	2,4-Dichlorophenoxyacetic acid
ABA	Abscisic acid
CK	Cytokinin
E	Embryogenic
GA	Gibberellin
GA_3	Gibberellic acid
IAA	Indole-3-acetic acid
IBA	Indole-3-butyric acid
NAA	Naphthalene acetic acid
NE	Nonembryogenic
PGR	Plant growth regulator
PAT	Polar auxin transport
SE	Somatic embryogenesis
TIBA	2,3,4-Triiodobenzoic acid

1
Introduction

Plant hormones play a determinant role in practically every developmental process studied to date in plants, somatic embryogenesis (SE) being no exception. Substances classified as plant hormones are of organic nature and act at very low concentrations. Whether these compounds should be named hormones or not, taking into consideration the properties of the corresponding compounds in animal physiology, has been a topic of some debate during the past few years. The term plant growth regulator (PGR) has been proposed as an alternative that matches more precisely the characteristics of these substances, but has the disadvantage that it has been used to name synthetic substances of this class (for details refer to Davis 1995). In this chapter, the term plant hormone will be used to define the endogenous and naturally occurring substances in the tissues, while the expression PGR will refer to those exogenously added compounds, usually of synthetic origin.

Most studies on regulation of SE have focused on one or another of the several stages in which this process has been divided. The first one involves the induction stage, in which somatic tissues acquire, directly (without a dedifferentiation step) or indirectly (by dedifferentiating tissues already differentiated, usually involving a callus phase), embryogenic (E) competence. This stage is followed by the expression of SE, in which the competent cells or proembryos start developing, after receiving the proper stimulus, passing through the phases characteristic of zygotic embryo development, i.e., globular, heart-shaped and torpedo-shaped stages in dicots, globular, scutellar (transition) and coleoptilar stages in monocots, and globular, early cotyledonary and late cotyledonary embryos in conifers (Jiménez 2001). Finally, during maturation, somatic embryos prepare themselves for germination, by desiccating and accumulating reserves.

SE is a very complex developmental process that shares similar characteristics, mainly in morphology and anatomy, within the same group of plants (monocots, dicots, gymnosperms), but which, at the same time, differs in the requirements needed to induce and govern its determination and progression. This complexity has impeded fully understanding the biochemistry and physiology of SE, the role of plant hormones and PGRs included. Therefore, in spite of the large amount of research conducted on the involvement of plant hormones, but especially of PGRs during SE, the way they interact with the cells and tissues to render an observed response is still not clear. Therefore, for specific genotypes, trial-and-error experiments to establish the proper culture conditions and media, especially the type and level of PGRs to be used, are nowadays still common practice (Huang et al. 2004; Zhang et al. 2005).

The aim of this review is to summarize relevant and recent findings related to the involvement of plant hormones and PGRs in the determination and progression of SE. Whenever possible, review works will be cited to avoid mentioning very large amounts of literature, although specific references, not always of recent publication, related to relevant findings will be also employed.

2 Effect of PGRs on SE

2.1 PGRs as Inducers of SE

The PGR composition of the culture medium is of prime importance to achieve the desired morphogenic reaction. In several culture systems, such as in *Arachis hypogaea* seedlings (Victor et al. 1999), *Juglans regia* embryonic axes (Fernández et al. 2000) and sunflower zygotic embryos (Thomas et al. 2004), the morphogenic pathway can be oriented through either shoot organogenesis or SE, only by modifying the PGR composition of the culture medium.

Among individual groups of PGRs, auxins are the most routinely used agents to mediate the transition from somatic to E cells. In more than 80% of 124 recently published protocols, induction of SE required the presence of auxins alone, or in combination with cytokinins (CKs) (Gaj 2004). The auxin most frequently used to initiate in vitro SE is the synthetic auxin 2,4-dichlorophenoxyacetic acid (2,4-D), also known for its herbicide activity. Naphthalene acetic acid (NAA), indole-3-acetic acid (IAA), indole-3-butyric acid (IBA), Picloram and Dicamba are used to a lesser extent (Raemakers et al. 1995).

The mode by which 2,4-D efficiently induces E competence remains unclear. On the one hand, 2,4-D could regulate SE through its strong auxinic activity, either directly or indirectly, by influencing the endogenous metabolism of other phytohormones (Sect. 5). On the other hand, 2,4-D could act as a strong stressor leading to SE, considered by some authors as an extreme stress response of cultured plant cells (reviewed by Fehér et al. 2003). This hypothesis is supported by the fact that several stress treatments can trigger SE (reviewed by Gaj 2004).

In most instances in which CKs induced SE, they were added to the culture medium together with auxins (Gaj et al. 2004). However, in some cases, the addition of CKs as the sole source of PGR is sufficient to generate somatic embryos (Bronner et al. 1994; Iantcheva et al. 1999). The most commonly used CKs in culture are N^6-benzylaminopurine, kinetin, zeatin and, more recently, thidiazuron.

There are only a few reports of abscisic acid (ABA) acting as an effective inducer of SE, in most cases by producing somatic embryos directly on the surface of the explants (Bell et al. 1993; Charrière and Hahne 1998). In a relatively recent work, Nishiwaki et al. (2000) reported the formation of somatic embryos from carrot seedlings cultured on a medium containing ABA as the sole source of PGRs. In this work, the number of embryos induced per number of seedlings was dependent on the seedling age as well as on the ABA concentration in the medium.

The effect of exogenously applied gibberellins (GAs) on induction of SE is highly variable in different species or tissues. For example, exogenous gibberellic acid (GA_3), the most commonly used synthetic GA, stimulated embryogenesis in both chickpea immature cotyledon cultures (Hita et al. 1997) and *Medicago sativa* tissue cultures (Rudus̈ et al. 2002), whereas it was detrimental to somatic embryo formation in geranium hypocotyl cultures (Hutchinson et al. 1997) and in *Citrus* ovule callus (Kochba et al. 1978). In carrot, the effect of GA on SE is also controversial. Tokuji and Kuriyama (2003) reported that GA_3 inhibited embryogenesis at the globular stage, while uniconazole, a GA biosynthesis inhibitor, promoted secondary embryogenesis when embryos were induced directly from carrot hypocotyl segments. In contrast, Mitsuhashi et al. (2003) observed that exogenous treatment with uniconazole caused a reduction of both the number of the developed embryos and the size of the torpedo-shaped embryos. These abnormalities in the latter case were prevented by GA_1 or GA_4 application.

There are several reports that support the inhibiting effect of external application of ethylene on induction of SE (reviewed by Minocha and Minocha 1995; Nomura and Komamine 1995; Thorpe 2000), while others indicate a neutral role (Roustan et al. 1994).

2.2 PGRs on Progression of SE

In most protocols in which auxins act as an efficient inducer of SE, development of somatic embryos is achieved by reducing or removing auxin from the culture medium. To explain this result, it was proposed that continuous exposition of explants to high exogenous auxin levels interferes with the polar auxin gradient that is normally established during embryogenesis, preventing the correct apical–basal embryo patterning (Schiavone and Cooke 1987; Liu et al. 1993).

The importance of polar auxin transport (PAT) in embryo morphogenesis was demonstrated by treating different stages of carrot somatic embryos with the PAT inhibitors 2,3,4-triiodobenzoic acid (TIBA) and *N*-(1-naphthyl)phthalamic acid (Schiavone and Cooke 1987; Cooke et al. 1993). Both inhibitors blocked the ability of somatic embryos to undergo morphogenic transitions to the subsequent stages. In a more recent experiment, Tokuji and Kuriyama (2003) treated carrot hypocotyls, in which SE was directly induced with a 24-h pulse of 2,4-D, with TIBA and 2,4,6-trichlorophenoxyacetic acid, another inhibitor of PAT, and found inhibition in the development of the somatic embryos, but not in the frequency of SE.

As will be pointed out later in this review (Sect. 3.2), formation of an auxin gradient appears to be necessary to establish bilateral symmetry during the initial steps of embryogenesis, a requisite for further development of the embryos (Schiavone and Cooke 1987; Liu et al. 1993; Fischer and Neuhaus 1996).

An exogenous supply of CKs during the expression phase has produced ambiguous results. While some reports indicate an inductive role of CKs during progression of SE (e.g., Fujimura and Komamine 1975; Sagare et al. 2000), the contrary has also been described (Li and Demarly 1996). As mentioned later, CKs seem to play an important role in cell division, rather than in embryo differentiation (Danin et al. 1993).

Addition of GAs during progression of SE has also shown confusing outcomes. On one hand, it stimulated embryo development in chickpea, *Iris germanica* and *M. sativa*, while on the other hand, it inhibited this event in carrot, mandarin, orange and anise (Rudus̈ et al. 2002 and references therein). Later in the progress of SE, it was observed that in some species normally showing dormancy, adding GA_3 promotes germination and conversion of somatic embryos into plants (reviewed by Gaj 2004).

Maybe the most relevant effect of ABA during progression of SE has been reported for conifers. In this plant group, development of somatic embryos has to be stimulated by exogenous addition of ABA (reviewed by Dong and Dunstan 2000; Stasolla et al. 2002). A more general effect of ABA has been observed during maturation of somatic embryos in numerous species, especially, but not restricted to, conifers (Mauri and Manzanera 2004; Sharma et al. 2004). Indeed, similarly to the effect produced by the natural increase of endogenous ABA in zygotic embryos, the addition of ABA into the culture medium induces a reduction in precocious germination and an increase in the number of mature somatic embryos. However, an extensive duration of the treatment could influence negatively conversion of mature embryos into plantlets (von Arnold et al. 2002).

Again, there is a limited number of reports on the effect of addition of ethylene on the progression of SE. In one of them, Roustan et al. (1994) observed an arrest in embryo development only when ethylene was applied during the first 7 days in the expression stage of carrot SE. When this compound was included after that moment, no effect was evident.

3 Is E Competence of Explants Determined by Endogenous Hormones?

3.1 The Situation in Donor Explants

Some attempts have been made to associate the endogenous hormone contents of donor plant tissues, on one hand, and of callus or cell suspension cultures derived from them, on the other, with their E competence.

Concerning the donor tissues, there are several works that report differences among responsive and unresponsive explants, and thus support the participation of endogenous plant hormones on E competence. In this way, a correlation between higher endogenous IAA concentrations and an increased E response was reported in leaves of alfalfa (Ivanova et al. 1994), *Pennisetum purpureum* (Rajasekaran et al. 1987a) and *Dactylis glomerata* (Wenck et al. 1988), as well as in immature zygotic embryos of wheat (Kopertekh and Butenko 1995) and sunflower (Charrière et al. 1999; Thomas et al. 2002).

A relationship between the levels of ABA and the E competence of the initial explants was reported in the aforementioned work of Kopertekh and Butenko (1995) and also by Jiménez and Bangerth (2001a), both in wheat. The latter authors suggest that the effect of ABA on competence might occur by reducing precocious germination and then indirectly favoring callus formation. Similarly, Rajasekaran et al. (1987b) found higher concentrations of ABA in E than in nonembryogenic (NE) leaf sections of *P. purpureum*, while Ivanova et al. (1994) found the opposite in equivalent tissues of *M. falcata*. Additional support for the positive role of endogenous ABA in determining E competence of the donor tissues derives from the results of Senger et al. (2001), working with *Nicotiana plumbaginifolia*. They reported that both transgenic plants that overexpress an anti-ABA single-chain variable fragment antibody and mutants that have a defect in the ABA synthesis rate exhibit abnormal morphogenesis at preglobular embryoid formation. This phenotype could be reversed by simple exogenous ABA application.

Concerning CKs, lower levels of total CKs were observed in competent tissues in leaves of *P. purpureum* (Rajasekaran et al. 1987a) and *D. glomerata* (Wenck et al. 1988), as well as in immature zygotic embryos of wheat (Kopertekh and Butenko 1995), than in their noncompetent ones. Sometimes the factor to be considered is not the pattern of total CKs, but the levels of individual members of this group of plant hormones. For example, even though Centeno et al. (1997) did not find differences in the total amounts of CKs between competent and noncompetent genotypes of *Coryllus avellana*, they reported differences in the contents of the individual CKs evaluated.

Supporting the positive role of CKs during induction of SE, Tokuji and Kuriyama (2003) reported that purine riboside, an anti-CK, severely inhibited SE from epidermal cells of carrot. This effect was counteracted by the simultaneous application of zeatin riboside, suggesting that CKs are involved in the very early stages of SE, such as the formation of E cell clumps. These findings support the concept, mentioned before, that CKs have a role in cell division rather than in embryo differentiation (Danin et al. 1993).

Regarding GAs, there are contrasting reports about the role played by endogenous levels of this hormone in donor explants. There are several publications that do not show differences in GA levels in genotypes differing in their E competence (Jiménez and Bangerth 2001a, b), an indication of the minor role these compounds might play during this phase. Supporting these findings, Rajasekaran et al. (1987a) observed that paclobutrazol, an inhibitor of GA synthesis, did not alter E nature of *P. purpureum* explants. However, other works suggest a negative role of endogenous GAs on E competence (Hutchinson et al. 1997).

Further evidence against a relationship between endogenous plant hormones in donor explants and their E competence comes from studies in maize (Jiménez and Bangerth 2001b) and asparagus (Limanton-Grevet et al. 2000) genotypes in which it was not possible to identify differences in the endogenous hormone contents in genotypes having different E capacity. Additionally, immature zygotic embryos of barley that contained variable IAA and GA levels displayed a similar degree of competence (Jimenez and Bangerth 2001c).

3.2
The Situation in Cultures with Distinct E Capacity

Some divergences have also been reported when endogenous hormone contents were evaluated in callus and cell suspension cultures varying in their degree of E capacity. Most works conducted with this purpose support the occurrence of higher auxin levels in E than in NE cultures (reviewed by Jiménez 2001). However, in other works, no differences in the endogenous auxin contents could be established between E and NE cultures (Besse et al. 1992; Michalczuk et al. 1992a). It was postulated that high endogenous auxin contents help to set up the auxin gradient necessary to establish bilateral symmetry during zygotic and SE (Schiavone and Cooke 1987; Liu et al. 1993; Fischer and Neuhaus 1996).

Higher levels of total CKs have been reported in NE than in E callus of *P. purpureum* (Rajasekaran et al. 1987a) and of *M. arborea* (Pintos et al. 2002). In addition, similarly to the aforementioned report of Centeno et al. (1997), Guiderdoni et al. (1995) found differences in the contents of individual CKs between E and NE callus cultures, in sugarcane. However, in spite of the previous reports, some researchers argue that CK levels are probably more related to the growth of the callus cultures than to the E competence (reviewed by Jiménez 2001).

A similar scene to the one described for auxins is found in ABA: even though the majority of publications support higher levels of this plant hormone in E than in NE cultures (reviewed by Jimenez 2001; Nakagawa et al. 2001), the contrary was reported for *Hevea brasiliensis* (Etienne et al. 1993) and alfalfa (Ivanova et al. 1994) cultures. A completely ambiguous situation is found in GAs, where higher levels of this hormone were found in E than in NE cultures in some works, while the contrary was found in others, whereas no differences were reported in some other publications (reviewed by Jimenez 2001).

Since ethylene quantification within the tissues is a very difficult task, indirect information about the role endogenous contents of this plant hormone might play in determination of the E potential of cultures comes from experiments using inhibitors of ethylene synthesis and action. Most works on this subject indicate that ethylene plays a negative role on induction of SE (reviewed by Thorpe 2000). However, in a carrot line in which ethylene promoted SE, the use of inhibitors of biosynthesis also slightly inhibited SE (Nissen 1994).

Because spatial information is lost during global quantification of endogenous hormones, researchers have tried to obtain a precise localization of plant hormones within the tissues. This can only be achieved with the help of in situ techniques, such as immunolocalization. Whereas analytical methods for quantifying plant hormones have been strongly improved during recent years, in situ specific detection of these compounds has been more difficult. As an example, despite auxin having proven to be a difficult molecule to localize in tissues, being highly diffusible and occurring in both active and inactive (conjugated) forms (Normanly and Bartel 1999), successful immunohistochemical localization of IAA has been recently reported (Moctezuma 1999; Moctezuma and Feldman 1999; Aloni et al. 2003), including a report during early phases of SE (Thomas et al. 2002).

4 How Do Endogenous Hormone Contents Evolve in the Progress of SE?

Several studies aiming to evaluate the way endogenous hormone concentrations change during development of SE, specifically after expression has been induced, have been carried out. In some of them, the initial stages of embryo development have been analyzed, i.e., before the first morphological changes had occurred, but when biochemical and physiological determination of embryo development has already started (Dodeman and Ducreux 1996). The other group of studies focused on the later phases of embryo development, when it is possible to synchronize and separate the different embryo stages through a series of steps of sieving and centrifugation (reviewed by Osuga et al. 1999; Sharma 1999). Synchronization of E cultures allows a more accurate estimation of the hormone status in each phase of embryo development.

Endogenous contents of most hormones remained steady or showed only minor changes during the first 7 days after 2,4-D had been eliminated from the medium in carrot E cultures (Fujimura and Komamine 1979; Michalczuk et al. 1992a; Jiménez et al. 2005); only increased contents of the polyamines putrescine, spermidine and spermine have been, to the best of our knowledge, reported (Feinberg et al. 1984). In citrus E cultures, in which expression of SE was triggered by a stimulus other than reducing the auxin content in the medium, auxin and CKs accumulated within the first 5 days after sucrose had been replaced by glycerol in the culture medium, the triggering factor, while the levels of ABA and GAs remained stable (Jiménez et al. 2001).

When studying evolution of endogenous hormones in SE after the first morphological changes had occurred, Michalczuk et al. (1992a) reported that auxin levels decline steadily after the globular stage in all subsequent stages of embryo development. Additional information, in their case for ABA, was provided by Kamada and Harada (1981). They found that, after remaining low during the first 7 days of culture in the absence of 2,4-D, the concentrations of ABA increased during further development of carrot somatic embryos until day 10, and then decreased. Similarly, Rajasekaran et al. (1982) found that ABA levels in hybrid grapevine somatic embryos decreased from the globular to the mature stage. The role of endogenous ABA has been more evident during the latter stages of embryo development, especially during maturation and germination. In this sense, Kermode et al. (1989) and, recently, Prewein et al. (2004) related an increase in germination to a reduction in ABA content in the tissues, while Finkelstein et al. (1985) related the beginning of germination to a change in the sensitivity of the tissues to this plant hormone.

Information regarding GA content during the final phases of somatic embryo development originates from two early works (Noma et al. 1982; Takeno et al. 1983). In the first one, polar and less polar GA contents were compared during this phase and lower levels of polar and higher levels of less polar GAs were found, while in the second, a reduction in the levels of free and highly soluble GA-like substances on a dry weight basis was observed during embryo development.

Endogenous ethylene increased at day 1 after transferring somatic embryos of white spruce into the maturation medium, and then declined transiently and increased again gradually, in the second half of the culture period (Kong and Yeung 1994). Concerning polyamines, very recently, Minocha et al. (2004) found a correlation in the relationship of several members of this plant hormone group with the developmental stage of red spruce somatic embryos.

5
PGRs Acting on Endogenous Hormones During SE

The mode of action of PGRs involves modulation of endogenous plant hormone concentrations, among other effects, a process that may occur directly, through synthesis of enzymes, or indirectly, with the intervention of effectors (Thorpe 2000; Gaspar et al. 2003; Gazzarrini and McCourt 2003). An exogenous PGR can, positively or negatively, modulate internal concentrations of plant hormones belonging to the same as well as to other groups.

Examples of exogenous PGRs modulating levels of endogenous hormones of the same group in SE include the accumulation of endogenous IAA in soybean hypocotyl explants after treatment with the synthetic auxins NAA and IBA (Liu et al. 1998). Also, using gas chromatography/mass spectrometry, Michalczuk et al. (1992a, b) showed that carrot cells treated with 2,4-D accumulate large amounts of endogenous IAA during SE. Further evidence in this sense is provided by the increase in the IAA levels observed in alfalfa leaf protoplasts cultured in the presence of 2,4-D (Pasternak et al. 2002). Moreover, Ceccarelli et al. (2002) found that two variant cell lines of carrot, capable of growing in high concentrations of 2,4-D and that showed disturbances in embryogenesis, raised the level of free IAA in response to the high exogenous auxin concentration.

Modulation of endogenous hormone levels by exogenous PGRs belonging to a different group has also been documented during SE. For example, application of ABA to immature zygotic sunflower embryos increased levels of endogenous IAA (Charrière et al. 1999). Moreover, high levels of exogenous ABA decreased ethylene contents during maturation of somatic embryos of white spruce (reviewed by Stasolla et al. 2002). There is also evidence, from a very early work, for 2,4-D regulating the rate of polar and less polar GAs (Noma et al. 1982).

It is suggested that the mechanism by which thidiazuron induces SE in peanut involves modulation of endogenous levels of auxin and CKs (Murthy et al. 1995). Moreover, the impairment in progression of embryo development caused by 2,4-D in carrot might be related to the increase in ethylene synthesis caused by the high levels of exogenous auxin (Minocha and Minocha 1995). Concerning polyamines, it has been observed that exogenous auxins suppressed the activity of two polyamine biosynthetic enzymes in carrot cultures, the effect of 2,4-D and IAA being distinct (Feinberg et al. 1984). Interaction of polyamines with other hormones has been reviewed by Kakkar and Sawhney (2002).

6
Concluding Remarks

Even though there are several factors that induce and govern SE in plants, the evidence available indicates that plant hormones, in response to the exogenous PGRs applied, or acting independently, in those few systems in which PGRs are not necessary for this process to occur, play a significant role. Together with the concentration of individual hormones, the interaction between members of different groups and the sensitivity/responsiveness of the tissues and cells (a factor not covered in this review) seem to condition the responses observed (Dudits et al. 1995; Thorpe 2000).

More than 10 years ago, when the first review articles involving quantification of endogenous hormones were published, it was postulated that knowing the endogenous hormone contents and their relation to the E competence of the explants would permit the induction and expression of SE in recalcitrant genotypes. That would take place through amendments to the culture medium, with substances that may mimic the inductive condition (supplying a deficiency or counteracting an excess) (Merkle et al. 1995). However, even though several works characterizing hormone status in responsive genotypes and E cultures have been published, most new publications defining adequate conditions to induce and allow progress of SE are still based on trial and error, as indicated at the beginning of the present review (Sect. 1).

Despite the progress achieved during the last few years in understanding the mechanisms involved in hormonal signaling of SE, there are still many aspects that are not fully understood and need to be studied in more detail. Progress is currently being achieved in comprehending the molecular responses that PGRs and plant hormones generate, mainly in gene expression (Thomas and Jiménez, this volume). It is to be expected that this alternative way to study hormonal regulation of SE will bring new insights on the subject.

References

Aloni R, Schwalm K, Langhans M, Ullrich CI (2003) Gradual shifts in sites of free-auxin production during leaf-primordium development and their role in vascular differentiation and leaf morphogenesis in *Arabidopsis*. Planta 216:841–853

Bell LM, Trigiano RN, Conger BV (1993) Relationship of abscisic acid to somatic embryogenesis in *Dactylis glomerata*. Environ Exp Bot 33:495–499

Besse I, Verdeil JL, Duval Y, Sotta B, Maldiney R, Miginiac E (1992) Oil palm (*Elaeis guineensis* Jacq.) clonal fidelity: endogenous cytokinins and indoleacetic acid in embryogenic callus cultures. J Exp Bot 43:983–989

Bronner R, Jeannin G, Hahne G (1994) Early cellular events during organogenesis and somatic embryogenesis induced on immature zygotic embryos of sunflower (*Helianthus annuus*). Can J Bot 72:239–248

Ceccarelli N, Mondin A, Lorenzi R, Picciarelli P, Lo Schiavo F (2002) The metabolic basis for 2,4-D resistance in two variant cell lines of carrot. Funct Plant Biol 29:575–583

Centeno ML, Rodríguez R, Berros B, Rodríguez A (1997) Endogenous hormonal content and somatic embryogenic capacity of *Corylus avellana* L. cotyledons. Plant Cell Rep 17:139–144

Charrière F, Hahne G (1998) Induction of embryogenesis versus caulogenesis on in vitro cultured sunflower (*Helianthus annuus* L.) immature zygotic embryos: role of plant growth regulators. Plant Sci 137:63–71

Charrière F, Sotta B, Miginiac É, Hahne G (1999) Induction of adventitious shoots or somatic embryos on in vitro cultured zygotic embryos of *Helianthus annuus*: variation of endogenous hormone levels. Plant Physiol Biochem 37:751–757

Cooke TJ, Racusen RH, Cohen JD (1993) The role of auxin in plant embryogenesis. Plant Cell 5:1494–1495

Danin M, Upfold SJ, Levin N, Nadel BL, Altman A, van Staden J (1993) Polyamines and cytokinins in celery embryogenic cell cultures. Plant Growth Regul 12:245–254

Davies PJ (1995) The plant hormones: their nature, occurrence, and functions. In: Davies PJ (ed) Plant hormones: physiology, biochemistry and molecular biology, 2nd edn. Kluwer, Dordrecht, pp 1–12

Dodeman VL, Ducreux G (1996) Total protein pattern expression during induction and development of carrot somatic embryos. Plant Sci 120:57–69

Dong J-Z, Dunstan DI (2000) Molecular biology of somatic embryogenesis in conifers. In: Jain SM, Minocha SC (eds) Molecular biology of woody plants, vol 1. Kluwer, Dordrecht, pp 51–87

Dudits D, Györgyey J, Bögre L, Bakó L (1995) Molecular biology of somatic embryogenesis. In: Thorpe TA (ed) In vitro embryogenesis in plants. Kluwer, Dordrecht, pp 276–308

Etienne H, Sotta B, Montoro P, Miginiac E, Carron MP (1993) Relation between exogenous growth regulators and endogenous indole-3-acetic acid and abscisic acid in the expression of somatic embryogenesis in *Hevea brasiliensis* (Müll. Arg.). Plant Sci 88:91–96

Fehér A, Pasternak TP, Dudits D (2003) Transition of somatic plant cells to an embryogenic state. Plant Cell Tissue Organ Cult 74:201–228

Feinberg AA, Choi JH, Lubich WP, Sung ZR (1984) Developmental regulation of polyamine metabolism in growth and differentiation of carrot culture. Planta 162:532–539

Fernández H, Pérez C, Sánchez-Tamés R (2000) Modulation of the morphogenic potential of the embryonic axis of *Juglans regia* by cultural conditions. Plant Growth Regul 30:125–131

Finkelstein RR, Tenbarge KM, Shumway JE, Crouch ML (1985) Role of ABA in maturation of rapeseed embryos. Plant Physiol 78:630–636

Fischer C, Neuhaus G (1996) Influence of auxin on the establishment of bilateral symmetry in monocots. Plant J 9:659–669

Fujimura T, Komamine A (1975) Effects of various growth regulators on the embryogenesis in a carrot cell suspension culture. Plant Sci Lett 5:359–364

Fujimura T, Komamine A (1979) Involvement of endogenous auxin in somatic embryogenesis in a carrot cell suspension culture. Z Pflanzenphysiol 95:13–19

Gaj MD (2004) Factors influencing somatic embryogenesis induction and plant regeneration with particular reference to *Arabidopsis thaliana* (L.) Heynh. Plant Growth Regul 43:27–47

Gaspar T, Kevers C, Faivre-Rampant O, Crèvecoeur M, Penel C, Greppin H, Dommes J (2003) Changing concepts in plant hormone action. In Vitro Cell Dev Biol Plant 39:85–105

Gazzarrini S, McCourt P (2003) Cross-talk in plant hormone signaling: what arabidopsis mutants are telling us. Ann Bot 91:605–612

Guiderdoni E, Mérot B, Eksomtramage T, Paulet F, Feldmann P, Glaszmann JC (1995) Somatic embryogenesis in sugarcane (*Saccharum* species). In: Bajaj YPS (ed) Somatic embryogenesis and synthetic seed I. Biotechnology in agriculture and forestry, vol 31. Springer, Berlin Heidelberg New York, pp 92–113

Hita O, Lafarga C, Guerra H (1997) Somatic embryogenesis from chickpea (*Cicer arietinum* L.) immature cotyledons: the effect of zeatin, gibberellic acid and indole-3-butyric acid. Acta Physiol Plant 19:333–338

Huang LVT, Takamura T, Tanaka M (2004) Callus formation and plant regeneration from callus through somatic embryo structures in *Cymbidium* orchid. Plant Sci 166:1443–1449

Hutchinson MJ, KrishnaRaj S, Saxena PK (1997) Inhibitory effect of GA_3 on the development of thidiazuron-induced somatic embryogenesis in geranium (*Pelargonium* x *hortorum* Bailey) hypocotyl cultures. Plant Cell Rep 16:435–438

Iantcheva A, Vlahova M, Bakalova E, Kondorosi E, Elliott MC, Atanassov A (1999) Regeneration of diploid annual medics via direct somatic embryogenesis promoted by thidiazuron and benzylaminopurine. Plant Cell Rep 18:904–910

Ivanova A, Velcheva M, Denchev P, Atanassov A, Van Onckelen HA (1994) Endogenous hormone levels during direct somatic embryogenesis in *Medicago falcata*. Physiol Plant 92:85–89

Jiménez VM (2001) Regulation of in vitro somatic embryogenesis with emphasis on the role of endogenous hormones. Rev Bras Fisiol Veg 13:196–223

Jiménez VM, Bangerth F (2001a) Endogenous hormone levels in initial explants and in embryogenic and non-embryogenic callus cultures of competent and non-competent wheat genotypes. Plant Cell Tissue Organ Cult 67:37–46

Jiménez VM, Bangerth F (2001b) Hormonal status of maize initial explants and of the embryogenic and non-embryogenic callus cultures derived from them as related to morphogenesis in vitro. Plant Sci 160:247–257

Jiménez VM, Bangerth F (2001c) In vitro culture and endogenous hormone levels in immature zygotic embryos, endosperm and callus cultures of normal and high-lysine barley genotypes. J Appl Bot 75:1–7

Jiménez VM, Guevara E, Herrera J, Bangerth F (2001) Endogenous hormone levels in habituated nucellar *Citrus* callus during the initial stages of regeneration. Plant Cell Rep 20:92–100

Jiménez VM, Guevara E, Herrera J, Bangerth F (2005) Evolution of endogenous hormone concentration in embryogenic cultures of carrot during early expression of somatic embryogenesis. Plant Cell Rep 23:567–572

Kakkar RK, Sawhney VK (2002) Polyamine research in plants—a changing perspective. Physiol Plant 116:281–292

Kamada H, Harada H (1981) Changes in the endogenous level and effects of abscisic acid during somatic embryogenesis of *Daucus carota* L. Plant Cell Physiol 22:1423–1429

Kermode AR, Dumbroff EB, Bewley JD (1989) The role of maturation drying in the transition from seed development to germination. VII. Effects of partial and complete desiccation on abscisic acid levels and sensitivity in *Ricinus communis* L. seeds. J Exp Bot 40:303–313

Kochba J, Spiegel-Roy P, Neumann H, Saad S (1978) Stimulation of embryogenesis in citrus ovular callus by ABA, Ethephon, CCC and Alar and its suppression by GA_3. Z Pflanzenphysiol 89:427–432

Kong L, Yeung EC (1994) Effects of ethylene and ethylene inhibitors on white spruce somatic embryo maturation. Plant Sci 104:71–80

Kopertekh LG, Butenko RG (1995) Naturally occurring phytohormones in wheat explants as related to wheat morphogenesis in vitro. Russ J Plant Physiol 42:488–491

Li XQ, Demarly Y (1996) Somatic embryogenesis and plant regeneration in *Medicago suffruticosa*. Plant Cell Tissue Organ Cult 44:79–81

Limanton-Grevet A, Sotta B, Brown S, Jullien M (2000) Analysis of habituated embryogenic lines in *Asparagus officinalis* L.: growth characteristics, hormone content and ploidy level of calli and regenerated plants. Plant Sci 160:15–26

Liu C-M, Xu ZH, Chua N-H (1993) Auxin polar transport is essential for the establishment of bilateral symmetry during early plant embryogenesis. Plant Cell 5:621–630

Liu Z-H, Wang W-C, Yen Y-S (1998) Effect of hormone treatment on root formation and endogenous indole-3-acetic acid and polyamine levels of *Glycine max* cultivated in vitro. Bot Bull Acad Sin 39:113–118

Mauri PV, Manzanera JA (2004) Effect of abscisic acid and stratification on somatic embryo maturation and germination of holm oak (*Quercus ilex* L.). In Vitro Cell Dev Biol Plant 40:495–498

Merkle SA, Parrott WA, Flinn BS (1995) Morphogenic aspects of somatic embryogenesis. In: Thorpe TA (ed) In vitro embryogenesis in plants. Kluwer, Dordrecht, pp 155–203

Michalczuk L, Cooke TJ, Cohen JD (1992a) Auxin levels at different stages of carrot embryogenesis. Phytochemistry 31:1097–1103

Michalczuk L, Ribnicky DM, Cooke TJ, Cohen JD (1992b) Regulation of indole-3-acetic acid biosynthetic pathways in carrot cell cultures. Plant Physiol 100:1346–1353

Minocha SC, Minocha R (1995) Role of polyamines in somatic embryogenesis. In: Bajaj YPS (ed) Somatic embryogenesis and synthetic seed I. Biotechnology in agriculture and forestry, vol 30. Springer, Berlin Heidelberg New York, pp 53–70

Minocha R, Minocha SC, Long S (2004) Polyamines and their biosynthetic enzymes during somatic embryo development in red spruce (*Picea rubens* Sarg.). In Vitro Cell Dev Biol Plant 40:572–580

Mitsuhashi W, Toyomasu T, Masui H, Katho T, Nakaminami K, Kashiwagi Y, Akutsu M, Kenmoku H, Sassa T, Yamaguchi S, Kamiya Y, Kamada H (2003) Gibberellin is essentially required for carrot (*Daucus carota* L.) somatic embryogenesis: dynamic regulation of gibberellin 3-oxidase gene expression. Biosci Biotechnol Biochem 67:2438–2447

Moctezuma E (1999) Changes in auxin patterns in developing gynophores of the peanut plant (*Arachis hypogaea* L.). Ann Bot 83:235–242

Moctezuma E, Feldman LJ (1999) Auxin redistributes upwards in graviresponding gynophores of the peanut plant. Planta 209:180–186

Murthy BNS, Murch SJ, Saxena PK (1995) Thidiazuron-induced somatic embryogenesis in intact seedlings of peanut (*Arachis hypogaea*): endogenous growth regulator levels and significance of cotyledons. Physiol Plant 94:268–276

Nakagawa H, Saijyo T, Yamauchi N, Shigyo M, Kako S, Ito A (2001) Effects of sugars and abscisic acid on somatic embryogenesis from melon (*Cucumis melo* L.) expanded cotyledon. Sci Hortic 90:85–92

Nishiwaki M, Fujino K, Koda Y, Masuda K, Kikuta Y (2000) Somatic embryogenesis induced by the simple application of abscisic acid to carrot (*Daucus carota* L.) seedlings in culture. Planta 211:756–759

Nissen P (1994) Stimulation of somatic embryogenesis in carrot by ethylene: Effects of modulators of ethylene biosynthesis and action. Physiol Plant 92:397–403

Noma M, Huber J, Ernst D, Pharis RP (1982) Quantitation of gibberellins and the metabolism of [^{3}H]gibberellin A_1, during somatic embryogenesis in carrot and anise cell cultures. Planta 155:369–376

Nomura K, Komamine A (1995) Physiological and biochemical aspects of somatic embryogenesis. In: Thorpe TA (ed) In vitro embryogenesis in plants. Kluwer, Dordrecht, pp 249–265

Normanly J, Bartel B (1999) Redundancy as a way of life—IAA metabolism. Curr Opin Plant Biol 2:207–213

Osuga K, Masuda H, Komamine A (1999) Synchronization of somatic embryogenesis at high frequency using carrot suspension cultures: model systems and application in plant development. Methods Cell Sci 21:129–140

Pasternak TP, Prinsen E, Ayaydin F, Miskolczi P, Potters G, Asard H, Van Onckelen HA, Dudits D, Fehér A (2002) The role of auxin, pH, and stress in the activation of embryogenic cell division in leaf protoplast-derived cells of alfalfa. Plant Physiol 129:1807–1819

Pintos B, Martín JP, Centeno ML, Villalobos N, Guerra H, Martín L (2002) Endogenous cytokinin levels in embryogenic and non-embryogenic calli of *Medicago arborea* L. Plant Sci 163:955–960

Prewein C, Vagner M, Wilhelm E (2004) Changes in water status and proline and abscisic acid concentrations in developing somatic embryos of pedunculate oak (*Quercus robur*) during maturation and germination. Tree Physiol 24:1251–1257

Raemakers CJJM, Jacobsen E, Visser RGF (1995) Secondary somatic embryogenesis and applications in plant breeding. Euphytica 81:93–107

Rajasekaran K, Vine J, Mullins MG (1982) Dormancy in somatic embryos and seeds of *Vitis*: changes in endogenous abscisic acid during embryogeny and germination. Planta 154:139–144

Rajasekaran K, Hein MB, Vasil IK (1987a) Endogenous abscisic acid and indole-3-acetic acid and somatic embryogenesis in cultured leaf explants of *Pennisetum purpureum* Schum. Plant Physiol 84:47–51

Rajasekaran K, Hein MB, Davis GC, Carnes MG, Vasil IK (1987b) Endogenous growth regulators in leaves and tissue cultures of *Pennisetum purpureum* Schum. J Plant Physiol 130:12–25

Roustan J-P, Latché A, Fallot J (1994) Role of ethylene on induction and expression of carrot somatic embryogenesis: relationship with polyamine metabolism. Plant Sci 103:223–229

Ruduś I, Kępczyńska E, Kępczyński J (2002) Regulation of *Medicago sativa* L. somatic embryogenesis by gibberellins. Plant Growth Regul 36:91–95

Sagare AP, Lee YL, Lin TC, Chen CC, Tsay HS (2000) Cytokinin-induced somatic embryogenesis and plant regeneration in *Corydalis yanhusuo* (Fumariaceae)—a medicinal plant. Plant Sci 160:139–147

Schiavone FM, Cooke TJ (1987) Unusual patterns of somatic embryogenesis in the domesticated carrot: developmental effects of exogenous auxins and auxin transport inhibitors. Cell Differ 21:53–62

Senger S, Mock H-P, Conrad U, Manteuffel R (2001) Immunomodulation of ABA function affects early events in somatic embryo development. Plant Cell Rep 20:112–120

Sharma AK (1999) Synchronization in plant cells—an introduction. Methods Cell Sci 21:73–78

Sharma P, Pandey S, Bhattacharya A, Nagar PK, Ahuja PS (2004) ABA associated biochemical changes during somatic embryo development in *Camellia sinensis* (L.) O. Kuntze. J Plant Physiol 161:1269–1276

Stasolla C, Kong L, Yeung EC, Thorpe TA (2002) Maturation of somatic embryos in conifers: morphogenesis, physiology, biochemistry, and molecular biology. In Vitro Cell Dev Biol Plant 38:93–105

Takeno K, Koshioka M, Pharis RP, Rajasekaran K, Mullins MG (1983) Endogenous gibberellin-like substances in somatic embryos of grape (*Vitis vinifera* x *Vitis rupestris*) in relation to embryogenesis and the chilling requirement for subsequent development of mature embryos. Plant Physiol 73:803–808

Thomas C, Bronner R, Molinier J, Prinsen E, van Onckelen H, Hahne G (2002) Immunocytochemical localization of indole-3-acetic acid during induction of somatic embryogenesis in cultured sunflower embryos. Planta 215:577–583

Thomas C, Meyer D, Himber C, Steinmetz A (2004) Spatial expression of a sunflower SERK gene during induction of somatic embryogenesis and shoot organogenesis. Plant Physiol Biochem 42:35–42

Thorpe TA (2000) Somatic embryogenesis: morphogenesis, physiology, biochemistry and molecular biology. Korean J Plant Tissue Cult 27:245–258

Tokuji Y, Kuriyama K (2003) Involvement of gibberellin and cytokinin in the formation of embryogenic cell clumps in carrot (*Daucus carota*). J Plant Physiol 160:133–141

Victor JMR, Murch SJ, KrishnaRaj S, Saxena PK (1999) Somatic embryogenesis and organogenesis in peanut: The role of thidiazuron and N^6-benzylaminopurine in the induction of plant morphogenesis. Plant Growth Regul 28:9–15

von Arnold S, Sabala I, Bozhkov P, Kyachok J, Filonova L (2002) Developmental pathways of somatic embryogenesis. Plant Cell Tissue Organ Cult 69:233–249
Wenck AR, Conger BV, Trigiano RN, Sams CE (1988) Inhibition of somatic embryogenesis in orchardgrass by endogenous cytokinins. Plant Physiol 88:990–992
Zhang Q, Chen J, Henny RJ (2005) Direct somatic embryogenesis and plant regeneration from leaf, petiole and stem explants of golden pothos. Plant Cell Rep 23:587–595

Plant Cell Monogr (2)
A. Mujib · J. Šamaj: Somatic Embryogenesis
DOI 10.1007/7089_032/Published online: 9 December 2005

Somatic Embryogenesis of Pine Species: From Functional Genomics to Plantation Forestry

Hely Häggman[1] (✉) · Jaana Vuosku[1] · Tytti Sarjala[2] · Anne Jokela[1] · Karoliina Niemi[3]

[1]Department of Biology, University of Oulu, P.O. Box 3000, 90014 University of Oulu, Finland
hely.haggman@oulu.fi

[2]Parkano Research Unit, Finnish Forest Research Institute, 39700 Parkano, Finland

[3]Department of Applied Biology, University of Helsinki, P.O. Box 27, 00014 University of Helsinki, Finland

Abstract Several economically important tree species belong to the genus *Pinus* and many of them form the ecological base of forest ecosystems. Pine wood is an important raw material for the forest industry and many of the pine species have been involved in conventional tree improvement programmes. A lot of effort has been made in the development of vegetative propagation methods, especially somatic embryogenesis, in order to rapidly gain the benefits of traditional breeding to be utilized in reforestation. The economically relevant clonal plantation forestry presumes effective mass-propagation systems with high-quality somatic embryo plants. Today this is feasible only for *Pinus banksiana* Lamb., *P. taeda* L. and *P. radiata* D. Don. The recent progress in somatic embryo production and the challenges in functional genomics have increased the understanding of pine zygotic embryo development, leading to improved protocols for somatic embryogenesis. Therefore, clonal plantation forestry might become a reality for more pine species in the coming years. This chapter highlights the recent challenges in the functional genomics of pine embryogenesis. Possibilities for molecular breeding or utilization of somatic embryo plants in conventional breeding and in clonal plantations in line for sustainable forestry are also covered. The importance of cryopreservation for elite genotype preservation and as a storage method during progeny testing is discussed, as well as the use of ectomycorrhizal fungi during somatic embryo conversion in vitro and acclimatization ex vitro.

1 Introduction

Several economically important tree species belong to the genus *Pinus* of the class Pinaceae and many of them form the ecological base of forest ecosystems. Pine wood is an important raw material for pulp production, saw-timber and the furniture industry. During recent decades the extraction of timber from managed or semi-natural tree plantations, instead of natural woodland, has been considered as sustainable forestry. In Europe these plantations are mainly composed of different species and various genotypes whereas in America and New Zealand, for example, plantations are composed of single species and a few genotypes. All natural and managed forest areas have a role in forest biodiversity and conservation. To maintain the present forest biodiversity levels all forests should be managed in an ecologically sustainable way. Today, forests are managed for many different purposes including wood production, recreation, ecological and cultural values, and biodiversity, as well as soil and groundwater protection. This brings new challenges to forest management and silviculture. On the other hand, plantation forestry might help to conserve the natural forests especially in developing countries, in which the majority of wood for construction and fires is supplied by natural forests (Walter 2004).

Many of the economically important pine species have been involved in conventional tree improvement programmes. Quite a lot of effort has been put into the development of vegetative propagation methods, especially somatic embryogenesis, in order to rapidly gain the benefits of traditional breeding to be utilized in reforestation. The economically relevant clonal plantation forestry presumes effective mass-propagation systems with high-quality somatic embryo plants. Today this is feasible only for three pine species: *Pinus banksiana* Lamb. (Park 2002), *P. taeda* L. and *P. radiata* D. Don (Smith et al. 1994; Handley et al. 1995; Sutton 2002; Attree 2004). The general obstacles in root production, conversion and acclimatization to ex vitro that hinder any technological outcomes in several pine species could be relieved by inoculation with specific ectomycorrhizal fungi (Niemi et al. 2004). In general, the recent progress in somatic embryo production (e.g. Pullman et al. 2003a) and the challenges in functional genomics have increased our understanding of pine zygotic embryo development, leading to improved protocols for somatic embryogenesis. Therefore, the economically relevant clonal plantation forestry might become a reality for more pine species in the coming years.

Recently, the potential for molecular breeding has also been considered. The classical tree-breeding work in pines is hindered by long life cycles and long generation intervals. Sexual or somatic hybridization may be limited by the sterility of the descents and the genetic barrier between the species. Overcoming this genetic barrier is only possible via genetic transformation. Important future approaches are considered to be the reduction of generation time, production of sterile trees, resistance to pest or fungal diseases and properties of the wood, especially lignin engineering (Peña and Séguin 2001; Diouf 2003). Today, however, the number of stably transformed pine species is limited and the potential practical applications will only be reached in the future. To achieve these goals and/or to apply the technology to conventional tree breeding, it is essential that individual genotypes are conserved during progeny testing in the field. During recent years, cryopreservation protocols have been developed for embryogenic cultures of several pine species.

The present chapter highlights the recent challenges in the functional genomics of pine embryogenesis. The possibilities for molecular breeding, or utilization of somatic embryo plants in conventional breeding or in clonal plantations in line for sustainable forestry, are also covered. The importance of cryopreservation for elite genotype preservation and as a storage method during progeny testing will be discussed, together with the role of ectomycorrhizal fungi within somatic embryo maturation and conversion in vitro and during ex vitro acclimatization.

2 The Present State of the Art in Pine Somatic Embryogenesis

Somatic embryogenesis is a process in which specific somatic cells are genetically reprogrammed towards the embryogenic pathway. Somatic embryo development of pine species encompasses four distinct phases, initiation (Figs. 1a, b), proliferation, maturation (Fig. 1d) and conversion, i.e. germination (Fig. 1e) and subsequent acclimatization ex vitro (Fig. 1g), which are induced by changes of the culture medium composition. After a successful initiation, the embryogenic potential of the proliferating embryogenic mass is maintained on the medium with high concentration of both auxin and cytokinin. Removal of these plant growth regulators is a prerequisite for the development of somatic embryos. During maturation, storage substances are accumulated and somatic embryos differentiate, desiccate and reduce their metabolic activity. These changes are induced by exogenous abscisic acid (ABA) and increased osmolality due to exogenous polyethylene glycol (PEG), sugars or increased gel strength of the medium. For germination, mature somatic embryos are usually cultivated on the medium without exogenous plant growth regulators and with lower concentrations of nutrients and sugar, which induces utilization of storage compounds in embryos. Germination and subsequent root elongation in vitro are critical phases for later acclimatization to ex vitro conditions in a greenhouse.

In pine species immature megagametophytes containing immature zygotic embryos have been the most responsive explants for the initiation of somatic embryogenesis (Handley et al. 1995; Häggman et al. 1999; Percy et al. 2000; Pullman et al. 2003b; Miguell et al. 2004; Niskanen et al. 2004). Somatic embryogenesis from mature zygotic embryos (Tang et al. 2001a; Malabadi et al. 2002) and vegetative shoot apices (Malabadi and van Staden 2005) have also been documented. Likewise, somatic organogenesis from mature zygotic embryos has been regarded as an alternative for somatic embryogenesis in *P. taeda* (Tang and Guo 2001; Tang et al. 2001c; Tang et al. 2004). The whole developmental process from initiation to conversion has succeeded in several pine species, including *P. banksiana* (Park 2002), *P. kesiya* Royle ex. Gord (Malabadi et al. 2002), *P. monticola* Dougl. (Percy et al. 2000), *P. patula* Schede et Deppe

Fig. 1 Somatic embryogenesis of *P. sylvestris*. **a** Developing green cone shortly after meiosis. **b** Initiation of somatic embryogenesis using immature embryos surrounded by megagametophytes and proliferation of embryogenic cell mass. **c** Option for cryopreservation of the germplasm. **d** Maturation of somatic embryos. **e** Conversion of somatic embryo plants. **f** Inoculation with specific ECM fungus improves root development. **g** Acclimatization to ex vitro conditions in a greenhouse

(Jones and van Staden 1995), *P. pinaster* Soland., non Ait. (Lelu et al. 1999; Miguel et al. 2004), *P. radiata* (Sutton 1999; Attree et al. 2004), *P. strobus* L. (Garin et al. 1998; Klimaszewska et al. 2001; Park 2002), *P. sylvestris* L. (Häggman et al. 1999; Lelu et al. 1999) and *P. taeda* (Handley et al. 1995; Sutton 2002; Attree 2004). Recently, a number of selection programmes have been started, predominantly by private forest companies, to test pine embryogenic clones (reviewed by Cyr and Klimaszewska 2002), and for *P. radiata* and *P. taeda* commercial production has dramatically increased (Sutton 2002; Attree et al. 2004).

Recently, somatic embryo production has been improved by achievements in functional genomics and physiology during pine zygotic embryogenesis, as well as by optimization work at the tissue culture media level and during acclimatization. Regardless of the developments, the application of somatic embryogenesis for most pine species is still limited, which is mainly due to low, cell line- and family-dependent initiation frequency and an inability of initiated cultures to become stable during proliferation. Furthermore, the inability of somatic embryos to fully mature results in low germination frequency and subsequently poor acclimatization ex vitro (Garin et al. 1998; Häggman et al. 1999; Pullman et al. 2003b; Miguell et al. 2004; Niskanen et al. 2004; Malabadi and van Staden 2005).

3
Functional Genomics of Pine Embryogenesis

All pine species have 12 pairs of chromosomes with essentially similar morphology (Sax and Sax 1933). The genome size is large, but there is variation between the pine species (Bogunic et al. 2003). Pine genomes are known to contain highly repeated DNA sequences (Kriebel 1985) and to harbour large complex gene families (Kinlaw and Neale 1997). However, the isozyme profiles of pines show less evidence for large gene families than is apparent from Southern hybridizations (Perry and Furnier 1996). Expressed sequence tag (EST) projects also suggest that the number of expressed gene family members may not be very high. On the other hand, the number of related non-expressed pseudo-genes is higher than in many other plant groups (Komulainen et al. 2003).

Recently, a programme on the functional genomics of *P. taeda* zygotic and somatic embryogenesis has been commenced (Cairney et al. 2003), and the development of a 10 000-clone *P. taeda* cDNA array enriched in sequences expressed in embryogenesis is in progress. Due to the success of heterologous hybridization in conifers (Van Zyl et al. 2002), this microarray will serve as a general pine cDNA allowing high-throughput gene expression analyses. Komulainen et al. (2003) found that the EST-based genetic maps between *P. sylvestris* and *P. taeda* are largely colinear. What is more, a comparative karyotypic analysis of four pine species suggested that the degree of chromosomal differentiation among species is very low (Hizume et al. 2002).

EST microarrays for *P. taeda* have been utilized in several gene expression studies of spruce species (Van Zyl et al. 2002, 2003; Stasolla et al. 2003, 2004). Generally, in somatic embryogenesis of *Picea abies* L. Karst. gene expression is upregulated during transition from proembryogenic masses to embryos, downregulated during early embryogeny and upregulated again at the onset of late embryogeny (Van Zyl et al. 2003). In *Arabidopsis* several regulatory genes responsible for embryo development have been identified by mutant analysis (Jurgens 2001), but in conifers the long generation interval makes the selection of embryo-specific mutants practically impossible. However, genotypes deviating from the normal embryo pattern formation and exhibiting developmental arrest at specific stages represent a tool for studying signalling and gene regulation during embryogenesis in conifers (Van Zyl et al. 2003). In *Picea abies* a comparison between transcript profiles of normal and developmentally arrested embryogenic lines showed that the early phases of normal embryo development were characterized by a precise pattern of gene expression. Several of these genes encoded proteins that are involved in carbohydrate metabolism, detoxification processes and methionine synthesis and utilization (Stasolla et al. 2004). Stasolla et al. (2003) compared the transcript profiles of stage-specific *Picea glauca* (Moench) Voss somatic embryos matured with or without PEG and found that several genes involved in the formation of the embryo body plan and in the control of the shoot and root apical meristems were up-regulated after PEG treatments. They also observed changes in the transcript levels of the genes involved in sucrose catabolism, nitrogen assimilation and utilization.

Preliminary studies on the molecular mechanisms regulating the phases of pine somatic embryogenesis have revealed several genes with differentially regulated expression between somatic and zygotic embryos. In *P. taeda*, gene expression patterns for 326 differentially expressed cDNA fragments were determined across the sequence of somatic and zygotic embryo development (Cairney 1999). Bishop-Hurley et al. (2003) compared gene expression in embryogenic and non-embryogenic tissues of *P. radiata* and identified six gene families that were preferentially expressed during somatic embryo development in vitro. These gene families include a cytochrome P450 enzyme and four putative extracellular proteins: germin, β-expansin, cellulase and 21-kDA protein precursor.

The understanding of embryogenesis has been increased due to the challenges in functional genomics at the genome, transcriptome and proteome levels. The identified conifer genes that are differentially expressed during embryogenesis are homologous to angiosperm seed storage protein genes (Dong and Dunstan 2000), *lea* genes (Dòng and Dunstan 1997), *KNOTTED1*-like homeobox gene (Hjortswang et al. 2002), *HD-GL2* homeobox gene family (Ingouff et al. 2003), *VP1/ABI3* gene family, and $p34^{cdc2}$ protein kinase (Footitt et al. 2003). This suggests that despite the differences in certain aspects of gymnosperm and angiosperm embryogenesis, the genes central to embryogenesis will exhibit a high degree of conservation. Germin-like proteins (GLPs) have also been identified in all plant species examined to date (Khuri et al. 2001). The GLPs have been reported to express in the embryogenic cell cultures of *P. caribea* L. and *P. radiata* (Domon et al. 1995; Bishop-Hurley et al. 2003). Preliminary observations suggest that the gymnosperm GLP *PcGER1* gene is unique in the pine genome (Neutelings et al. 1998), which contrasts with the broad divergence of GLPs among the angiosperms. In our own studies on polyamine biosynthesis in *P. sylvestris* the arginine decarboxylase (*ADC*) gene expression and enzyme activity increased during zygotic embryo development, and the *ADC* mRNA transcripts were localized in specific dividing cells of the shoot meristems of the late embryos (unpublished results). In *P. taeda*, the transcript of an aquaglyceroporin gene, *PtNIP1;1*, was found to be abundant in immature zygotic and somatic embryos, and the gene was expressed preferentially in suspensor tissues (Ciavatta et al. 2001). Ciavatta et al. (2002) suggest that this preferential expression in suspensors was due to specific elements of the putative *PtNIP1;1* promoter.

The rapid increase in the availability of EST sequences has opened new prospects for analysing embryogenesis in conifers. Some pine genes, which are activated or expressed differentially during embryogenesis, have now been isolated. However, the accurate mechanism controlling gene expression and the detailed roles of the genes in directing embryogenesis are not clearly understood. In biological systems, information flow goes from DNA to RNA and further to protein and to metabolites. This means that large-scale protein analyses are needed to complement the data derived from transcriptome analysis. Protein arrays and specific antibodies will be generated and used for the functional characterization of woody plant systems (Cánovas et al. 2004). The precise localization of mRNAs and proteins in embryogenic cells and tissues will provide new insights into the organization of metabolic pathways during pine embryo development. Subsequently, because post-translational factors are functionally important in the cell, metabolome-level studies will be of great importance in gaining a comprehensive view of pine embryogenesis.

4 From Conventional Breeding Towards Molecular Breeding

Generally, vegetative propagation is an important tool for achieving significant credits for both conventional tree breeding and propagation of genetically improved material. By in vitro propagation it is possible to realize additional gain due to the potential exploitation of non-additive genetic variation, to increase homogeneity of the material and to compensate potential shortage of improved seeds from seed orchards. The credits for progeny testing and selection of genotypes for the next generations will also be achieved by testing vegetatively propagated material under various environmental conditions. Somatic embryogenesis is expected to have positive effects on both tree breeding and propagation of conventionally improved pine material. However, for the pine species that have been studied so far, practical applications have been hindered particularly by genotype-dependent initiation, uneven maturation and low germination frequency. Although a lot of prospects have been linked to molecular breeding of coniferous species, problems in both the vegetative propagation and the production of genetically transformed material still limit the biotechnological applications of several pine species. Recent advances in somatic embryogenesis have certainly brought these prospects closer to reality (as reviewed by Merkle and Dean 2000).

The first stably transformed coniferous species, *Larix decidua* Miller, was produced through *Agrobacterium rhizogenes*-mediated genetic transformation (Huang et al. 1991). Since then, *A. rhizogenes* has been considered as a potential tool for rooting recalcitrant woody plants including pine species (as reviewed by Häggman and Aronen 2000). The next stably transformed coniferous species, *Picea glauca*, was achieved by application of particle bombardment technology (Bommineni et al. 1993; Ellis et al. 1993). However, the first report of a stably transformed pine species, *P. radiata*, was published by the New Zealand group of Wagner and co-workers in 1997, which was followed by Walter and co-workers one year later in 1998, i.e. ten years later than the first report on stably transformed hardwood, *Populus alba* × *P. grandidentata* (Fillatti et al. 1987). Later on, both direct gene transfer by particle bombardment, which in most cases means Biolistic® transformation, and *Agrobacterium tumefaciens*-mediated transformation were applied to pine species. *Agrobacterium*-mediated transformation has also been developed as an alternative to Biolistic® transformation for conifers. The advantages of *Agrobacterium*-mediated genetic transformation compared with the Biolistic® method are a lower average copy number, less fragmentation of the transgenes and precise gene integration (Kumar and Faldung 2001, and as reviewed by Walter 2004). Controversially, some papers indicate high integration of vector backbone sequences in plants like rice, tomato, grape and potato after *Agrobacterium*-mediated transformation (Hanson et al. 1999; Kim et al. 2003; Rommens et al. 2004).

At present, regeneration of transgenic pines has been reported via *Agrobacterium*-mediated transformation from organogenic (Tang et al. 2001b) and embryogenic (Tang and Tian 2003) material of *P. taeda*, as well as from embryogenic cultures of *P. strobus* (Levee et al. 1999). In addition to the pioneering work of Wagner et al. (1997) and Walter et al. (1998), transgenic pines via Biolistic® transformation have been produced from embryogenic tissues of *P. radiata* by Bishop-Hurley et al. (2001). Overall, the list of transgenic pines derived from material in tissue culture is still short, only three species. This does not mean that these would be the only pine species which have been targets of genetic transformation, but it might rather reflect the effort put into the development of transformation protocols or severe difficulties in regeneration. As an example, we have studied genetic transformation of organogenic material (Aronen et al. 1994, 1995, 1996; Aronen and Häggman 1995) and embryogenic cultures of *P. sylvestris* (Häggman and Aronen 1998) using both *Agrobacterium*-mediated gene transfer and Biolistic® transformation, but failed to produce transgenic pines mostly due to difficulties in regeneration. In the case of embryogenic cultures, especially slow growth of the cultures together with prolonged antibiotic selection have prevented regeneration. Another example is the work of Wenck et al. (1999), who transformed embryogenic cultures of two coniferous species, *Picea abies* and *Pinus taeda*, using disarmed *Agrobacterium* helper strains to which either a constitutively expressed *virG* or extra copies of *virG* and *virB* were added. Transformation efficiencies in *Picea abies* and *Pi-*

nus taeda increased 1000- and 10-fold, respectively, but regeneration of stably transformed somatic embryo plants was successful only in *Picea abies*.

In species recalcitrant for in vitro propagation, such as *P. sylvestris*, development of protocols without tissue culture would be of great value. One possibility might be to use pollen, which is a natural carrier of genetic material and as such a good target for foreign gene delivery. Transgenic tobacco plants have been regenerated successfully by applying transformed pollen in conventional pollinations (van der Leede-Plegt et al. 1995) or through tissue culture (Stoger et al. 1995). Protocols for the genetic transformation of pine pollen, resulting in transient expression of reporter genes, have been published for *P. banksiana*, *P. contorta* Dougl. ex Loud (Hay et al. 1994), *P. aristata* Engelm., *P. griffithii* McClell, *P. monticola* (Fernando et al. 2000), *P. pinaster* (Martinussen et al. 1995) and *P. sylvestris* (Häggman et al. 1997). For *P. sylvestris* pollen, we developed the particle bombardment protocol and the necessary dehydration–storage protocol for bombarded pollen that is compatible with the conventional crossing technique (Häggman et al. 1997; Aronen et al. 1998). Furthermore, we reported on the production of transgenic plants via the use of transformed pollen in controlled crossings (Aronen et al. 2003). The frequency of transgenic progenies is, however, still low but might be improved by increasing the efficiency of progeny screening. Another option might be to combine the method with the existing somatic embryogenesis protocol for *P. sylvestris*. This means that after controlled pollinations with bombarded pollen, the immature embryos surrounded by the immature megagametophyte could be dissected from the developing cones to initiate somatic embryogenesis.

So far, most of the research on pine species has focused on the development of genetic transformation protocols, and the traits transferred to pine species are listed in Table 1 (reporter genes: β-glucuronidase gene *uidA* or green fluorescent protein gene *gfp*; selectable marker genes: neomycin phosphotransferase *nptII* or hygromycin phosphotransferase *hph*). Considering other traits, there are only two reports that might have feasible options for plantation forestry. Bishop-Hurley et al. (2001) transferred the *bar* gene, which confers herbicide resistance into *P. radiata*, and found that transgenic plants spray tested with Buster (glufosinate) survived with minor or no damage to their needles. Tang and Tian (2003) reported on the integration of the synthetic *Bacillus thuringiensis CRY1Ac* coding sequence, i.e. a modified δ-endotoxin gene to *P. taeda*, and subsequently, in feeding bioassays, they demonstrated an increased resistance to the lepidopteran larvae *Dendrolimus punctatus* Walker and *Crypyothelea formosicola* Staud.

It is clear that the genetic improvement or molecular breeding of all forest crops that utilize genetic transformation techniques is today at an early stage, and forest trees can still be regarded as undomesticated wild trees for the majority of our wood product needs. However, there is a global shift to-

Table 1 The target material, genetic transformation method and transferred genes used in the production of stably transformed pine species

Pinus species	Target material	Transformation method	Gene	Reference
P. radiata	Embryogenic	Biolistic®	*uidA, nptII*	Walter et al. 1998 Wagner et al. 1987
	Embryogenic	Biolistic®	*uidA, nptII, bar, germin*	Bishop-Hurley et al. 2001
P. strobus	Embryogenic	Agrobacterium	*uidA, gfp, nptII*	Levee et al. 1999
P. sylvestris	Pollen	Biolistic®	*uidA*	Aronen et al. 2003
P. taeda	Organogenic	Agrobacterium	*uidA, hph*	Tang et al. 2001b
	Mature, zygotic embryos	Biolistic®	*cry1Ac, nptII*	Tang & Tian 2003

wards tree plantations to meet the increasing need for fibre and to maximize both growth and yield. In this context, the potential of genetically modified tree crops will also be evaluated. At present, the most important approaches include the reduction of generation time, production of sterile trees, resistance to pest or fungal diseases and evaluation of the properties of the wood, especially lignin engineering (Peña and Séguin 2001; Diouf 2003). In addition to these practical goals, a transgenic approach has been widely used as a tool in tree and plant physiology, ecology, genetics and molecular biology (as reviewed by Herschbach and Kopriva 2002). So far, the first and only report in which a transgenic approach has been used to study pine embryogenesis was from Bishop-Hurley et al. (2001), who introduced a specific germin cDNA into *P. radiata* embryogenic cultures.

Biosafety issues of transgenic plants have recently been discussed in several reviews (e.g. Walter 2004) and it has been emphasized, for instance, by the establishment of a Europe-wide, web-based, public-access database (www.versailles.inra.fr/europe/gmorescom) to enhance communication regarding biosafety research. In short, environmental concerns about transgenic technology in plants have arisen from the possibility of not only vertical but also horizontal gene flow, the possible undesirable effects of the transgenes or traits and their possible effect on non-target organisms. All pine species are wind-pollinated, characterized by long life cycles and many of them are the key species of their ecosystems. Therefore, the recognition of the unexpected (e.g. epistatic or pleiotrophic) effects of the transgenes as well as other biosafety concerns have to be considered seriously. However, as also pointed out by Walter (2004), the potential risks or benefits of the genetic modification technology should be discussed in comparison with the risks or benefits of not using this technology.

5 Cryopreservation of Embryogenic Cultures of Pines

Cryopreservation, i.e. storage of material in liquid nitrogen at – 196 °C, represents the only safe and cost-effective option for long-term conservation of plant germplasm (as reviewed by Engelmann 2004). In pine species the recent progress in somatic embryogenesis, the production of genetically modified plants (Table 1) and the efforts towards plantation forestry have emphasized the need for germplasm conservation with functional cryopreservation protocols (Häggman et al. 2000, 2001; Park 2002).

Reliable long-term maintenance of embryogenic cultures requires that the cultures are stored using cryopreservation techniques. It is well known that in conifers the embryogenic cell lines may remain stable for years, the growth and embryogenic potential may vary with time or they may be lost after some months of sub-culturing (as reviewed by Häggman et al. 2000). Recently, it has also been proposed that cryopreservation could be used for cryoselection, i.e. for selection of material with specific properties (Engelmann 2004). In this way it could be used as a tool to "rejuvenate" the cultures with decreasing proliferation capacities (Engelmann 2004), which might be of great value especially for pine species. A protocol for the cryopreservation of conifer embryogenic tissue was first developed by Kartha et al. (1988) and it is still used with minor modifications in conifers including both *Picea* and *Pinus* species. Most of the cryopreservation protocols developed for specific pine species follow the classical cryopreservation techniques that involve the potential pre-treatment of the material and a slow cooling down to a defined pre-freezing temperature, followed by rapid immersion in liquid nitrogen. The material has to be re-warmed fast to avoid the phenomenon of re-crystallization, i.e. re-formation of large and damaging ice crystals by melting ice. This method has been successfully applied with some modifications to several pine species including *P. taeda* (Gupta et al. 1987), *P. caribaea* (Lainé et al. 1992), *P. radiata* (Hargreaves and Smith 1992; Hargreaves et al. 2002), *P. pinaster* (Bercetche and Páques 1995), *P. sylvestris* (Häggman et al. 1998), *P. patula* (Ford et al. 2000) and *P. roxburghii* (Mathur et al. 2003).

New vitrification-based cryopreservation techniques rely on cell dehydration prior to freezing, e.g. by exposure of samples to concentrated cryoprotective medium (Engelmann 1997). Compared with the classical techniques, these new techniques are simpler and have been adopted really quickly for several plant species. At present, in vegetatively propagated species, vitrification-based protocols have been employed almost exclusively (Engelmann 2004). Recently, Touchell et al. (2002) reported in *Picea mariana* the first successful preservation of a coniferous embryogenic culture using a vitrification-based protocol. However, it has not yet been employed with any pine species.

The combining of a clonal forestry strategy with conventional breeding is dependent on cryopreservation. The most important factor in conifer propagation via somatic embryogenesis is the opportunity to cryostore embryogenic lines (Fig. 1c) when the trees are tested in the field. In this way, it is possible to circumvent physiological maturation and hence increase propagation potential. The trees that turn out to be genetically superior in the field may be propagated consistently from the cryogenic storage. Furthermore, as pointed out by Park (2002), sufficient quantities of tested clones can be maintained indefinitely in liquid nitrogen by repeating cycles of cryopreservation, thawing, proliferation and re-cryopreservation. In conclusion, the protocols have to be reliable during the prolonged storage times to ensure genetic stability.

The potential aberrations in genetic stability during cryopreservation might be due to the generally used mutagenic chemical dimethyl sulphoxide (DMSO) in cryoprotectant mixtures (e.g. Häggman et al. 2000), prolonged sub-culturing (DeVerno et al. 1999) and especially in pine species the genetic integrity of clonal lines (Häggman et al. 2000; Park et al. 2002). In pines, megagametophytes may contain multiple archegonia indicating their capability of producing multiple genotypes (e.g. Becwar et al. 1991; Häggman et al. 1998) and the possibility that the subsequently cryopreserved clones may contain mixed genotypes. According to Park et al. (2002), this might be circumvented by re-initiating the cryopreserved clones from mature somatic embryos. This has been achieved from *P. strobus* and *P. banksiana* but at a lower rate (Park et al. 2002). These results emphasize the importance of monitoring the genetic fidelity of cryopreserved material both in vitro and ex vitro at multiple levels. Molecular markers have been used in a few cases. In *Picea glauca*, the genetic stability of randomly selected clones was evaluated by randomly amplified polymorphic DNA (RAPD) fingerprints (De Verno et al. 1999). Variant banding patterns were found in two clones out of six for in vitro culture 12 months after thawing and in plants regenerated from aberrant somatic embryos. De Verno et al. (1999) emphasized the importance of avoiding prolonged sub-culturing as well as the selection of somatic embryos with normal morphology. To our knowledge, the only pine species evaluated by RAPD fingerprints after reestablishment from cryogenic storage is *P. sylvestris* (Häggman et al. 1998), but no variation was found when cryopreserved cultures were compared with unfrozen ones. Overall, molecular markers can be used to detect genetic changes that are not readily expressed as morphological or physiological variations of the phenotype. However, they should preferably be used together with other approaches such as morphological and cytological observations (Fourré et al. 1997). Tsai and Hubscher (2004) pointed out the need to consider additional quality control issues, ranging from the soundness of liquid nitrogen Dewar flasks and cryogenic temperature monitoring to the security of storage facilities and remote backup collections.

6
Obstacles in Conversion and Acclimatization of Pine Somatic Embryos: Do We Need Symbiotic Ectomycorrhizal Fungi?

In nature, all pine species live in mutualistic interaction with specific ectomycorrhizal (ECM) fungi that colonize the roots of the host plant. In ECM symbiosis, the fungal partner increases plant nutrition by increasing the surface that absorbs nutrients and by enabling the use of organic forms of nutrients. Water and nutrients taken up by the fungus are exchanged for carbohydrates derived from the host plant (Smith and Read 1997). To date, genes encoding for nitrate and ammonium transporters have been characterized in an ECM fungus *Hebeloma cylindrosporum* Romagnesi often associated with *P. pinaster* (Jargeat et al. 2003; Javelle et al. 2003), and genes encoding for a general amino acid permease have been characterized in both *H. cylindrosporum* and *Amanita muscaria* (L. ex. Fr.) Pers. (Nehls et al. 1999; Wipf et al. 2002). Furthermore, phosphate, potassium, sulphate and micronutrient transporters were recently identified from a collection of ESTs in *H. cylindrosporum* (Lambilliotte et al. 2004).

The presence of compatible ECM fungi in the pine root system results in dramatic changes in root morphology. Lateral root formation is induced (Tranvan et al. 2000; Niemi et al. 2002, 2005), and furthermore, the tips of short roots may undergo dichotomous branching (Smith and Read 1997). In mature ectomycorrhizas, short roots of the host plant are covered by a hyphal mantle, and a highly branched structure called a Hartig net is formed as the fungus penetrates between epidermal and cortical cells (Smith and Read 1997). The formation of ECM symbiosis causes inhibition in root hair proliferation and external hyphae replace root hairs for absorbing water and nutrients from the soil (Béguiristain and Lapeyrie 1997; Ditengou et al. 2000). The necessity of ECM symbiosis to coniferous species has resulted in attempts to apply ECM fungi in root formation of vegetatively propagated material. Inoculation of the plant cuttings with specific ECM fungi has resulted in a higher rooting frequency, higher number of roots per shoot, and improved root growth of several recalcitrant coniferous species, including pines. However, interaction during root formation has been highly dependent on the plant and fungus genotypes (reviewed by Niemi et al. 2004).

In somatic embryogenesis, successful germination and subsequent growth of the root system are prerequisites for acclimatization to the conditions ex vitro in a greenhouse. However, somatic embryo germination is often poor, and roots elongate and branch slowly or not at all (e.g. Jones and van Staden 1995; Häggman et al. 1999; Niemi and Häggman 2002; Miguel et al. 2004). In nature, germinated seedlings become colonized immediately by mycorrhizal fungi, resulting in better growth of the root system and plant adaptation to the conditions in the soil. Therefore, inoculation with specific ECM fungi might be a potential tool to improve conversion of mature somatic embryos.

So far, there have only been four reports on specific ECM fungi affecting the conversion of mature somatic embryos (Sasa and Krogstrup 1991; Piola et al. 1995; Díez et al. 2000; Niemi and Häggman 2002), one of which is on a pine species (Niemi and Häggman 2002). In our study, four out of five cell lines of *P. sylvestris* increased their germination frequency as a result of inoculation with the ECM fungus *Pisolithus tinctorius* (Pers.) Coker and Couch. Positive responses were observed when the fungal mycelium and germinating embryos were far enough apart to avoid physical contact (Fig. 1f). In contrast, when placed in physical contact, the fungus grew aggressively over the whole embryo. This imbalance between symbiotic partners was probably due to the relatively high concentration of nutrients and sugar in the germination medium. Subsequent inoculation of the germinated somatic embryos with the same fungus on a medium with low nutrient and sugar concentrations resulted in extensive root elongation, root branching and finally mycorrhiza formation (Figs. 2a–c) (Niemi and Häggman 2002). Successful root development and mycorrhiza formation was also observed when somatic embryo plants of *Larix* × *eurolepis* were inoculated with specific ECM fungi (Piola et al. 1995), whereas in *Picea sitchensis* (Bong.) Carr. only a slight or no increase was observed in the growth due to mycorrhiza formation (Sasa and Krogstrop 1991). These results indicate that positive interaction between a somatic embryo and ECM fungus is highly dependent on the developmental phase of the somatic embryo, the fungal and plant genotype, and the composition of the medium.

Similarly, acclimatization of rooted cuttings to the conditions ex vitro was improved in the presence of a specific ECM fungus (Supriyanto and Rohr 1994; Normand et al. 1996). This was also the case with somatic embryo plants

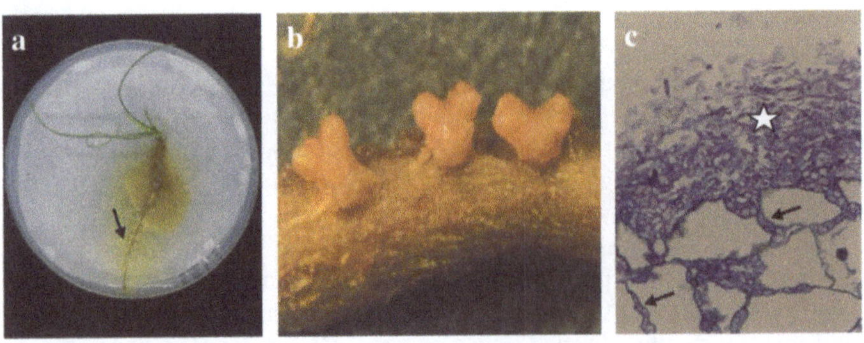

Fig. 2 Ectomycorrhizal symbiosis between *P. sylvestris* somatic embryo plant and *Pisolithus tinctorius* in vitro. **a** An elongated main root of somatic embryo plant and dichotomously branched short roots covered by fungal hyphae (*arrow*). **b** Dichotomously branched mycorrhizal short roots stained red with Ponceau S. **c** Cross section of an ectomycorrhizal short root. Hyphal mantle over the short root (*star*); Hartig net formed by the fungus between epidermal and cortical cells (*arrows*)

of *P. sylvestris* inoculated ex vitro with *Pisolithus tinctorius*. Depending on the plant cell line, better adaptation was observed as either an increased survival rate or increased shoot and root growth. *Pisolithus tinctorius* formed neither hyphal mantle nor Hartig net in the root system, which shows that the plant may benefit from the specific ECM fungus even without mycorrhiza formation (Niemi and Häggman 2002).

Regardless of the necessity for ECM interaction of pines in nature, hardly any attention has been paid to its potential use in somatic embryogenesis. Studies with Scots pine (Niemi and Häggman 2002) and three other tree species (Sasa and Krogstrup 1991; Piola et al. 1995; Díez et al. 2000) clearly show that inoculation with specific ECM fungi is a potential tool for improving both the germination of mature somatic embryos and the acclimatization process of somatic embryo plants. However, since the reactions are highly specific and dependent on the genotypes of both symbiotic partners, it is important to test several fungus strain–plant cell line interactions before any larger-scale use.

7
Concluding Remarks

Pine species are globally important woody species with a wide distribution. During recent decades, plantation forestry has generally been considered as sustainable forestry. Somatic embryogenesis is expected to be a potential mass-scale technology that would allow the production of vegetatively propagated pine clones for reforestation and breeding purposes. Recent achievements in functional genomics, especially in zygotic embryogenesis and physiological outcomes, have improved somatic embryo production. Development of cryopreservation protocols for several pine species have also contributed to practical and tree breeding applications. Nevertheless, there are still obstacles in somatic embryogenesis, e.g. in proper root formation, and certainly more attention should be paid to the potential of natural symbiotic ECM fungi at the germination and acclimatization stages. The progress in somatic embryogenesis has also opened the door to molecular breeding using the transgenic approach. However, this approach is in its infancy and the years to come will show how this technology will be adopted. It is certain that this development has to be based on sustainable forestry.

Acknowledgements We wish to thank Prof. James Graham from the Citrus Research and Education Center, University of Florida, for valuable comments on the manuscript and Mr. Jouko Lehto from the Finnish Forest Research Institute, Punkaharju Research Station, for the photos in Figs. 1 and 2a. We acknowledge the research funding from the Academy of Finland (grants 105214 to H.H., 202415 to K.N. and 53440 to T.S.) and from the Finnish Cultural Foundation (a grant to K.N.).

References

Aronen T, Häggman H, Hohtola A (1994) Transient β-glucuronidase expression in Scots pine tissues derived from mature trees. Can J For Res 24:2006–2011

Aronen T, Hohtola A, Laukkanen H, Häggman H (1995) Seasonal changes in the transient expression of a 35S CaMV-GUS gene construct introduced into Scots pine buds. Tree Physiol 15:65–70

Aronen T, Häggman H (1995) Differences in *Agrobacterium* infections in silver birch and Scots pine. Eur J Forest Pathol 25:197–213

Aronen T, Häggman H, Salonen M (1996) Rooting of Scots pine fascicular shoots by *Agrobacterium rhizogenes*. For Genet 3:15–24

Aronen T, Nikkanen T, Häggman H (1998) Compatibility of different pollination techniques with microprojectile bombardment of Norway spruce and Scots pine pollen. Can J For Res 28:79–86

Aronen TS, Nikkanen TO, Häggman HM (2003) The production of transgenic Scots pine (*Pinus sylvestris* L.) via the application of transformed pollen in controlled crossings. Transgenic Res 12:375–378

Attree SM, Denchev P, Kong L, Lobatcheva I, Folk R, Lawrence B (2004) Developing a commercial somatic embryogenesis (SE) production platform for conifers. In: Proceedings of the 2004 IUFRO joint conference of division 2. Forest genetics and tree breeding in the age of genomics: progress and future. 1–5 Nov 2004, Charleston, SC, p 142

Becwar MR, Blush TD, Brown DW, Chesick EE (1991) Multiple paternal genotypes in embryogenic tissue derived from individual immature loblolly pine seeds. Plant Cell Tissue Organ Cult 26:37–44

Bercetche J, Pâques M (1995) Somatic embryogenesis in maritime pine (*Pinus pinaster*). In: Jain SM, Gupta PK, Newton RJ (eds) Somatic embryogenesis in woody plants, vol 3. Kluwer, Dordrecht, pp 221–242

Béguiristain T, Lapeyrie F (1997) Host plant stimulates hypaphorine accumulation in *Pisolithus tinctorius* hyphae during ectomycorrhizal infection while excreted fungal hypaphorine controls root hair development. New Phytol 136:525–532

Bishop-Hurley S, Gardner RC, Walter C (2003) Isolation and molecular characterization of genes expressed during somatic embryo development in *Pinus radiata*. Plant Cell Tissue Organ Cult 74:267–281

Bishop-Hurley S, Zabkievicz L, Grace L, Gardner RC, Wagner A, Walter C (2001) Conifer genetic engineering: transgenic *Pinus radiata* (D. Don) and *Picea abies* (karst) plants are resistant to the herbicide Buster. Plant Cell Rep 20:235–243

Bogunic F, Muratovic E, Brown SC, Siljak-Yakovlev S (2003) Genome size and base composition of five *Pinus* species from the Balkan region. Plant Cell Rep 22:59–63

Bommineni VR, Chibbar RN, Datla RSS, Tsang EWT (1993) Transformation of white spruce (*Picea galuca*) somatic embryos by microprojectile bombardment. Plant Cell Rep 13:17–23

Cairney J, Buell R, Pullman J, Quackenbush J (2003) Genomics of embryogenesis in loblolly pine. In: Abstracts of the IUFRO tree biotechnology meeting, Umeå, Sweden, 7-12 June 2003

Cairney J, Xu N, Pullman GS, Ciavatta VT, Johns B (1999) Natural and somatic embryo development in loblolly pine: gene expression studies using differential display and DNA arrays. Appl Biochem Biotechnol 77–79:5–17

Cánovas FM, Dumas-Gaudot E, Recorbet G, Jorrin J, Mock H-P, Rossignol M (2004) Plant proteome analysis. Proteomics 4:285–298

Ciavatta VT, Egertsdotter U, Clapham D, von Arnold S, Cairney J (2002) A promoter from the loblolly pine PtNIP1;1 gene directs expression in an early-embryogenesis and suspensor-specific fashion. Planta 215:694–698

Ciavatta VT, Morillon R, Pullman GS, Chrispeels MJ, Cairney J (2001) An aquaglyceroporin is abundantly expressed early in the development of the suspensor and the embryo proper of loblolly pine. Plant Physiol 127:1556–1567

Cyr DR, Klimaszewska K (2002) Conifer somatic embryogenesis: II. Applications. Dendrobiologia 48:41–49

DeVerno LL, Park YS, Bonga JM, Barrett JD (1999) Somaclonal variation in cryopreserved embryogenic clones of white spruce (*Picea glauca* (Moench) Voss). Plant Cell Rep 18:239–261

Díez J, Manjon JL, Kovács GM, Celestino C, Toribio M (2000) Mycorrhization of in vitro plants from somatic embryos of cork oak (*Quercus suber* L.). Appl Soil Ecol 15:119–123

Diouf D (2003) Genetic transformation of forest trees. Afr J Biotechnol 2:328–333

Ditengou FA, Béguiristain T, Lapeyrie F (2000) Root hair elongation is inhibited by hypaphorine, the indole alkaloid from the ectomycorrhizal fungus *Pisolithus tinctorius*, and restored by indole-3-acetic acid. Planta 211:722–728

Domon J-M, Dumas B, Lainé E, Meyer Y, Alain D, David H (1995) Three glycosylated polypeptides secreted by several embryogenic cell cultures of pine show highly specific serological affinity to antibodies directed against the wheat germin apoprotein monomer. Plant Physiol 108:141–148

Dong J-Z, Dunstan D (1997) Characterization of cDNAs representing five abscisic acid-responsive genes associated with somatic embryogenesis in *Picea glauca*, and their responses to abscisic acid stereostructure. Planta 203:448–453

Dong J-Z, Dunstan DI (2000) Molecular biology of somatic embryogenesis in conifers. In: Jain SM, Minocha SC (eds) Molecular biology of woody plants, vol 1. Kluwer, Dordrecht, pp 51–87

Ellis DD, McCabe DE, McInnis S, Ramachandran R, Russell DR, Wallace KM, Martinelli BJ, Roberts DR, Raffa KF, McCown BH (1993) Stable transformation of *Picea glauca* by particle bombardment. Biotechnology (NY) 11:84–89

Engelmann F (1997) Importance of desiccation for cryopreservation of recalcitrant seed and vegetatively propagated species. Plant Genet Resour Newsl 112:9–18

Engelmann F (2004) Plant cryopreservation: progress and prospects. In Vitro Cell Dev Biol Plant 40:427–433

Fernando DD, Owens JN, Misra S (2000) Transient gene expression in pine pollen tubes following particle bombardment. Plant Cell Rep 19:224–228

Fillatti JJ, Selmer J, McCown B, Haissig B, Comai L (1987) *Agrobacterium*-mediated transformation and regeneration of Populus. Mol Gen Genet 206:192–199

Ford CS, Jone NB, van Staden J (2000) Cryopreservation and plant regeneration from somatic embryos of *Pinus patula*. Plant Cell Rep 19:610–615

Footitt S, Ingouff M, Clapham D, von Arnold S (2003) Expression of the viviparous 1 (*Pavp1*) and p34^{cdc2} protein kinase (*cdc2Pa*) genes during somatic embryogenesis in Norway spruce (*Picea abies* [L.] Karst). J Exp Bot 54:1711–1719

Fourré JL, Berger P, Niquet L, André P (1997) Somatic embryogenesis and somaclonal variation in Norway spruce: morphogenetic, cytogenetic and molecular approaches. Theor Appl Genet 94:159–169

Garin E, Isabel N, Plourde A (1998) Screening of large numbers of seed families of *Pinus strobus* L. for somatic embryogenesis from immature and mature zygotic embryos. Plant Cell Rep 18:37–43

Gupta PK, Durzan DJ, Finkle BJ (1987) Somatic polyembryogenesis in embryogenic cell masses of *Picea abies* (Norway spruce) and *Pinus taeda* (loblolly pine) after thawing from liquid nitrogen. Can J For Res 17:1130–1134

Häggman H, Aronen T (1998) Transgene expression in regenerating cotyledons and embryogenic cultures of Scots pine. J Exp Bot 49:1147–1156

Häggman H, Aronen T (2000) *Agrobacterium rhizogenes* for rooting recalcitrant woody plants. In: Jain SM, Minocha SC (eds) Molecular biology of woody plants, vol 2. Kluwer, Dordrecht, pp 47–78

Häggman HM, Aronen TS, Nikkanen TO (1997) Gene transfer by particle bombardment to Norway spruce and Scots pine pollen. Can J For Res 27:928–935

Häggman HM, Aronen TS, Ryynänen LA (2000) Cryopreservation of embryogenic cultures of conifers. In: Jain SM, Gupta PK, Newton RJ (eds) Somatic embryogenesis in woody plants, vol 6. Kluwer, Dordrecht, pp 707–728

Häggman H, Ryynänen L, Aronen T (2001) Cryopreservation of forest tree germplasm. In: Sorvari S, Karhu S, Kanervo E, Pihakaski S. Proceedings of the 4th international symposium on in vitro culture and horticultural breeding. Acta Hortic 560:121–124

Häggman HM, Ryynänen LA, Aronen TS, Krajnakova J (1998) Cryopreservation of embryogenic cultures of Scots pine. Plant Cell Tissue Organ Cult 54:45–53

Häggman H, Jokela A, Krajnakova J, Kauppi A, Niemi K, Aronen T (1999) Somatic embryogenesis of Scots pine: cold treatment and characteristics of explants affecting induction. J Exp Bot 50:1769–1778

Handley LW, Becwar MR, Chesick EE, Coke JE, Godbey AP, Rutter MR (1995) Research and development of commercial tissue culture systems in loblolly pine. Tappi J 78:169–175

Hanson B, Engler D, Moy Y, Newman B, Ralston E, Gutterson N (1999) A simple method to enrich an *Agrobacterium*-transformed population for plants containing only T-DNA sequences. Plant J 19:727–734

Hargreaves CL, Grace LJ, Holden DG (2002) Nurse culture for efficient recovery of cryopreserved *Pinus radiata* D. Don embryogenic cell lines. Plant Cell Rep 21:40–45

Hargreaves C, Smith DR (1992) Cryopreservation of *Pinus radiata* embryogenic tissue. International Plant Propagators' Society combined proceedings 42:327–333

Hay I, Lachance D, von Aderkas P, Charest PJ (1994) Transient chimeric gene expression in pollen of five conifer species following microparticle bombardment. Can J For Res 24:2417–2423

Herschbach C, Kopriva S (2002) Transgenic trees as tools in tree and plant physiology. Trees 16:250–261

Hizume M, Shibata F, Matsusaki Y, Garajova Z (2002) Chromosome identification and comparative karyotypic analyses of four Pinus species. Theor Appl Genet 105:491–497

Hjortswang HI, Sundås Larsson A, Bharathan G, Bozhkov PV, von Arnold S, Vahala T (2002) KNOTTED1-like homeobox genes of a gymnosperm, Norway spruce, expressed during somatic embryogenesis. Plant Physiol Biochem 40:837–843

Huang Y, Diner AM, Karnosky DF (1991) *Agrobacterium rhizogenes*-mediated genetic transformation and regeneration of a conifer *Larix decidua*. In Vitro Cell Dev Biol Plant 27:201–207

Ingouff M, Farbos I, Wiweger M, von Arnold S (2003) The molecular characterization of *PaHB2*, a homeobox gene of the *HD-GL2* family expressed during embryo development in Norway spruce. J Exp Bot 54:1343–1350

Jargeat P, Rekangalt D, Verner MC, Gay G, Debaud JC, Marmeisse R, Fraissinet-Tachet L (2003) Characterization and expression analyses of a nitrate transporter and nitrite reductase genes, two members of a gene cluster for nitrate assimilation from the symbiotic basidiomycete *Hebeloma cylindrosporum*. Curr Genet 43:199–205

Javelle A, Morel M, Rodriguez-Pastrana BR, Andre B, Marini MA, Brun A, Chalot M (2003) Molecular characterization, function and regulation of ammonium transportes (Amt) and ammonium-metabolizing enzymes (GS, NADP-GDH) in the ectomycorrhizal fungus *Hebeloma cylindrosporum*. Mol Plant Microbe Interact 17:202–215

Jones NB, van Staden J (1995) Plantlet production from somatic embryos of *Pinus patula*. J Plant Physiol 145:519–525

Jurgens G (2001) Apical-basal pattern of embryo formation in *Arabidopsis* embryogenesis. EMBO J 20:3609–3616

Kartha K, Fowke L, Leung N, Caswell K, Hakman I (1988) Induction of somatic embryos and plantlets from cryopreserved cell cultures of white spruce (*Picea glauca*). J Plant Physiol 132:529–539

Khuri S, Bakker FT, Dunwell JM (2001) Phylogeny, function and evolution of the cupins, a structurally conserved, functionally diverse superfamily of proteins. Mol Biol Evol 18:593–605

Kim SR, Lee J, Jun SH, Park S, Kang HG, Kwon S, An G (2003) Transgene structures in T-DNA inserted rice plants. Plant Mol Biol 52:761–773

Kinlaw CS, Neale DB (1997) Complex gene families in pine genomes. Trends Plant Sci 2:356–359

Klimaszewska K, Park YS, Overton C, Maceacheron I, Bonga JM (2001) Optimized somatic embryogenesis in *Pinus strobus* L. In Vitro Cell Dev Biol Plant 37:392–399

Komulainen P, Brown GR, Mikkonen M, Karhu A, García-Gil MR, O'Malley D, Lee B, Neale DB, Savolainen O (2003) Comparing EST-based genetic maps between *Pinus sylvestris* and *Pinus taeda*. Theor Appl Genet 107:667–678

Kriebel HB (1985) DNA sequence components of *Pinus strobus* nuclear genome. Can J For Res 15:1–4

Kumar S, Fladung M (2001) Gene stability in transgenic aspen (*Populus*). II. Molecular characterization of variable expression of transgene in wild and hybrid aspen. Planta 213:731–740

Lainé E, Pascale B, David A (1992) Recovery of plants from cryopreserved embryogenic cell suspensions of *Pinus caribaea*. Plant Cell Rep 11:295–298

Lambilliotte R, Cooke R, Samson D, Fizames C, Gaymard F, Plassard C, Tatry MV, Berger C, Laudie M, Legeai F, Karsenty E, Delseny M, Zimmerman S, Sentenac H (2004) Large-scale identification of genes in the fungus *Hebeloma cylindrosporum* paves the way to molecular analyses of ectomycorrhizal symbiosis. New Phytol 164:505–513

Lelu MA, Bastien C, Drugeault A, Gouez LM, Klimaszewska K (1999) Somatic embryogenesis and plantlet development in *Pinus sylvestris* and *Pinus pinaster* on the medium with and without growth regulators. Physiol Plantarum 105:719–728

Levee V, Garin E, Klimaszewska K, Seguin A (1999) Stable genetic transformation of white pine (*Pinus strobus* L.) after cocultivation of embryogenic tissues with *Agrobacterium tumefaciens*. Mol Breed 5:429–440

Malabadi RB, Choudhury H, Tandon P (2002) Plant regeneration via somatic embryogenesis in *Pinus kesiya* (Royle ex. Grod.). Appl Biol Res 4:1–10

Malabadi RB, van Staden J (2005) Somatic embryogenesis from vegetative shoot apices of mature trees of *Pinus patula*. Tree Physiol 25:11–16

Martinussen I, Bate N, Weterings K, Junttila O, Twell D (1995) Analysis of gene regulation in growing pollen tubes of angiosperms and gymnosperm species using microprojectile bombardment. Physiol Plantarum 93:445–450

Mathur G, Alkutkar VA, Nadgauda RS (2003) Cryopreservation of embryogenic culture of *Pinus roxburghii*. Biol Plantarum 46:205–210

Merkle SA, Dean JFD (2000) Forest tree biotechnology. Curr Opin Biotechnol 11:298–302
Miguel C, Goncalves S, Tereso S, Marum L, Maroco J, Oliveira MM (2004) Somatic embryogenesis from 20 open-pollinated families of Portuguese plus trees of maritime pine. Plant Cell Tissue Organ Cult 76:121–130
Nehls U, Kleber R, Wiese J, Hampp R (1999) Isolation and characterization of a general amino acid permease from the ectomycorrhizal fungus *Amanita muscaria*. New Phytol 144:343–349
Neutelings G, Domon J, Membre N, Bernier F, Meyer Y, David A, David H (1998) Characterization of a germin-like protein gene expressed in somatic and zygotic embryos of pine (*Pinus caribaea* Morelet). Plant Mol Biol 38:1179–1190
Niemi K, Häggman H (2002) *Pisolithus tinctorius* promotes germination and forms mycorrhizal structures in Scots pine somatic embryos in vitro. Mycorrhiza 12:263–267
Niemi K, Julkunen-Tiitto R, Tegelberg R, Häggman H (2005) Light sources with different spectra affect root and mycorrhiza formation of Scots pine *in vitro*. Tree Physiol 25:123–128
Niemi K, Scagel C, Häggman H (2004) Application of ectomycorrhizal fungi in vegetative propagation of conifers. Plant Cell Tissue Organ Cult 78:83–91
Niemi K, Vuorinen T, Ernstsen A, Häggman H (2002) Ectomycorrhizal fungi and exogenous auxins influence root and mycorrhiza formation of Scots pine hypocotyl cuttings in vitro. Tree Physiol 22:1231–1239
Niskanen AM, Lu J, Seitz S, Keinonen K, von Weissenberg K, Pappinen A (2004) Effect of parent genotype on somatic embryogenesis of Scots pine (*Pinus sylvestris*). Tree Physiol 24:1259–1265
Normand L, Bärtschi H, Debaud JC, Gay G (1996) Rooting and acclimation of micropropagated cuttings of *Pinus pinaster* and *Pinus sylvestris* are enhanced by the ectomycorrhizal fungus *Hebeloma cylindrosporum*. Physiol Plantarum 98:759–766
Park YS (2002) Implementation of conifer somatic embryogenesis in clonal forestry: technical requirements and deployment considerations. Ann For Sci 59:651–656
Peña L, Séguin A (2001) Recent advances in the genetic transformation of trees. Trends Biotechnol 12:500–506
Percy RE, Klimaszewska K, Cyr DR (2000) Evaluation of somatic embryogenesis for clonal propagation of western white pine. Can J For Res 30:1867–1876
Perry DJ, Furnier G (1996) *Pinus banksiana* has at least seven expressed alcohol dehydrogenase genes in two linked groups. Proc Natl Acad Sci USA 93:13020–13023
Piola F, Rohr R, von Aderkas P (1995) Controlled mycorrhizal initiation as a means to improve root development in somatic embryo plantlets of hybrid larch. Physiol Plantarum 95:575–580
Pullman GS, Johnson S, Peter G, Cairney J, Xu N (2003a) Improving loblolly pine somatic embryo maturation: comparison of somatic and zygotic embryo morphology, germination, and gene expression. Plant Cell Rep 21:747–758
Pullman GS, Namjoshi K, Zhang Y (2003b) Somatic embryogenesis in loblolly pine (*Pinus taeda* L.): improving culture initiation with abscisic acid and silver nitrate. Plant Cell Rep 22:85–95
Rommens CM, Humara JM, Ye J, Yan H, Richael C, Zhang L, Perry R, Swords K (2004) Crop improvement through modification of the plant's own genome. Plant Physiol 135:421–431
Sasa M, Krogstrup P (1991) Ectomycorrhizal formation in plantlets derived from somatic embryos of Sitka spruce. Scand J For Res 6:129–136
Sax K, Sax HJ (1933) Chromosome number and morphology in the conifers. J Arnold Arbor 14:356–375

Smith S, Read D (eds) (1997) Mycorrhizal symbiosis, 2nd edn. Academic Press, San Diego
Smith DR, Warr A, Grace L, Walter C, Hargreaves CL (1994) Somatic embryogenesis joins the plantation forestry revolution in New Zealand. In: Proceedings of the TAPPI 1994 biological sciences symposium, pp 19–24
Stasolla C, Bozhkov P, Tzu-Ming C, van Zyl L, Egertsdotter U, Suarez MF, Craig D, Wolfinger RD, von Arnold S, Sederoff RR (2004) Variation in transcript abundance during somatic embryogenesis in gymnosperms. Tree Physiol 24:1073–1085
Stasolla C, van Zyl L, Egertsdotter U, Craig D, Liu W, Sederoff RR (2003) The effects of polyethylene glycol on gene expression of developing white spruce somatic embryos. Plant Physiol 131:49–60
Stoger E, Fink C, Pfosser M, Heberle-Bors E (1995) Plant transformation by particle bombardment of embryogenic pollen. Plant Cell Rep 14:273–278
Supriyanto M, Rohr R (1994) In vitro regeneration of plantlets of Scots pine (*Pinus sylvestris*) with mycorrhizal roots from subcultured callus initiated from needle adventitious buds. Can J Bot 72:1144–1150
Sutton B (1999) The need for planted forests and an example of radiata pine. New Forest 17:95–109
Sutton B (2002) Commercial delivery of genetic improvement to conifer plantations using somatic embryogenesis. Ann For Sci 59:657–661
Tang W, Guo Z (2001) In vitro propagation of loblolly pine via direct somatic organogenesis from mature cotyledons and hypocotyls. Plant Growth Regul 33:25–31
Tang W, Guo Z, Ouyang F (2001a) Plant regeneration from embryogenic cultures initiated from mature loblolly pine zygotic embryos. In Vitro Cell Dev Biol Plant 37:558–563
Tang W, Harris LC, Outhavong V, Newton RJ (2004) Antioxidants enhance in vitro plant regeneration by inhibiting the accumulation of peroxidase in Virginia pine (*Pinus virginia* Mill.). Plant Cell Rep 22:871–877
Tang W, Sederoff R, Whetten R (2001b) Regeneration of transgenic loblolly pine (*Pinus taeda* L.) from zygotic embryos transformed with *Agrobacterium tumefaciens*. Planta 213:981–989
Tang W, Tian Y (2003) Transgenic loblolly pine (*Pinus taeda* L.) plants expressing a modified delta-endotoxin gene from *Bacillus thuringiensis* with enhanced resistance to *Dendrolimus punctatus* Walker and *Crypyothelea formosicola* Staud. J Exp Bot 54:835–844
Tang W, Whetten R, Sederoff R (2001c) Genotypic control of high-frequency adventitious shoot regeneration via somatic organogenesis in loblolly pine. Plant Sci 161:267–272
Touchell DH, Chiang VL, Tsai CJ (2002) Cryopreservation of embryogenic cultures of *Picea mariana* (black spruce) using vitrification. Plant Cell Rep 21:118–124
Tranvan H, Habricot Y, Jeannette E, Gay G, Sotta B (2000) Dynamics of symbiotic establishment between an IAA-overproducing mutant of the ectomycorrhizal fungus *Hebeloma cylindrosporum* and *Pinus pinaster*. Tree Physiol 20:123–129
Tsai CJ, Hubscher SL (2004) Cryopreservation in *Populus* functional genomics. New Phytol 164:73–81
van der Leede-Plegt LM, van de Ven BCE, Schilder M, Franken J, van Tunen A (1995) Development of a pollen-mediated transformation method for *Nicotiana glutinosa*. Transgenic Res 4:77–86
van Zyl L, Bozhkov PV, Clapham DH, Sederoff RR, von Arnold S (2003) Up, down and up again is a signature global gene expression pattern at the beginning of gymnosperm embryogenesis. Gene Expr Patterns 3:83–91
van Zyl L, von Arnold S, Bozhkov P, Chen Y, Egertsdotter U, MacKay J, Sederoff RR, Shen J, Zelena L, Clapham DH (2002) Heterologous array analysis in Pinaceae: hy-

bridization of *Pinus taeda* cDNA arrays with cDNA from needles and embryogenic cultures of *P. taeda, P. sylvestris* or *Picea abies*. Comp Funct Genomics 3:306–318

Wagner A, Moody J, Grace L, Walter C (1997) Stable transformation of *Pinus radiata* based on selection with Hygromycin B. NZ J For Sci 27:280–288

Walter C (2004) Genetic engineering in conifer forestry: technical and social considerations. In Vitro Cell Dev Biol Plant 40:434–441

Walter C, Grace LJ, Wagner A, White DWR, Walden A, Donaldson SS, Hinton H, Gardner RC, Smith DR (1998) Stable transformation and regeneration of transgenic plants of *Pinus radiata* D. Don. Plant Cell Rep 17:460–468

Wenck A, Quinna M, Whetten R, Pullman GS, Sederoff RR (1999) High-efficiency Agrobacterium-mediated transformation of Norway spruce (*Picea abies*) and loblolly pine (*Pinus taeda*). Plant Mol Biol 39:407–416

Wipf D, Benjdia M, Tegeder M, Fommer WB (2002) Characterization of a general amino acid permease from *Hebeloma cylindrosporum*. FEBS Lett 528:119–124

Plant Cell Monogr (2)
A. Mujib · J. Šamaj: Somatic Embryogenesis
DOI 10.1007/7089_023/Published online: 30 November 2005

Somatic Embryogenesis in *Pinus nigra* Arn.: Some Physiological, Structural and Molecular Aspects

Terezia Salaj (✉) · Jana Moravcikova · Jan Salaj

Institute of Plant Genetics and Biotechnology, Slovak Academy of Sciences, Akademicka 2, P.O. Box 39A, 950 07 Nitra 1, Slovak Republic
nrgrtesa@savba.sk

Abstract In this chapter we summarize different aspects of *Pinus nigra* somatic embryogenesis including initiation, maintenance and maturation on both solid and liquid media, and evaluation of the role of plant growth regulators and composition of the media in this process. In addition, the establishment and maintenance of *Pinus nigra* suspension cultures is described. The experiments on genetic transformation of *Pinus nigra* embryogenic tissue with reporter *uid*A and selection *npt* II genes are also reported and discussed here.

1 Introduction

In the genus *Pinus*, somatic embryogenesis was initiated mostly from immature zygotic embryos prior to cotyledon development. This initiation pattern has been described for *Pinus strobus* (Finer et al. 1989; Klimaszewska and Smith 1997; Kaul 1995), *Pinus taeda* (Becwar et al. 1990), *Pinus caribaea* (Laine and David 1990), *Pinus nigra* (Salajova and Salaj 1992; Salajova at al. 1995), *Pinus elliottii* (Newton et al. 1995), *Pinus lambertiana* (Gupta 1995), *Pinus patula* (Jones and van Staden 1995), *Pinus pinaster* (Bercetche and Paques 1995; Lelu et al. 1999; Miguel et al. 2004), *Pinus radiata* (Chandler and Young 1995), *Pinus sylvestris* (Keinonen-Mettälä et al. 1996; Häggman et al. 1999), *Pinus roxburghii* (Mathur et al. 2000; Arya et al. 2000) and *Pinus monticola* (Percy et al. 2000). The embryogenic tissue usually extruded from megagametophyte explants. These extrusions can be considered to be an indicator of positive explant response although they did not always lead to tissue proliferation. Only about 20% of extrusions produced proliferating embryogenic tissues in *Pinus monticola* (Percy et al. 2000). The initiation is dependent on the concentration of growth regulators in the culture medium. Recently, Klimaszewska et al. (2001) have found that exposure of explants to a medium with a lower concentration of plant growth regulators significantly improved the initiation of somatic embryogenesis in *Pinus strobus*. Embryogenic tissues have also been occasionally initiated in the absence of plant growth regulators in the medium, as was the case with *Pinus sylvestris* and *P. pinaster* (Lelu et al. 1999).

The embryogenic tissues of *Pinus* can be maintained in long-term culture by regular transfers to new medium every 2–3 weeks. However, this long-term cultivation of embryogenic tissues holds the risk of genetic changes as well as loss of maturation and regeneration ability of somatic embryos. In order to prevent these undesirable events, the method of cryopreservation has been developed and applied for several *Pinus* species. Embryogenic tissues of *Pinus taeda* (Gupta et al. 1987), *Pinus caribaea* (Laine et al. 1992), *Pinus sylvestris* (Häggman et al. 1998), *Pinus patula* (Ford et al. 2000), *Pinus roxburghii* (Mathur et al. 2003) and *Pinus pinaster* (Marum et al. 2004) have been successfully cryopreserved in liquid nitrogen. After cryostorage most of the cell lines recovered and showed growth on a proliferation medium. Importantly, the reestablished cultures maintained an embryogenic potential similar to non-frozen cultures. Moreover, the RAPD assays suggested that the cryostorage treatment preserved the genetic fidelity of embryogenic cultures in *Pinus sylvestris* (Häggman et al. 1998).

2
Initiation of Embryogenic Tissues

Initiation of somatic embryogenesis (embryogenic tissues) in *Pinus nigra* Arn. occurred from immature zygotic embryos enclosed in megagametophytes dissected from immature seeds (Salajová et al. 1995). The presence of the megagametophyte is an important requirement for successful initiation. During the period of collection and placing of the explants on the medium, the immature zygotic embryos are of microscopic size and their cultivation is very problematic. The specific effect of megagametophytes for the initiation has not been determined. It is very likely that the surrounding megagametophyte supplies nutrients to the zygotic embryos and may also serve as a support preventing the embryo from mechanical damage. The culture of megagametophytes as explants also holds the risk of multiple paternal genotype initiation owing to the polyembryony in a single pine seed (Becwar et al. 1991). Observations of isolated zygotic embryos under the dissecting microscope as well as a histological study revealed that the responsive immature embryos yielding the embryogenic tissues were at the pre-cotyledonary stage of development (Fig. 1). Using zygotic embryos in later developmental stages resulted in only very low or no initiation frequencies. The ability of zygotic embryos to produce embryogenic tissue diminished completely when they matured (Salajova et al. 1995). Zygotic embryos, under the typical climatic conditions of Slovakia, reach the proper developmental stage for somatic embryogenesis initiation in the middle of June or in the second half of June. Therefore, explant collection has been restricted to this part of the year.

DCR (Gupta and Durzan 1985) was used as the initiation medium. The basal medium was supplemented with 2,4-D (2 $mg\,l^{-1}$), BA (0.5 $mg\,l^{-1}$), enzy-

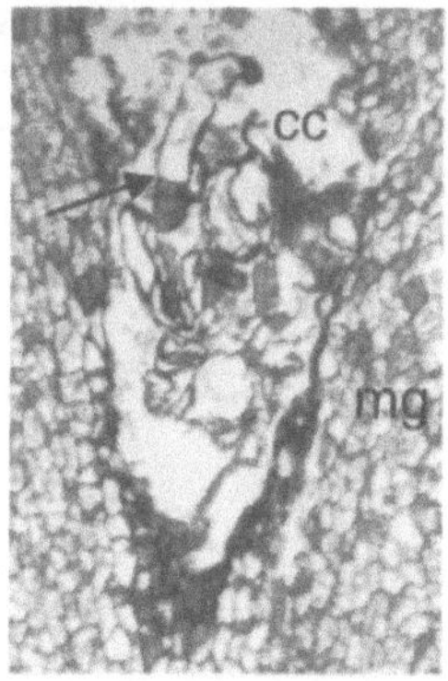

Fig. 1 Section through megagametophyte (*mg*) with corrosion cavity (*cc*) and zygotic embryo (*arrow*) at the responsive stage

matic caseinhydrolysate (500 mg l^{-1}), glutamine (50 mg l^{-1}), sucrose (20 g l^{-1}) and gelified with Phytagel (3 g l^{-1}).

The embryogenic tissue proliferated from the micropylar pole of the megagametophytes. The proliferation occurred soon after the placing of the explants on the culture medium (usually within 4–16 days) and the majority of explants formed embryogenic tissue within six weeks of cultivation. Initiation frequencies differed from year to year and reached values of 1.53 to 24.1% (Table 1).

Growth regulators play an important role in the initiation of somatic embryogenesis. In our experiments, we have used the same combination and concentrations of 2,4-D (2 mg l^{-1}) and BA (0.5 mg l^{-1}) for several years. In later experiments, we tested the effect of different combinations of BA; 2,4-D and NAA in a range of concentrations between 0 and 2 mg l^{-1} on the embryogenic tissue initiation with the aim of improving the initiation frequencies.

These experiments were repeated twice, in the years 2000 and 2001. Different initiation frequencies have been obtained although no profound improvement has been achieved (Table 2). The most productive response (5.68% initiation frequency) was obtained with an equal mixture of 2,4-D (1 mg l^{-1}) and BA (1 mg l^{-1}). Embryogenic tissue initiation occurred also on media containing sole growth regulators reaching values from 1.13 to 5% in the year 2000 or relatively high frequencies from 7.14 to 9.09% in the year 2001. In the year 2001, a higher total initiation frequency was achieved (5.54%) in comparison to the previous year (3.06% in 2000). In both cases explants were collected in the middle of June but the zygotic embryos were in a more progressive developmental stage owing to the extreme climatic conditions of the year 2000. Some of the developing zygotic embryos were at the early or late cotyledonary stage in 2000 while all observed zygotic embryos were at the precotyledonary developmental stage in 2001.

Table 1 Initiation of embryogenic tissues in *P. nigra* megagametophytes dissected from immature seeds (from Salajova et al. 1999, with permission of Elsevier)

Date	Number of explants	% of initiation
1994		
June 20	253	24.11
June 24	237	16.01
1997		
June 20	77 [a]	2.59
June 25	120	1.66
1998		
June 18	240	5.41
June 25	130	1.53

[a] The number of dissected explants was 171, however, 94 explants were contaminated during the culture.

Table 2 The effect of growth regulators on the initiation frequency of embryogenic tissues from immature *P. nigra* zygotic embryos (from Salaj and Salaj 2005, with permission of Biologia Plantarum)

Growth regulators ($mg\,ml^{-1}$)	Year 2000 Number of explants	Initiation frequency (%)	Year 2001 Number of explants	Initiation frequency (%)
2, 4-D (2.0) BA (0.5)	88	2.27	81	7.40
2, 4-D (0.5) BA (2.0)	88	4.54	99	1.01
2, 4-D (2.0) BA (2.0)	88	5.68	84	8.33
BA (1.0)	88	1.13	83	3.61
2, 4-D (1.0)	80	1.25	84	7.14
PGR-free	80	0.00	79	3.89
NAA (2.0) BA (0.5)	84	1.19	77	6.49
NAA (0.5) BA (2.0)	80	8.75	93	7.52
NAA (2.0) BA (2.0)	88	2.27	102	5.88
NAA (1.0)	80	5.00	88	9.09

In *Pinus nigra* Arn., attempts were focused on the initiation of somatic embryogenesis from mature zygotic embryos. Zygotic embryos were isolated from surface sterilized mature seeds and pretreated with BA (0, 1, 5, 10, 20 mg l^{-1}), and subsequently cultured on DCR basal medium containing 2,4-D (2 mg l^{-1}) and BA (0.5 mg l^{-1}) or different concentrations of both of these growth regulators (0.5, 1 and 2 mg l^{-1} applied simultaneously). The cultured mature zygotic embryos showed a similar response pattern regardless of the composition of the medium. Very early soft callus-like tissue was formed on the radicula pole without further proliferation. During the culture period white, filamentous structures appeared mainly on cotyledons and formed soft white tissue resembling embryogenic tissue. Microscopic investigations using the squash preparation method showed the presence of long cells but not bipolar somatic embryos visible in embryogenic tissue initiated from immature zygotic embryos.

The developmental stage of the zygotic embryo is the critical factor determining the frequency of embryogenic tissue initiation. In *Pinus*, the immature zygotic embryos, mostly in the precotyledonary stage of development, were the superior source of explants for embryogenic tissue initiation (Finer et al. 1990; Laine and David 1990; Keinonen-Mettälä 1996; Klimaszewska et al. 2001; Miguel et al. 2004). The origin of somatic embryos in pine was traced to the suspensor region (Becwar et al. 1990). Diverse criteria have been used to characterize developing zygotic embryos suitable for embryogenic initiation. In loblolly pine, a culture of zygotic embryos smaller than 0.3 mm led to embryogenic tissue formation (Becwar et al. 1988). Recent experiments have showed that an extension of the "initiation window" for some *Pinus* species is also possible. In *Pinus strobus*, late cotyledonary embryos formed embryonal masses (Klimaszewska and Smith 1997). Our attempts to initiate embryogenic tissue from mature zygotic embryos were unsuccessful. The soft white tissue formed early on cultured mature zygotic embryos resembled embryogenic tissue, but somatic embryos with bipolar organization were not observed.

3
Proliferation and Maintenance of Embryogenic Tissues

Embryogenic tissues were cultured for long-term maintenance on a DCR medium containing 2,4-D and BA. Although the initiation frequencies were relatively low, evaluation of the cell lines three months after initiation showed survival rates reached 94.11%. Growing embryogenic tissue was regularly transferred to fresh media at 2–3 week intervals. Four months after initiation all the initiated cell lines were investigated under a light microscope using the squash preparation method. These embryogenic tissues contained heavy stained meristematic cell groups, long vacuolized single cells and bipolar somatic embryos composed of embryonal and suspensor parts resembling their zygotic counterparts in seeds. Detailed microscopic observations revealed differences in the cytological features of somatic embryos among cell lines. According to the cytological organization of the somatic embryos the cell lines were categorized into three groups. Cell lines containing bipolar somatic embryos composed of a tightly packed meristematic "head" with a regular outline and long vacuolized suspensor cells often arranged into bundles were categorized as group one (Fig. 2). In some cell lines the embryonal part consisted of loosely con-

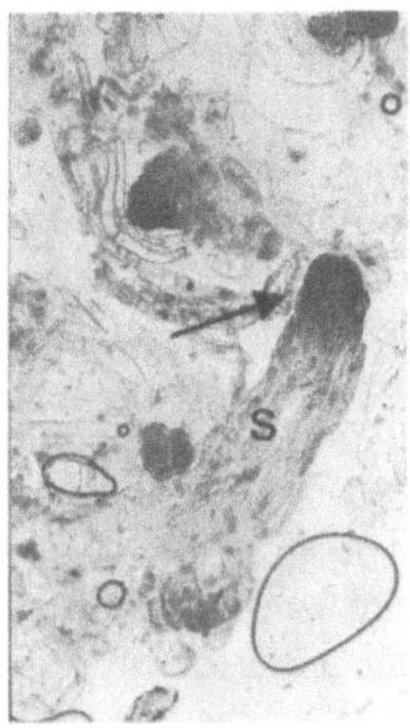

Fig. 2 Well-developed somatic embryo (*arrow*) with long suspensor cells (*s*) categorized as group one (cell line E103)

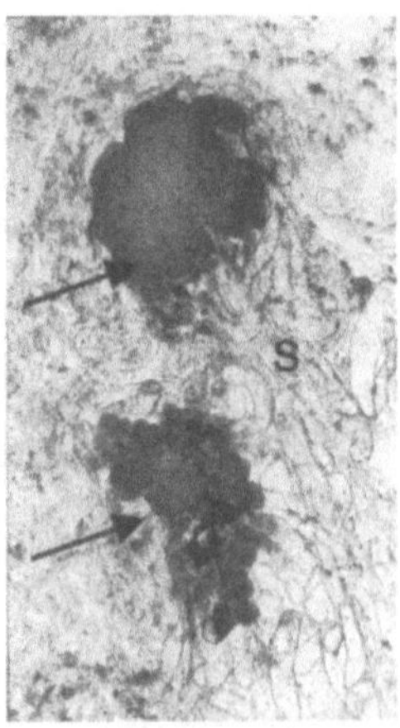

Fig. 3 Somatic embryos belonging to group two have loosely packed meristematic cells (*arrows*) and irregular shape with few suspensor (*s*) cells (cell line E55)

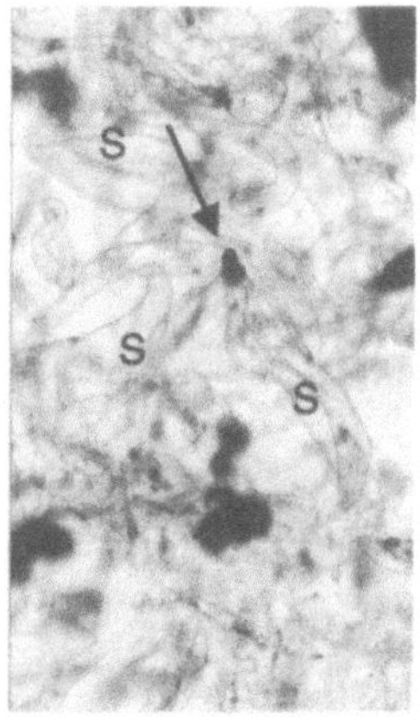

Fig. 4 Somatic embryos of group three are less developed, with a few meristematic cells (*arrow*) and many long suspensor (*s*) cells (cell line E49)

nected meristematic cells with an irregular outline. The long vacuolized suspensor cells were attached without forming into bundles. These cell lines were categorized as group two (Fig. 3). In the cell lines categorized as group three, aggregates of meristematic cells were mostly present. The only structures resembling somatic embryos were the occasional meristematic cells connected with 1–2 long vacuolized cells (Fig. 4). In the embryogenic cell lines group two dominated over the other two groups (Salaj and Salaj 2005).

4 Somatic Embryo Maturation

4.1 Maturation on a Medium with Maltose

Early somatic embryos developed to the cotyledonary stage in embryogenic cultures using medium containing abscisic acid (0.1–10.0 mg l^{-1}). However, embryo development was infrequent and irregular, and finally no plantlet regeneration occurred (Salajova and Salaj 1992).

Improved somatic embryo maturation has been achieved by using maltose and abscisic acid (ABA) simultaneously in the maturation medium. Cotyledonary somatic embryos developed in the presence of ABA (25 mg l^{-1}) and maltose. The process was strongly dependent on the cell line and maltose concentration (Table 3). The tested cell lines (E7, E15, E16) showed different responses. Precotyledonary somatic embryos developed in all of the three tested cell lines, but further development was restricted to the lines E15

Table 3 The effect of maltose on the maturation of *P. nigra* somatic embryos (from Salajova et al. 1999 with permission of Elsevier)

Maltose concentration	Precotyledonary somatic embryos			Cotyledonary somatic embryos		
	E7	E15	E16	E7	E15	E16
3%	19 (7.74)	89 (4.78)	19 (2.20)	-	10 (1.60)	-
6%	20 (1.94)	121 (7.49)	14 (1.33)	-	48 (4.70)	-
9%	15 (2.05)	135 (9.06)	6 (1.70)	-	66 (8.07)	10 (1.67)

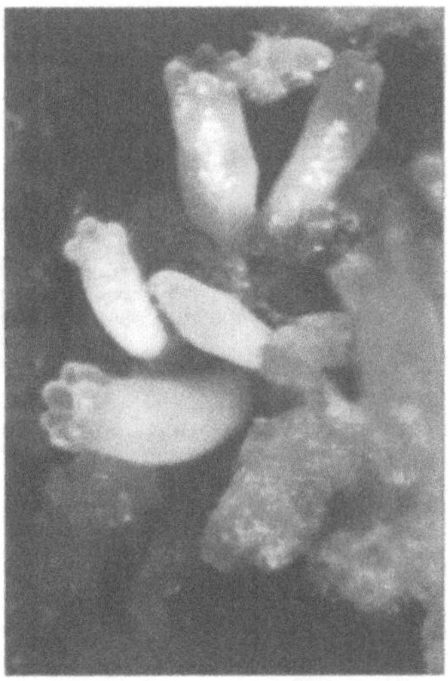

Fig. 5 Cotyledonary somatic embryos of the cell line E15 after 9 weeks of culture

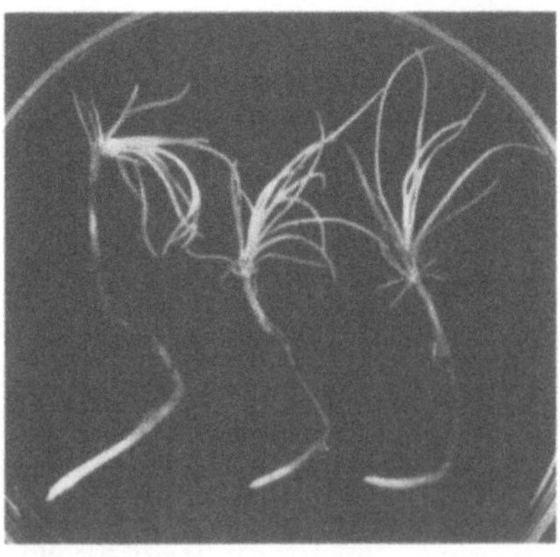

Fig. 6 Regenerated plantlets before placing into soil

Fig. 7 Plantlets of *P. nigra* growing in the soil for five months

(Fig. 5) and E16. Higher concentrations of maltose (6 or 9%) were beneficial for maturation in both cell lines, although cotyledonary somatic embryos appeared only in the presence of 9% maltose using the E16 line. In cell line E15, the cotyledonary somatic embryos germinated and developed plantlets (Fig. 6) that were successfully transferred to the soil (Fig. 7).

4.2
Maturation on a Medium with PEG

PEG has been reported to affect somatic embryo maturation in conifers, therefore the effect of polyethylene glycol was also tested in our experiments. The promoting effect of PEG is considered to be a consequence of the induced osmotic stress (Attree et al. 1991). Gene expression studies have confirmed maximum β-coniferine transcript accumulation after the combined ABA and PEG treatment, suggesting an influence of PEG on the "quality" of somatic embryos (Leal et al. 1995). Although in the mentioned species a positive effect of PEG on the quantity and quality of somatic embryos was demonstrated, Klimaszewska and Smith (1997) showed that in *Pinus strobus* PEG was not equally effective.

In our experimental system PEG-4000 was not effective for somatic embryo maturation. Application of PEG-4000 in different concentrations and its combination with sucrose as a carbon source resulted in very limited maturation. The tested cell lines responded to the maturation medium by forming precotyledonary somatic embryos. Despite their numbers being high in cell lines E15 and E16 they did not develop beyond this stage. Exceptionally, cotyledonary somatic embryos were present in low numbers in cell line E15, but they soon degenerated without further development. On the basis of these results we can state that PEG-4000 treatments combined with sucrose had no positive effect on somatic embryo maturation. The process stopped at the precotyledonary stage of development without further formation of cotyledonary somatic embryos capable of germination and plantlet production.

4.3 Structural Features and Somatic Embryo Maturation

Our results indicate that the morphological organization of somatic embryos plays a role in the maturation capacity (Table 4). Somatic embryos with a well-organized embryonal "head" developed into cotyledonary somatic embryos capable of germination and plantlet regeneration. The less organized somatic embryos (categorized as group two) developed mainly to the precotyledonary stage and formed abnormal structures. Development was very limited in the cell lines containing somatic embryos categorized as group three.

Table 4 Somatic embryo maturation in different cell lines of *P. nigra*. Mean number of developing somatic embryos calculated per 1 g of fresh mass inoculum (from Salaj and Salaj 2005, with permission of Biologia Plantarum)

Cell line	Precotyledonary somatic embryos	Cotyledonary somatic embryos	Germination frequencies (%)
E 19 ***	5.0 ± 1.32	–	–
E 27 ***	7.0 ± 2.21	–	–
E 34 **	78.0 ± 16.31	abnormal	–
E 42 *	59.0 ± 10.39	33.0 ± 7.95	41.9
E 43 **	25.9 ± 7.94	abnormal	–
E 47 **	33.0 ± 6.14	abnormal	–
E 49 ***	4.0 ± 0.86	–	–
E 50 **	80.0 ± 12.89	50.0 ± 9.32	39.95
E 52 **	39.0 ± 10.66	abnormal	–
E 57 ***	59.0 ± 22.12	–	–
E 103 *	122.0 ± 10.73	54.0 ± 3.99	47.86
E 104 *	135.0 ± 4.08	42.0 ± 4.79	42.53
E 106 **	32.0 ± 6.47	abnormal	–
E 113 **	24.0 ± 3.03	abnormal	–
E 114 **	51.0 ± 10.93	abnormal	–

* Cell lines categorized as group 1 ** Cell lines categorized as group 2 *** Cell lines categorized as group 3

5 Establishment and Characterization of *P. nigra* Suspension Cultures

The embryogenic tissues of conifers are also able to grow and produce plantlets in liquid suspension cultures. Therefore, suspension cultures provide a reliable experimental model system for the study of growth parameters, nutrient uptake and/or maturation of somatic embryos (Krogstrup 1990; Find et al. 1998; Gorbatenko and Hakman 2001).

Suspension cultures have been established also from embryogenic tissues of *Pinus nigra* (Salaj et al. 2003b). The embryogenic tissues (0.5 g, 1.0 g, 2.5 g) of 15 selected cell lines were resuspended in a liquid medium with a regular change of the medium every two weeks. The sedimented cell volume (SCV) was used as a nondestructive quantitative parameter of growth.

Embryogenic tissues of *P. nigra* were able to grow in suspension cultures although their growth parameters were influenced by the initial tissue weight used for the establishment of suspension culture (Table 5). An initial tissue weight of 0.5 g was not sufficient for the establishment of culture, and most of the cell lines failed to grow. Out of the 15 cell lines tested, only four grew in the liquid medium with minimal SCV. Better results were obtained using an initial tissue weight of 1.0 g or 2.5 g. Relatively large differences in SCV

Table 5 Growth of different *P. nigra* cell lines in suspension culture

Cell lines	Inoculum 0.5 g	(se)	1.0 g	(se)	2.5 g	(se)
E 42	9.33	0.99	20.52	0.65	9.22	1.26
E 43	1.00	–	15.87	2.8	14.7	0.8
E 47	0	–	0	–	3.71	0.21
E 49	1.50	–	17.0	1.0	16.92	0.55
E 50	0	–	0	–	2.14	–
E 78	0	–	contamin.	–	1.6	–
E 80	0	–	contamin.	–	10.27	0.89
E 98	0	–	0	–	1.0	–
E 103	0	–	0	–	2.37	–
E 104	2.75	0.26	15.28	0.34	10.5	0.55
E 106	0	–	23.50	0.51	17.5	1.94
E 114	0	–	16.50	1.41	1.0	–
E 115	0	–	16.83	4.6	10.42	1.68
E 127	0	–	6.16	0.72	0.91	–
E 130	0	–	1.0	–	0.5	–

were observed among the cell lines. The morphology of the somatic embryos was not profoundly affected by culture in the liquid medium. Several cell lines (E42, E104, E140, E146, E157) were selected for maturation capacity testing. These cell lines were characterized by the presence of somatic embryos able to form plantlets on solid media. The maturation of cell lines was very limited. After one month in the liquid maturation medium (DCR medium containing 9% maltose and 25 mg l^{-1} ABA or DCR supplemented with 7.5% PEG-4000, 3% maltose and 25 mg l^{-1} ABA), the embryonal part enlarged as a result of cell division. Such embryos stopped their growth and necrotized at this developmental stage. Medium exchange did not induce further maturation of somatic embryos.

6 Genetic Transformation of *P. nigra* Embryogenic Tissues

Conifer embryogenic tissues are often the targets of genetic transformation experiments using *Agrobacterium*-mediated gene transfer or direct transformation by biolistic bombardment (Minocha and Minocha 1999). Transient and stable genetic transformation via particle bombardment has been achieved in *Pinus pinea* (Humara et al. 1998), *P. sylvestris* (Häggman and Aronen 1998) and *Pinus radiata* (Walter et al. 1998).

Embryogenic cell lines E 103 and E 104 were selected for the genetic transformation of *Pinus nigra*. The plasmid pCW 122 (Walter et al. 1994) carried the GUS-intron reporter gene under the control of the double CaMV 35S promoter and the *npt II* selection gene driven by the single 35S promoter.

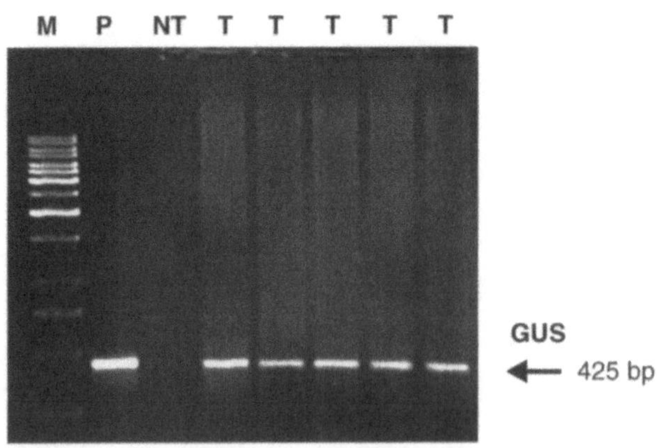

Fig. 8 PCR analysis of transformed *P. nigra* embryogenic tissues using primers specific for *uid*A (*gus*) gene. M—1 kb DNA ladder (Fermentas), P—plasmid pCW 122, NT—control, non-transformed tissue, T—transformed tissue of five sub-lines E104

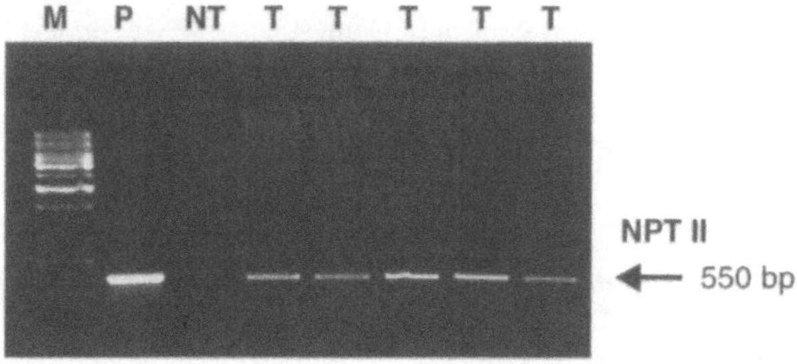

Fig. 9 PCR analysis of transformants using primers specific for the *npt II* selection gene

Tissue regeneration on the selection medium containing 20 mg l^{-1} geneticin G 418 was observed 10–12 weeks after transformation by particle bombardment. Under geneticin selection, the cell line E104 produced five resistant sub-lines, while cell line E103 showed no regeneration. The expression of foreign genes was confirmed by the ability to form embryogenic callus in the presence of geneticin, and by histochemical assays revealing GUS activity. Each of the five geneticin resistant sub-lines of the cell line E104 showed extensive GUS activity in embryogenic tissue, which was concentrated mainly in the meristematic embryonal cells of the head of the somatic embryo. PCR analysis of the five selected sub-lines showed the presence of T-DNA (Figs. 8 and 9) using *GUS* and *npt II*—specific primer pairs. Thus, these regenerated sub-lines of line E104 were confirmed to be transformants (Salaj et al. 2003a).

7
Conclusions and Future Prospects

The results reported for *Pinus nigra* somatic embryogenesis demonstrate that factors such as the developmental stage of immature zygotic embryos and the composition of the cultivation media play an important role in the initiation frequency of somatic embryogenesis. Although the initiation frequency was relatively low, the survival of initiated cell lines reached 94%. The maturation of selected lines of somatic embryos (categorized as group one) can by improved using maltose and ABA simultaneously. In addition, these cell lines were also able to grow and produce somatic embryos in suspension cultures, as well as regenerate transgenic embryogenic tissue.

Recent research has been focused on the improvement of the quality and uniformity of the developing somatic embryos at various stages in their development, and on long-term embryo storage by cryopreservation. In addition, we would also like to improve some selected traits (e.g. ornamental features) using genetic transformation of *Pinus nigra* and other conifers.

Acknowledgements Support from the Slovak Grant Agency VEGA, project No. 2/5022/25, is greatly appreciated.

References

Arya S, Kalia RK, Arya ID (2000) Induction of somatic embryogenesis in *Pinus roxburghii* Sarg. Plant Cell Rep 19:775–780

Attree SM, Moore D, Sawhney LC, Fowke LC (1991) Enhanced maturation and desiccation tolerance of white spruce [*Picea glauca* (Moench) Voss] somatic embryos: effects of a non-plasmolysing water stress and abscisic acid. Ann Bot 68:519–525

Becwar MR, Blush TD, Brown DW, Chesick EE (1991) Multiple paternal genotypes in embryogenic tissue derived from individual immature loblolly pine seeds. Plant Cell Tiss Org Cult 26:37–44

Becwar MR, Nagmani R, Wann SR (1990) Initiation of embryogenic cultures and somatic embryo development in loblolly pine (*Pinus taeda*). Can J For Res 2:810–817

Becwar MR, Wann SR, Johnson MA, Verhangen RP, Feier RP, Nagmani R (1988) Development and characterization of in vitro embryogenic systems in conifers. In: Ahuja MR (ed) Somatic cell genetics of woody plants. Kluwer Acad Publ, Dordrecht, p 1–18

Chandler SF, Young R (1995) Somatic embryogenesis in *Pinus radiata* Don. In: Jain SM, Gupta PK, Newton RJ (eds) Somatic embryogenesis in woody plants. vol 3, Gymnosperms. Kluwer Acad Publ, Dordrecht, p 243–256

Ford CS, Jones NB, van Staden J (2000) Cryopreservation and plant regeneration from somatic embryos of *Pinus patula*. Plant Cell Rep 19:610–615

Find JI, Norgaard JV, Krogstrup P (1998) Growth parameters, nutrient uptake and maturation capacity of two-lines of Norway spruce (*Picea abies*) in suspension culture. J Plant Physiol 152:510–517

Finer JJ, Kriebel HB, Becwar MR (1989) Initiation of embryogenic callus and suspension cultures of eastern white pine (*Pinus strobus* L.). Plant Cell Rep 8:203–206

Gorbatenko O, Hakman I (2001) Desiccation-tolerant somatic embryos of Norway spruce (*Picea abies*) can be produced in liquid cultures and regenerated into plantlets. Int J Plant Sci 162:1211–1218

Gupta PK (1995) Somatic embryogenesis in sugar pine (*Pinus lambertiana* Dougl.). In: Jain SM, Gupta PK, Newton RJ (eds) Somatic embryogenesis in woody plants. vol 3, Gymnosperms. Kluwer Acad Publ, Dordrecht, p 197–206

Gupta PK, Durzan DJ, Finkle BJ (1987) Somatic polyembryogenesis in embryogenic cell masses of *Picea abies* (Norway spruce) and *Pinus taeda* (loblolly pine) after thawing from liquid nitrogen. Can J For Res 17:1130–1133

Häggman H, Aronen T (1998) Transgene expression in regenerating cotyledons and embryogenic cultures of Scotch pine. J Exp Bot 49:1147–1156

Häggmann H, Jokele A, Krajňáková J, Kuppi A, Niemi K, Aronen T (1999) Somatic embryogenesis of Scotch pine: cold treatment and characteristics of explants affecting induction. J Exp Bot 50:1769–1778

Häggmann H, Ryynänen A, Aronen T, Krajňáková J (1998) Cryopreservation of embryogenic cultures of Scotch pine. Plant Cell Tiss Org Cult 54:45–53

Humara JM, Lopez M, Ordas RJ (1998) Modifying transient β-glucuronidase expression in pine species using introns. Plant Cell Tiss Org Cult 52:183–187

Jones NB, van Staden J (1995) Plantlet production from somatic embryos of *Pinus patula*. J Plant Sci 145:519–525

Kaul K (1995) Somatic embryogenesis in eastern white pine (*Pinus strobus* L.). In: Jain SM, Gupta PK, Newton RJ (eds) Somatic embryogenesis in woody plants. vol 3, Gymnosperms. Kluwer Acad Publ, Dordrecht, p 257–268

Keinonen-Mettala K, Jalonen P, Eurola P, von Arnold S, Weissenberg K (1996) Somatic embryogenesis of *Pinus sylvestris*. Scand J For Res 11:242–250

Klimaszewska K, Smith DR (1997) Maturation of somatic embryos of *Pinus strobus* is promoted by a high concentration of gellan gum. Physiol Plant 100:949–957

Krogstrup P (1990) Effect of culture densities on cell proliferation and regeneration from embryogenic cell suspensions of *Picea sitchensis*. Plant Sci 72:115–123

Laine E, Bade P, David A (1992) Recovery of plants from cryopreserved embryogenic cell suspensions of *Pinus caribaea*. Plant Cell Rep 11:295–298

Laine E, David A (1990) Somatic embryogenesis in immature embryos and protoplasts of *Pinus caribaea*. Plant Sci 69:215–224

Leal I, Mistra S, Attree SM, Fowke LC (1995) Effect of abscisic acid, osmoticum and desiccation on 11S storage protein gene expression in somatic embryos of white spruce. Plant Sci 106:121–128

Lelu M-A, Bastien C, Drugeault A, Gouey M-L, Klimaszewska K (1999) Somatic embryogenesis and plantlet development in *Pinus sylvstris* and *Pinus pinaster* on medium with and without growth regulators. Physiol Plant 105:719–728

Marum L, Estevaco C, Oliveira M, Amanico S, Rodrigues L, Miguel C (2004) Recovery of cryopreserved embryogenic cultures of maritime pine—effect of cryoprotectant and suspension density. CryoLetters 25:363–374

Mathur G, Alkutkar VA, Nadgauda RS (2003) Cryopreservation of embryogenic culture of *Pinus roxburghii*. Biol Plant 46:205–210

Mathur G, von Arnold S, Nadgauda R (2000) Studies on somatic embryogenesis from immature zygotic embryos of chir pine (*Pinus roxburghii* Sarg.). Curr Sci 79:999–1004

Miguel C, Goncalves S, Tereso S, Marum L, Maroco J, Oliveira M (2004) Somatic embryogenesis from 20 open-pollinated families of Portuguese plus trees of maritime pine. Plant Cell Tiss Org Cult 76:121–130

Minocha SC, Minocha R (1999) Genetic transformation in conifers. In: Jain SM, Gupta PK, Newton RJ (eds) Somatic embryogenesis in woody plants, vol 5. Kluwer Acad Publ, Dordrecht, p 291–312

Newton RJ, Marek-Swize KA, Magallenes-Cedeno ME, Dong S, Sen S, Jain SM (1995) Somatic embryogenesis in Slash pine (*Pinus elliottii* Engelm). In: Jain SM, Gupta PK, Newton RJ (eds) Somatic embryogenesis in woody plants. vol 3, Gymnosperms. Kluwer Acad Publ, Dordrecht, p 183–197

Percy RE, Klimaszewska K, Cyr DR (2000) Evaluation of somatic embryogenesis for clonal propagation of western white pine. Can J For Res 30:1867–1876

Salaj T, Salaj J (2005) Somatic embryogenesis in *Pinus nigra* Arn.: Embryogenic tissue initiation, maturation and regeneration ability of established cell lines. Biol Plant 49:333–339

Salaj T, Moravcikova J, Grec-Niquet L (2003a) Biolistic transformation of embryogenic tissues of *Pinus nigra* Arn. In: Gajdosova A, Libiakova G (eds) Plant Biotechnology: Progress and Developments. 5th Int Symp Recent Advances in Plant Biotechnology, Stara Lesna, Sept. 7–13, 2003, Slovakia, p 94

Salaj T, Salaj J, Kormutak A (2003b) Establishment and characterization of embryogenic suspension cultures of *Pinus nigra* Arn. In: COST 843—WG 2 Advanced propagation techniques. 4th Meeting in San Remo, Italy, Nov. 27–30, 2003, p 15–16

Salajova T, Salaj J (1992) Somatic embryogenesis in European black pine (*Pinus nigra* Arn.). Biol Plant 34:213–218

Salajova T, Salaj J, Kormutak A (1999) Initiation of embryogenic tissues and plantlet regeneration from somatic embryos of *Pinus nigra* Arn. Plant Sci 145:33–40

Salajova T, Salaj J, Jasik J, Kormutak A (1995) Somatic embryogenesis in *Pinus nigra* Arn. In: Jain SM, Gupta PK, Newton RJ (eds) Somatic embryogenesis in woody plants. vol 3, Gymnosperms. Kluwer Acad Publ, Dordrecht, p 207–220

Walter C, Grace LJ, Wagner A, White DWR, Walden AR, Donaldson SS, Hinton H, Gardner RC, Smith DR (1998) Stable transformation and regeneration of transgenic plants of *Pinus radiata*. Plant Cell Rep 17:460–468

Walter C, Smith DR, Connet MB, Grace L, White DRW (1994) A biolistic approach for the transfer and expression of gusA reporter gene in embryogenic cultures of *Pinus radiata*. Plant Cell Rep 14:69–74

Plant Cell Monogr (2)
A. Mujib · J. Šamaj: Somatic Embryogenesis
DOI 10.1007/7089_040/Published online: 30 November 2005

Mode of Action of Plant Hormones and Plant Growth Regulators During Induction of Somatic Embryogenesis: Molecular Aspects

Clément Thomas[1] (✉) · Víctor M. Jiménez[2]

[1]Plant Molecular Biology, CRP-Santé, Bâtiment modulaire,
84 Val Fleuri, 1526 Luxembourg, Luxembourg
clement.thomas@crp-sante.lu

[2]CIGRAS, Universidad de Costa Rica, 2060 San Pedro, Costa Rica
vjimenez@cariari.ucr.ac.cr

Abstract Plant hormones play critical roles in the establishment of somatic embryogenesis. During this process, somatic plant cells reverse their state of differentiation, acquire pluripotentiality and set up a new developmental program. The identification of the regulatory mechanisms that govern the key events of somatic embryogenesis requires molecular and genetic investigations. One critical issue is how plant hormones and growth regulators act to mediate somatic embryogenesis. Do they function as simple stimuli or participate directly, as central signals, in the reprogramming of the somatic cells towards an embryogenic fate? The latter scenario is now well supported by a number of studies that provide evidence of close interconnections between plant hormones and the molecular pathways that control somatic embryogenesis, including chromatin remodeling, gene expression patterning, reactivation of cell cycle and division and regulation of protein turnover. In this chapter we describe recent advances in the understanding of molecular and genetic mechanisms underlying the early stages of somatic embryogenesis. The roles and mode of action of plant hormones are especially emphasized.

Abbreviations

2,4-D	2,4-Dichlorophenoxyacetic acid
ABA	Abscisic acid
ABP1	Auxin binding protein 1
ARF	Auxin-response factors
aza-C	5-Azacytidine
BBM	*BABY BOOM*
BAP	Benzylaminopurine
CDK	Cyclin-dependent kinase
DD-RT PCR	Differential display reverse transcription polymerase chain reaction
ER	Endoplasmic reticulum
GA	Gibberellin
IAA	Indole-3-acetic acid
LEC	*LEAFY COTYLEDON*
NAA	Naphthalene acetic acid
PGR	Plant growth regulator

PH	Plant hormone
PKL	*PICKLE*
SE	Somatic embryogenesis
(*SERK*)	*SOMATIC EMBRYOGENESIS RECEPTOR-LIKE KINASE*
(*WUS*)	*WUSCHEL*

1 Introduction

Somatic embryogenesis (SE) has been observed to be induced by different factors (reviewed by Jiménez 2001; Fehér et al. 2003). Independently of the nature of the external stimulus, the establishment of SE necessarily involves profound changes at the molecular level, such as the coordinated expression of different sets of genes that drive the switch from the current vegetative growth pattern to an embryogenic development. Thus, the identification of the genes that trigger key phases of SE, i.e. cell dedifferentiation, cell cycle reentry and establishment of a new embryogenic fate, is highly desirable. Additionally, the elucidation of the signaling pathways by which plant cells remodel their gene expression program is central to understanding the regulation of the SE process.

As discussed in detail elsewhere (Jiménez and Thomas, this volume), plant growth regulators (PGRs) are among the external stimuli most often employed to induce SE and to regulate the further development of embryogenic tissues. There was some controversy as to whether PGRs/plant hormones (PHs) act only as stimuli or are more directly involved in the mechanisms that regulate gene expression (Gaspar et al. 2003). However, during the last few years, a large body of experimental data supports the view that PHs play a central role in the establishment of SE. The understanding of the underlying mechanisms of PH action requires investigation of hormone receptors, signal transduction pathways, and genetic programs that lead to the final cell response.

The first step in any event associated with a response to a hormone is a proper recognition by the target cells. This recognition normally involves receptors, which are proteins associated with the cell membranes or are located in the cytoplasm. Receptors have been identified and characterized for hormone groups such as ethylene (Chang et al. 1993; Schaller and Bleecker 1995) and cytokinins (Inoue et al. 2001; Ueguchi et al. 2001). These receptors activate a signal transduction pathway that either induces or inhibits cellular functions, or controls gene expression (reviewed by Kulaeva and Prokoptseva 2004).

In the case of auxins, although some auxin-binding proteins have been isolated, it is still uncertain whether they represent receptors for different auxin-mediated processes (Gaspar et al. 2003). To date, auxin binding protein 1 (ABP1) is the best-studied putative auxin receptor protein. ABP1 predominantly accumulates in the lumen of the endoplasmic reticulum (ER), an unusual location for a hormone receptor (Barbier-Brygoo et al. 1989; Inohara et al. 1989). However, there is evidence that ABP1 is active in auxin responsiveness at the surface of the plasma membrane although it carries the ER retention motif (Rück et al. 1993; Thiel et al. 1993; Leblanc et al. 1999; Steffens et al. 2001; Shimomura et al. 1989). ABP1 has been shown to mediate early auxin responses such as auxin-induced electrical responses (Rück et al. 1993; Thiel et al. 1993; Zimmermann et al. 1994; Bauly et al. 2000) and cell expansion (Jones et al. 1998; Chen et al. 2001a, b). However, its involvement in auxin-induced gene expression has not been proved yet. Although mechanisms responsible for auxin signal transduction from receptor to genome are still poorly known, significant progress has been achieved in auxin-regulated gene expression (reviewed by Hagen and Guilfoyle 2002).

The PGRs most widely used to induce and regulate in vitro SE are auxins and cytokinins. It has been observed that members from both hormone groups regulate the cell cycle and trigger cell divisions (Francis and Sorrell 2001), two very important factors that have been related to initiation of SE (Dudits et al. 1991, 1995; Yeung 1995). Recent data provide evidence that the elaboration and execution of developmental programs require a proper control of the cell cycle and division, indicating the regulators of the cell cycle machinery as key determinants of SE.

In addition to their influence on cell cycle progression, PGRs/PHs have been demonstrated to trigger substantial changes in chromatin structure and alteration of transcription that lead to the formation of either dedifferentiated callus tissues or somatic embryos (Dudits et al. 1995). Studies on the links between PH action and gene expression have resulted in the cloning of several genes responsive to auxins, cytokinins, or to both hormones. Although, the functions of a number of these genes remain unknown, others have obvious connections with the cell cycle or developmental processes including SE.

In this chapter we describe those findings related to cell division and changes in the pattern of gene expression during early stages of SE and we further highlight how hormonal signals are integrated into these processes.

2 Reactivation of Cell Cycle and Division

The reactivation of the cell cycle and division in differentiated cells is indispensable for the initiation of plant developmental processes, including SE (Dudits et al. 1995). The cell cycle is usually divided in four sequential phases: G1, S (DNA replication), G2 and M (cytokinesis). The basic control mechanisms that regulate progression through the cell cycle are remarkably well conserved during evolution and operate mainly at the G1–S and G2–M transitions (reviewed by Stals and Inzé 2001; De Veylder et al. 2003; Dewitte and Murray 2003). These two key check points depend on highly conserved serine/threonine kinases, named cyclin-dependent kinases (CDKs), and their associated regulatory subunits called cyclins.

The two mentioned groups of hormones, auxin and cytokinins, are generally sufficient to stimulate and sustain the in vitro proliferation of most plant cell types and have therefore been the best documented direct regulators of the cell cycle progression. The expression of genes related to the *cdc2* gene, which encodes the catalytic subunit of the key G1–S and G2–M regulator cdc2 protein kinase, is upregulated by auxin in alfalfa (Hirt et al. 1991), soybean (Miao et al. 1993) and tobacco pith explants (John et al. 1993). Although auxin enhances *cdc2* gene expression, cotreatment with cytokinin is absolutely required to induce a basic *cdc2* expression in tobacco pith explants (John et al. 1993), illustrating the synergic regulation exerted by both growth regulators on cell proliferation. The observation that most systems require only exogenously added auxin to resume cell division suggests that the rate of endogenous cytokinin synthesis is sufficient to sustain growth (del Pozo et al. 2005). In situ analysis of *cdc2* expression in plants such as *Arabidopsis* (Martinez et al. 1992; Hemerly et al. 1993) and soybean (Miao et al. 1993) revealed that *cdc2a* expression is not only associated with cell proliferation but also precedes it, suggesting that it reflects a state of competence to divide (Hemerly et al. 1993).

More recently, auxin has been shown to upregulate the expression of an alfalfa A2-type cyclin, whose promoter contains auxin-response-like elements (Roudier et al. 2003). In addition, auxin treatment of alfalfa plants affects the spatial expression pattern of this cyclin by shifting its expression from the phloem to the xylem poles, where lateral root formation is initiated in response to auxin. This auxin-regulated spatial cyclin expression illustrates another aspect of the complexity of hormonal regulation of the cell cycle in planta.

Cyclins D represent important connections between PHs and the cell cycle. Consistent with its regulatory function in the cell cycle progression, cyclin CycD3 is expressed in tissues having a high rate of cell divisions, including shoot meristems, young leaf primordia, axillary buds, procambium and vascular tissues of developing leaves (Riou-Khamlichi et al. 1999). The *CycD3* gene is highly responsive to cytokinin in both cell cultures and whole plants and is rapidly induced by cytokinin during the G1 phase of cells reentering the cell cycle. Constitutive expression of this cyclin in transgenic *Arabidopsis* plants leads to diverse disorders, e.g., extensive leaf curling and disorganized meristems, and, importantly, it renders callus growth independent of cytokinin application (Riou-Khamlichi et al. 1999). This demonstrates that cytokinins promote cell division by inducing the *CycD3* expression at the G1–S phase transition.

Cytokinins have also been reported to play a regulatory role at the mitotic control point of the G2–M transition. This is well illustrated by the observation that application of the cytokinin biosynthesis inhibitor lovastatin blocks cells of tobacco BY-2 in G2 (Laureys et al. 1998). This effect is nullified by the addition of an exogenous cytokinin, such as zeatin. This is in line with previous observations that cytokinins accumulate transiently during the G2–M phase (Redig et al. 1996) and that the removal of cytokinin from the culture medium leads to an arrest in G2 of tobacco suspension cell cultures, which accumulate inactive CDK complexes (Zhang et al. 1996). The kinase activity of the latter is restored by either addition of cytokinin or by tyrosine dephosphorylation, suggesting that the inactivation of CDK complexes under cytokinin deprivation is due to phosphorylation of regulatory residues of the CDK subunit.

Although auxin and cytokinins are generally considered as the main hormonal signals triggering cell cycle progression, others PHs with enhancing or inhibitory functions participate in the cell cycle control by modulating the transcriptional expression of different cell cycle genes. For example, gibberellin (GA) stimulates CDK and cyclin accumulation in a tissue-specific manner (Sauter 1997). Abscisic acid (ABA) induces a decrease in Cdc2a-like kinase activity by increasing the expression level of a CDK inhibitor gene, namely the *ICK1* gene, whose product interacts with Cdc2a and CycD3 (Wang et al. 1998). Although its mode of action is still unclear, jasmonic acid has been reported to block synchronized BY2 cells in both G1–S and G2–M transitions (Swiatek et al. 2002).

3
Reprogramming of the Gene Expression Pattern

The establishment of totipotency and the subsequent induction and development of somatic embryos require reprogramming the cultures. This is in part achieved by synthesis of new RNA molecules. Therefore, early inductive molecular events have been investigated by monitoring gene transcripts that are synthesized under the influence of external stimuli that trigger the embryogenic fate. Several examples of changes in the expression of genes related to initiation of SE have been reported. Here, we make reference only to those works in which the change in gene expression can be traced back to PGRs/PHs.

In an attempt to identify genes that switch on the SE program, researchers have employed systematic approaches aimed at comparing the population of transcripts expressed in embryogenic conditions with the population of the transcript expressed in nonembryogenic conditions. This was carried out using techniques such as differential complementary DNA library screening, differential display reverse transcription polymerase chain reaction (DD-RT PCR), cold plaque screening and more recently microarrays. The application of these techniques to the induction phase of SE has been complicated by the difficulty to identify and isolate embryogenic cells in the initial steps when no morphological changes are visible. However, improvements in in vitro culture systems and methods of molecular analysis have allowed progress in the exploration of the early phases of SE.

Nagata et al. (1994) isolated three auxin-regulated genes, *parA*, *parB* and *parC*, that are transiently expressed during the regaining of meristematic activity of tobacco mesophyll protoplasts. The corresponding transcripts were detected as early as 20 min after the beginning of incubation of protoplasts with auxin. Importantly, they were no longer detected after 48 h of culture when protoplasts started to divide, suggesting that they are specifically involved in the reentry into the plant cell cycle.

Kitamiya et al. (2000) isolated two carrot genes that are differentially expressed in hypocotyl cells induced to form somatic embryos by treatment with 2,4-dicholorophenoxyacetic acid (2,4-D) for 2 h. One of these genes, namely the *D. carota heat-shock protein 1* (*Dchsp-1*) gene, is related to low molecular weight heat-shock proteins and was found to be expressed during embryo development. The other gene has homology to the auxin-regulated genes, including *par A* (Takahashi et al. 1989), and thus was named *D. carota auxin-regulated gene 1* (*Dcarg-1*). Interestingly, there is a parallel relationship between the expression of *Dcarg-1* and the formation of somatic embryos. In addition, in contrast to *Dchsp-1*, *Dcarg-1* was not responsive to stress treatment and was not expressed during development of somatic embryos, implying that its function was not required for this process to occur.

Using DD-RT PCR, Yasuda et al. (2001) attempted to identify genes that are preferentially expressed during the early stages of auxin-induced carrot SE. Three transcripts that accumulate immediately after somatic cells divide to form cell clusters, but that do not accumulate or barely accumulate in nonembryogenic cell suspension cultures, were characterized. Although these genes represent potential key regulators of SE, a clear function has still not been attributed to them.

An important issue is how PH action on the gene expression level pattern is mediated. In a general view, the hormonal signal activates a signaling cascade that recruits specific transcription factors. These induce the expression of target genes, which in turn trigger the final response. Numerous genes have been described containing *cis*-acting elements in their promoter region that confer hormone responsiveness. Over the past 20 years, sequences that are upregulated or downregulated by PHs have been described for auxins (Guilfoyle et al. 1998; Ulmasov et al. 1999), ABA (Marcotte et al. 1992), GAs (Gubler and Jacobsen 1992) and ethylene (Meller et al. 1993).

The auxin-modulated gene expression system is based, at least in part, on two interacting protein families. The multifamily protein auxin-response factors (ARFs) can activate or repress target genes by directly binding to specific DNA sequences, i.e., auxin response elements (Ulmasov et al. 1999). In contrast, the auxin/indole-3-acetic acid (IAA) proteins do not bind to DNA directly but can inactivate ARF transcription factors by interacting with them through heterodimerization (Tiwari et al. 2001). Auxin exerts its regulation on gene expression through modulation of auxin/IAA protein turnover via a specialized branch of the ubiquitin–proteasome pathway (Worley et al. 2000; Gray et al. 2001; Dharmasiri and Estelle 2002). Such a PH-regulated proteolysis has been shown to be involved in many aspects of plant developmental processes including SE.

4 Chromatin Structure and DNA Methylation

A specific gene expression program is the result of the balance between the part of the genome that is transcribed, i.e., euchromatin, and the part that is repressed, i.e., heterochromatin. Many aspects of plant development, including embryonic and meristem development, flowering and seed formation, involve modifications of chromatin structure that affect the accessibility of target genes to regulatory factors that control their expression (reviewed by Li et al. 2002). Since maintaining the cellular differentiated state largely relies on chromatin-dependent gene silencing, the cellular dedifferentiation and the switch to a new embryogenic program necessarily involve important changes in chromatin structure.

Zhao et al. (2001) identified two distinct phases of chromatin decondensation during in vitro induced dedifferentiation of tobacco mesophyll cells. The first was independent of any hormonal treatment and was linked to the acquisition of pluripotentiality or dedifferentiation of cells. In contrast, the second phase of chromatin decondensation required auxin and cytokinin treatment and was linked to the reentry into the S phase.

Dynamic changes in chromatin structure are influenced by both posttranslational modifications of histone amino terminal tails and direct modifications of the DNA, such as methylation. The degree of DNA methylation has been reported to influence plant morphogenesis (reviewed by Li et al. 2002). The overexpression of an antisense DNA methyltransferase copy in transgenic tobacco plants provokes development disorders, including small leaves, short internodes and abnormal flower morphology (Nakano et al. 2000). The role of DNA methylation in early phases of SE has been recently addressed by Yamamoto et al. (2005) by investigating the effects of 5-azacytidine (aza-C), an inhibitor of DNA methylation, on the induction of direct carrot SE. Aza-C treatment totally inhibited the formation of embryogenic cell clumps from epidermal carrot cells. When applied during morphogenesis of embryos, aza-C downregulated the expression of *C-LEC1*, an important gene that participates in the embryonic program (Sect. 6). Additionally, in untreated cells, a DNA methyltransferase gene transcript transiently accumulated after auxin application but before the formation of embryogenic cell clumps, suggesting a direct role for DNA methylation in the establishment of embryogenic competence in carrot somatic cells.

Other chemical substances, such as the antibiotic kanamycin, have been observed to considerably modify the level of DNA methylation during plant in vitro culture (Bardini et al. 2003). In this case, DNA methylation is considered as a potential source of somaclonal variation, a phenomenon (often undesirable) observed in plant cell and tissue cultures (Caplan et al. 1998).

5 Some Key Regulators of the Vegetative-to-Embryogenic Transition

As already stated, different strategies have been used to identify genes that are differentially expressed during SE (Thomas 1993; Lin et al. 1996; Schmidt et al. 1997). Although several genes have been cloned, their function or functions often remain obscure. However, improvements in plant transformation protocols and the availability of new mutants allowed the characterization of genes that regulate the vegetative-to-embryogenic transition. The ectopic expression of these genes either enhances SE in in vitro cultures or even provokes spontaneous embryo formation on intact plants. One new challenge is to identify possible existing links between the PRG/pH and the genes that possibly influence the vegetative-to-embryogenic transition during SE.

5.1 The *SOMATIC EMBRYOGENESIS RECEPTOR-LIKE KINASE* Gene

Most of the molecular markers of SE identified to date are related to late stages of embryo development. However, one gene, encoding a leucine-rich-repeat receptor-like kinase, has been found to be specifically upregulated during the very precocious phases of the SE process. The *SOMATIC EMBRYOGENESIS RECEPTOR-LIKE KINASE* (*SERK*) gene was originally cloned from a carrot cell suspension culture where it was found to mark cells that are competent to form somatic embryos, i.e., cells in transition between the somatic and the embryogenic states (Schmidt et al. 1997). Using in situ hybridization, *DcSERK* expression was shown to first appear in single cells of embryogenic cultures induced with 2,4-D for 7 days. *DcSERK* expression continues until the 100-cell stage of the globular somatic embryo and then ceases. Interestingly, a similar *SERK* expression pattern was observed during early zygotic embryogenesis, suggesting that the same SERK signaling pathway is activated during both SE and zygotic embryogenesis (Schmidt et al. 1997).

Several homologs of the carrot *DcSERK* have been identified in monocots, e.g., maize (Baudino et al. 2001) and *Dactylis glomerata* (Somleva et al. 2000), and dicots, e.g., *Medicago truncatula* (Nolan et al. 2003), *Arabidopsis thaliana* (Hecht et al. 2001) and sunflower (Thomas et al. 2004). Plant genomes contain several *SERK* genes. As an example, the *Arabidopsis SERK* gene family comprises five members (Hecht et al. 2001). The expression of *AtSERK1*, the *Arabidopsis* gene most closely related to the carrot *DcSERK*, also marks embryogenic competent cells in culture. As in carrot, the level of *SERK* expression increases in response to the auxin treatment used to induce somatic embryos. Auxin-dependent *SERK* expression was also reported for the *M. truncatula MtSERK* gene, which is upregulated by the auxin naphthalene acetic acid (NAA) but not by the cytokinin benzylaminopurine (BAP; Nolan et al. 2003). However, addition of BAP to the culture medium potentiates NAA-induced *SERK* expression, possibly by stimulating endogenous auxin synthesis. In the direct SE system of sunflower, *SERK* transcripts specifically accumulate in the future morphogenic region of explants within the first few hours of culture. Although the only PGR supplied in the medium is a cytokinin, analysis of the endogenous PH content revealed that the internal IAA concentration transiently increases in explants during this early period (Thomas et al. 2002). A link between auxin and *SERK* expression is also suggested by the accumulation of *SERK* transcripts in plant tissues that contain high auxin levels, e.g., vascular tissue and leaf primordia (Hecht et al. 2001; Thomas et al. 2004). However, since *SERK* is not induced by auxin in all the cell explants or cell cultures, it is probably not an integral part of the auxin machinery or its expression requires other, still unknown, factors (Hecht et al. 2001).

Evidence that *AtSERK1* is not only a good marker of embryogenic competent cells in *Arabidopsis* but is also involved in the establishment of the embryogenic competence comes from ectopic overexpression of the *AtSERK1* gene in *Arabidopsis* (Hecht et al. 2001). Although during normal growth transgenic seedlings do not show any specific phenotype, their embryogenic capacity is considerably enhanced (approximately 4 times compared with the wild type) during in vitro culture. A similar increase in embryogenic competence is conferred by mutation in shoot apical meristem regulatory genes such as *AMP1*, *CLV1* and *CLV2* (Mordhorst et al. 1998). The higher *AtSERK1* expression level in *amp1* cultures, in comparison with that in wild-type cultures, suggests that one role of AMP1 could be to downregulate the expression of *AtSERK1* after germination (Hecht et al. 2001).

The identification of SERK-activating ligand(s) as well as the downstream targets of SERK is highly desirable to further characterize the function(s) of SERK in both zygotic embryogenesis and SE.

5.2 The *BABY BOOM* Gene

Another gene that potentially activates signal transduction pathways leading to the induction of embryo development from differentiated somatic cells is the *BABY BOOM* (*BBM*) gene (Boutilier et al. 2002). It was identified by a screening approach aimed at identifying genes differentially expressed during early phases of *Brassica napus* microspore embryogenesis. The *B. napus* microspore culture system relies on the ability of the vegetative cell of an immature pollen grain to develop into an embryo in response to high-temperature (above 25 °C) culture conditions (Custers et al. 1994).

The *BBM* gene encodes a protein that belongs to the AP2/ERF family, a plant-specific class of transcription factors that regulate several developmental processes, such as floral organ identity determination and control of leaf epidermal cell identity (reviewed by Riechmann and Meyerowitz 1998). It is preferentially expressed during embryo and seed development (Boutilier et al. 2002).

Overexpression of the *BBM* gene under the control of a constitutive promoter leads to the spontaneous formation of somatic embryos and cotyledon-like structures on different tissues of intact plants (Boutilier et al. 2002). Additionally, in vitro cultured explants, coming from *BBM*-overexpressing transgenic plants, display an enhanced capacity to regenerate through shoot organogenesis. This suggests that *BBM* plays a broader role in cell division and differentiation rather than being a specific element of the SE pathway.

Importantly, in contrast to *SERK*, ectopic expression of *BBM* is able to promote SE in the absence of exogenously applied PGR. It has been proposed that *BBM* could act by stimulating an increase of PH and/or increasing the cellular hormonal sensitivity (Boutilier et al. 2002). In that sense, Klucher et al. (1996) speculated that AP2/ERF domain proteins, being unique to plants, might have coevolved with plant-specific pathways such as PH signal transduction. Alternatively, it is also conceivable that the *BBM* product acts in a PH signaling pathway downstream of the hormone perception as previously shown for some other AP2/ERF domain proteins (Finkelstein et al. 1998; Menke et al. 1999; Gu et al. 2000; Banno et al. 2001; van der Fits and Memelink 2001).

5.3 The *LEAFY COTYLEDON* Genes

Arabidopsis mutants that display abnormalities in embryo development represent powerful tools to investigate the molecular pathways underlying SE. The *LEAFY COTYLEDON1* (*LEC1*) and *LEAFY COTYLEDON2* (*LEC2*) genes were identified originally as loss-of-function mutants showing defects in both embryo identity and seed maturation processes (Meinke et al. 1994; West et al. 1994). *Lec* embryos present a reduction in desiccation tolerance and do not accumulate normal storage materials. In addition, *lec* mutants exhibit other anatomical characteristics, including the presence of trichomes on cotyledons, which in *Arabidopsis* wild-type plants are specific to true leaves (Meinke et al. 1994; West et al. 1994; Stone et al. 2001). The pleiotropic effects of *lec* mutations pinpoint the *LEC* genes as central regulators of embryo and seed development.

Identification and analysis of the *Arabidopsis LEC1* and *LEC2* genes confirmed their regulatory roles in embryogenesis and provided significant insight into their functions. Both *LEC* genes encode seed-expressed transcriptional activators. *LEC1* encodes a protein related to the heme-activated protein 3 subunit of the CCAAT box-binding factor, a eukaryotic transcription factor (Lotan et al. 1998). *LEC2* encodes a protein that contains the plant-specific B3 domain (Stone et al. 2001), which is found in several plant transcription factors including ABA INSENSITIVE3 (Luerssen et al. 1998) and VIVIPAROUS1 (McCarty et al. 1989 and 1991).

Ectopic overexpression of either *lec1* or *lec2* results in the spontaneous formation of somatic embryos directly on the leaf surface, suggesting that *lec* genes play a role in conferring embryogenic competence to cells (Lotan et al. 1998; Stone et al. 2001). It also confers embryonic characteristics to seedlings. The expression of embryo-specific genes, such as those encoding cruciferin A, 2S storage protein and oleosin, in adult transgenic seedlings, confirms the activation and maintenance of embryo-specific programs in vegetative tissues. Interestingly, the *35S::LEC1* phenotype is relatively weak, i.e., only a few plants show sporadic embryo development, whereas the *35S::LEC2* phenotype is stronger and comparable to that observed for *35S::BBM* plants. The fact that the *BBM* and *LEC* genes exhibit similar putative functions as transcription factors, are both preferentially expressed in seeds, and confer to plants a similar phenotype when ectopically expressed suggests that they function in the same molecular pathway. However, *BBM* transcripts are present in *lec1* mutant seeds (Boutilier et al. 2002), indicating that the expression of *BBM* is not dependent per se on the presence of the *LEC1* protein. Thus, *BBM* could either function upstream of *LEC1* or operate in an *LEC1*-separated but overlapping pathway.

Recently, Yazawa et al. (2004) isolated a carrot functional homolog of *Arabidopsis LEC1*, as demonstrated by complementation experiments. In the SE system of carrot, the highest expression of *C-LEC1* was detected in cell clusters of 38–63 μm in diameter that were being cultured for induction of somatic embryos. Strikingly, cell clusters of this size are also those that are the most efficient for somatic embryo production (Satoh et al. 1986).

5.4 The *PICKLE* Gene

Another interesting *Arabidopsis* mutant is the *pickle* (*pkl*) mutant described by Ogas et al. (1997). At the opposite side of the *lec* phenotypes, a null mutation in the *PKL* gene induces embryonic characteristics in the roots of *Arabidopsis* seedlings, including accumulation of lipids and seed storage proteins normally found in seeds (Ogas et al. 1999; Rider et al. 2004). When excised and cultured on a medium lacking PGR, roots of *pkl* seedlings spontaneously develop somatic embryos. Exogenous application of GA is sufficient to suppress the mutant phenotype, whereas decreasing the level of GA in germinating seeds increases significantly the penetrance of the *pkl* root phenotype (Ogas et al. 1997). These observations suggest that *PKL* functions in a GA pathway that controls the switch of root cells from an embryonic to a vegetative fate. The *PKL* gene encodes a CHD3-chromatin remodeling factor, and thus is likely to function as a negative regulator of transcription of embryo-specific genes (Eshed et al. 1999; Ogas et al. 1999). This is supported by the observation that *LEC1* and *LEC2* expression levels are significantly higher in *pkl* than in wild-type seedlings. Thus, expression of embryonic traits in *pkl* seedlings is highly suspected to be a consequence of the failure to repress expression, in a GA-dependent manner, of the master regulators of embryogenic identity, such as the *LEC* genes, during germination (Rider et al. 2003). However, as noted by Henderson et al. (2004), data that demonstrate a direct link between *PKL* activity and GA are still missing and thus it could not be absolutely decided whether repression of *LEC1* is or is not a GA-dependent event. The observation that GA can act in the absence of *PKL* to repress expression of the *pkl* root phenotype (Ogas et al. 1997) demonstrates that there also exists a *PKL*-independent pathway by which GA represses expression of embryonic traits. This is consistent with the recent metabolic analysis that revealed that *pkl Arabidopsis* roots accumulate some but not all seed-specific metabolites (Rider et al. 2004).

5.5 The *WUSCHEL* Gene

Using a genetic approach to identify gain-of-function mutations that can promote embryogenic callus formation from *Arabidopsis* root explants, Zuo et al. (2002) identified a gene, *PAG6*, that was found to be identical to *WUSCHEL* (*WUS*), a gene previously characterized as a key regulator for specification of stem cell fate in floral and shoot meristems (Laux et al. 1996). *WUS* encodes a homeodomain protein and is expressed in a small group of cells, namely, the organizing center, below the shoot meristem central zone, which contains the stem cells (Mayer et al. 1998; Schoof et al. 2000).

Overexpression of *WUS* induces the formation of highly embryogenic callus in the presence of auxin (Zuo et al. 2002). In addition, ectopic overexpression of *WUS* in transgenic plants directly induces somatic embryos from various vegetative tissues independently of any external PGR treatment. Therefore, *WUS* appears to be able to trigger the vegetative-to-embryogenic transition, bypassing the auxin requirement or taking advantage of the endogenous auxin flux (Zuo et al. 2002).

Interestingly, *WUS* cannot reprogram the shoot apex towards SE when overexpressed under the control of meristem-specific promoters such as CLV1, ANT (Schoof et al. 2000), LFY, AP3 and AG (Lenhard et al. 2001;

Lohmann et al. 2001). This raises the possibility that some factors could favor one or the other *WUS* function (a shoot meristem or an embryo organizer). Gallois et al. (2004) addressed this possibility by studying the effects of ectopic expression of *WUS* in roots. In the absence of additional cues, *WUS* expression in the root induced shoot stem identity and leaf development indicating that *WUS* establishes stem cells with an intrinsic shoot identity. However, when *WUS* is coexpressed with *LEAFY*, which is a master regulator of floral development (Weigel et al. 1992), *WUS* induces the formation of floral tissues. Finally, when exogenous auxin is supplied, the expression of *WUS* leads to the development of somatic embryos. This elegant work demonstrates that although *WUS* expression specifies an intrinsic shoot activity (in the absence of additional cues) it also makes cells developmentally flexible and able to be directed to floral organ or embryo development, depending on additional cues.

6 Concluding Remarks

PGRs/PHs are largely used to elicit in vitro SE and are therefore suspected to play important roles in this process; however, the question of their exact function remains open. One difficulty in elucidating the role of PGRs/PHs in SE is that they are likely to be involved at different levels. Although they are very efficient stimuli, they also represent signaling molecules that are an integral part of the molecular pathways underlying SE. As exogenous stimuli, they can occasionally be replaced by other treatments, including stresses such as osmotic or heat shock (Jiménez and Thomas, this volume). In contrast, it becomes obvious that endogenous PHs play essential roles in directing crucial SE-related events, including reentry into the cell cycle and dedifferentiation and redifferentiation of somatic cells. Recent developments in the elucidation of modes of action of PHs have shown that they trigger profound modifications in cellular gene expression patterns both by influencing chromatin structure and DNA methylation and by a finer and more specific transcriptional regulation of target genes.

Recent data suggest that the cellular embryonic competence is "actively" repressed in postembryonic plant tissues by proteins such as AMP1 or PICKLE. Derepression, e.g., by null mutation in repressor genes, opens the way to SE. However, somatic embryo induction is only activated when local tissue/cellular conditions, such as a proper hormonal balance, are appropriate. This would explain why all cells of *pickle* or *amp1* mutants do not uniformly enter an embryonic developmental program. The observation that different mutations induce similar embryonic phenotypes in postembryonic plants reflects the complexity of SE and the possible existence of overlapping pathways triggering this developmental process.

In recent years functional genomics allowed identification of several potential candidate genes that may be responsible for the establishment of the SE program. Although the participation of these genes in the induction of SE in wild-type plants has not been proved yet, they represent very exciting tracks to pursue in the exploration of molecular pathways underlying SE.

References

Banno H, Ikeda Y, Niu QW, Chua NH (2001) Overexpression of Arabidopsis ESR1 induces initiation of shoot regeneration. Plant Cell 13:2609–2618

Barbier-Brygoo H, Ephritkhine G, Klämbt D, Ghislain M, Guern J (1989) Functional evidence for an auxin receptor at the plasmalemma of tobacco mesophyll protoplasts. Proc Natl Acad Sci USA 86:891–895

Bardini M, Labra M, Winfield M, Sala F (2003) Antibiotic-induced DNA methylation changes in calluses of Arabidopsis thaliana. Plant Cell Tissue Organ Cult 72:157–162

Baudino S, Hansen S, Brettschneider R, Hecht VF, Dresselhaus T, Lörz H, Dumas C, Rogowsky PM (2001) Molecular characterisation of two novel maize LRR receptor-like kinases, which belong to the *SERK* gene family. Planta 213:1–10

Bauly JM, Sealy IM, Macdonald H, Brearley J, Droge S, Hillmer S, Robinson DG, Venis MA, Blatt MR, Lazarus CM, Napier RM (2000) Overexpression of auxin-binding protein enhances the sensitivity of guard cells to auxin. Plant Physiol 124:1229–1238

Boutilier K, Offringa R, Sharma VK, Kieft H, Ouellet T, Zhang L, Hattori J, Liu CM, van Lammeren AA, Miki BL, Custers JB, van Lookeren Campagne MM (2002) Ectopic expression of *BABY BOOM* triggers a conversion from vegetative to embryonic growth. Plant Cell 14:1737–1749

Caplan A, Berger PH, Naderi M (1998) Phenotypic variation between transgenic plants: what is making gene expression unpredictable? In: Jain SM, Brar DS, Ahloowalia BS (eds) Somaclonal variation and induced mutations in crop improvement. Kluwer, Dordrecht, pp 539–562

Chang C, Kwok SF, Bleecker AB, Meyerowitz EM (1993) Arabidopsis ethylene-response gene *ETR1*: similarity of product to two-component regulators. Science 262:539–544

Chen JG, Shimomura S, Sitbon F, Sandberg G, Jones AM (2001a) The role of auxin-binding protein 1 in the expansion of tobacco leaf cells. Plant J 28:607–617

Chen JG, Ullah H, Young JC, Sussman MR, Jones AM (2001b) ABP1 is required for organized cell elongation and division in Arabidopsis embryogenesis. Genes Dev 15:902–911

Custers JBM, Cordewener JHG, Nöllen Y, Dons JJM, van Lookeren Campagne MM (1994) Temperature controls both gametophytic and sporophytic development in microspore cultures of Brassica napus. Plant Cell Rep 13:267–271

De Veylder L, Joubes J, Inze D (2003) Plant cell cycle transitions. Curr Opin Plant Biol 6:536–543

del Pozo JC, Lopez-Matas MA, Ramirez-Parra E, Gutierrez C (2005) Hormonal control of the plant cell cycle. Physiol Plant 123:173–183

Dewitte W, Murray JA (2003) The plant cell cycle. Annu Rev Plant Biol 54:235–264

Dharmasiri S, Estelle M (2002) The role of regulated protein degradation in auxin response. Plant Mol Biol 49:401–409

Dudits D, Bögre L, Györgyey J (1991) Molecular and cellular approaches to the analysis of plant embryo development from somatic cells in vitro. J Cell Sci 99:475–484

Dudits D, Györgyey J, Bögre L, Bakó L (1995) Molecular biology of somatic embryogenesis. In: Thorpe TA (ed) In Vitro embryogenesis in plants. Kluwer, Dordrecht, pp 276–308

Eshed Y, Baum SF, Bowman JL (1999) Distinct mechanisms promote polarity establisment in carpels of Arabidopsis. Cell 99:199–209

Fehér A, Pasternak T, Dudits D (2003) Transition of somatic plant cells to an embryogenic state. Plant Cell Tissue Organ Cult 74:201–228

Finkelstein RR, Wang ML, Lynch TJ, Rao S, Goodman HM (1998) The Arabidopsis abscisic acid response locus ABI4 encodes an APETALA 2 domain protein. Plant Cell 10:1043–1054

Francis D, Sorrell DA (2001) The interface between the cell cycle and plant growth regulators: a mini review. Plant Growth Regul 33:1–12

Gallois JL, Nora FR, Mizukami Y, Sablowski R (2004) *WUSCHEL* induces shoot stem cell activity and developmental plasticity in the root meristem. Genes Dev 18:375–380

Gaspar T, Kevers C, Faivre-Rampant O, Crèvecoeur M, Penel C, Greppin H, Dommes J (2003) Changing concepts in plant hormone action. In Vitro Cell Dev Biol Plant 39:85–105

Gray WM, Kepinski S, Rouse D, Leyser O, Estelle M (2001) Auxin regulates SCF(TIR1)-dependent degradation of AUX/IAA proteins. Nature 414:271–276

Gu YQ, Yang C, Thara VK, Zhou J, Martin GB (2000) Pti4 is induced by ethylene and salicylic acid, and its product is phosphorylated by the Pto kinase. Plant Cell 12:771–786

Gubler F, Jacobsen JV (1992) Gibberellin-responsive elements in the promoter of a barley high-pI alpha-amylase gene. Plant Cell 4:1435–1441

Guilfoyle T, Hagen G, Ulmasov T, Murfett J (1998) How does auxin turn on genes? Plant Physiol 118:341–347

Hagen G, Guilfoyle T (2002) Auxin-responsive gene expression: genes, promoters and regulatory factors. Plant Mol Biol 49:373–385

Hecht V, Vielle-Calzada JP, Hartog MV, Schmidt ED, Boutilier K, Grossniklaus U, de Vries SC (2001) The Arabidopsis *SOMATIC EMBRYOGENESIS RECEPTOR KINASE 1* gene is expressed in developing ovules and embryos and enhances embryogenic competence in culture. Plant Physiol 127:803–816

Hemerly AS, Ferreira P, de Almeida Engler J, Van Montagu M, Engler G, Inze D (1993) *cdc2a* expression in Arabidopsis is linked with competence for cell division. Plant Cell 5:1711–1723

Henderson JT, Li HC, Rider SD, Mordhorst AP, Romero-Severson J, Cheng JC, Robey J, Sung ZR, de Vries SC, Ogas J (2004) PICKLE acts throughout the plant to repress expression of embryonic traits and may play a role in gibberellin-dependent responses. Plant Physiol 134:995–1005

Hirt H, Pay A, Györgyey J, Bako L, Nemeth K, Bogre L, Schweyen RJ, Heberle-Bors E, Dudits D (1991) Complementation of a yeast cell cycle mutant by an alfalfa cDNA encoding a protein kinase homologous to p34cdc2. Proc Natl Acad Sci USA 88:1636–1640

Inohara N, Shimomura S, Fukui T, Futai M (1989) Auxin-binding protein located in the endoplasmic reticulum of maize shoots: molecular cloning and complete primary structure. Proc Natl Acad Sci USA 86:3564–3568

Inoue T, Higuchi M, Hashimoto Y, Seki M, Kobayashi M, Kato T, Tabata S, Shinozaki K, Kakimoto T (2001) Identification of CRE1 as a cytokinin receptor from Arabidopsis. Nature 409:1060–1063

Jiménez VM (2001) Regulation of in vitro somatic embryogenesis with emphasis on the role of endogenous hormones. Rev Bras Fisiol Veg 13:196–223

John PCL, Zhang K, Dongg C, Diederich L, Wightman F (1993) A p34cdc2 related protein in control of cell cycle progression, the switch between division and differentiation in tissue development and stimulation of division by auxin and cytokinin. Aust J Plant Physiol 20:503–526

Jones AM, Im KH, Savka MA, Wu MJ, DeWitt NG, Shillito R, Binns AN (1998) Auxin-dependent cell expansion mediated by overexpressed auxin-binding protein 1. Science 282:1114–1117

Kitamiya E, Suzuki S, Sano T, Nagata T (2000) Isolation of two genes that were induced upon the initiation of somatic embryogenesis on carrot hypocotyls by high concentrations of 2,4-D. Plant Cell Rep 19:551–557

Klucher KM, Chow H, Reiser L, Fischer RL (1996) The *AINTEGUMENTA* gene of Arabidopsis required for ovule and female gametophyte development is related to the floral homeotic gene APETALA2. Plant Cell 8:137–153

Kulaeva ON, Prokoptseva OS (2004) Recent advances in the study of mechanisms of action of phytohormones. Biochemistry (Moscow) 69:233–247

Laureys F, Dewitte W, Witters E, Van Montagu M, Inze D, Van Onckelen H (1998) Zeatin is indispensable for the G2-M transition in tobacco BY-2 cells. FEBS Lett 426:29–32

Laux T, Mayer KF, Berger J, Jurgens G (1996) The *WUSCHEL* gene is required for shoot and floral meristem integrity in Arabidopsis. Development 122:87–96

Leblanc N, David K, Grosclaude J, Pradier JM, Barbier-Brygoo H, Labiau S, Perrot-Rechenmann C (1999) A novel immunological approach establishes that the auxin-binding protein, Nt-abp1, is an element involved in auxin signaling at the plasma membrane. J Biol Chem 274:28314–28320

Lenhard M, Bohnert A, Jurgens G, Laux T (2001) Termination of stem cell maintenance in Arabidopsis floral meristems by interactions between WUSCHEL and AGAMOUS. Cell 105:805–814

Li G, Hall TC, Holmes-Davis R (2002) Plant chromatin: development and gene control. Bioassays 24:234–243

Lin X, Hwang GJ, Zimmerman JL (1996) Isolation and characterization of a diverse set of genes from carrot somatic embryos. Plant Physiol 112:1365–1374

Lohmann JU, Hong RL, Hobe M, Busch MA, Parcy F, Simon R, Weigel D (2001) A molecular link between stem cell regulation and floral patterning in Arabidopsis. Cell 105:793–803

Lotan T, Ohto M, Yee KM, West MA, Lo R, Kwong RW, Yamagishi K, Fischer RL, Goldberg RB, Harada JJ (1998) Arabidopsis LEAFY COTYLEDON1 is sufficient to induce embryo development in vegetative cells. Cell 93:1195–1205

Luerssen H, Kirik V, Herrmann P, Misera S (1998) FUSCA3 encodes a protein with a conserved VP1/AB13-like B3 domain which is of functional importance for the regulation of seed maturation in Arabidopsis thaliana. Plant J 15:755–764

Marcotte WR Jr, Guiltinan MJ, Quatrano RS (1992) ABA-regulated gene expression: cis-acting sequences and trans-acting factors. Biochem Soc Trans 20:93–97

Martinez MC, Jorgensen JE, Lawton MA, Lamb CJ, Doerner PW (1992) Spatial pattern of *cdc2* expression in relation to meristem activity and cell proliferation during plant development. Proc Natl Acad Sci USA 89:7360–7364

Mayer KF, Schoof H, Haecker A, Lenhard M, Jurgens G, Laux T (1998) Role of WUSCHEL in regulating stem cell fate in the Arabidopsis shoot meristem. Cell 95:805–815

McCarty DR, Carson CB, Stinard PS, Robertson DS (1989) Molecular analysis of *viviparous-1*: an abscisic acid-insensitive mutant of maize. Plant Cell 1:523–532

McCarty DR, Hattori T, Carson CB, Vasil V, Lazar M, Vasil IK (1991) The *viviparous-1* developmental gene of maize encodes a novel transcriptional activator. Cell 66:895–905

Meinke DW, Franzmann LH, Nickle TC, Yeung EC (1994) Leafy cotyledon mutants of Arabidopsis. Plant Cell 6:1049–1064

Meller Y, Sessa G, Eyal Y, Fluhr R (1993) DNA-protein interactions on a cis-DNA element essential for ethylene regulation. Plant Mol Biol 23:453–463

Menke FL, Champion A, Kijne JW, Memelink J (1999) A novel jasmonate- and elicitor-responsive element in the periwinkle secondary metabolite biosynthetic gene Str interacts with a jasmonate- and elicitor-inducible AP2-domain transcription factor, ORCA2. EMBO J 18:4455–4463

Miao GH, Hong Z, Verma DP (1993) Two functional soybean genes encoding p34cdc2 protein kinases are regulated by different plant developmental pathways. Proc Natl Acad Sci USA 90:943–947

Mordhorst AP, Voerman KJ, Hartog MV, Meijer EA, van Went J, Koornneef M, de Vries SC (1998) Somatic embryogenesis in Arabidopsis thaliana is facilitated by mutations in genes repressing meristematic cell divisions. Genetics 149:549–563

Nagata T, Ishida S, Hasezawa S, Takahashi Y (1994) Genes involved in the dedifferentiation of plant cells. Int J Dev Biol 38:321–327

Nakano Y, Steward N, Sekine M, Kusano T, Sano H (2000) A tobacco NtMET1 cDNA encoding a DNA methyltransferase: molecular characterization and abnormal phenotypes of transgenic tobacco plants. Plant Cell Physiol 41:448–457

Nolan KE, Irwanto RR, Rose RJ (2003) Auxin up-regulates *MtSERK1* expression in both root-forming and embryogenic cultures. Plant Physiol 133:218–230

Ogas J, Cheng JC, Sung ZR, Somerville C (1997) Cellular differentiation regulated by gibberellin in the Arabidopsis thaliana pickle mutant. Science 277:91–94

Ogas J, Kaufmann S, Henderson J, Somerville C (1999) PICKLE is a CHD3 chromatin-remodeling factor that regulates the transition from embryonic to vegetative development in Arabidopsis. Proc Natl Acad Sci USA 96:13839–13844

Redig P, Shaul O, Inze D, Van Montagu M, Van Onckelen H (1996) Levels of endogenous cytokinins, indole-3-acetic acid and abscisic acid during the cell cycle of synchronized tobacco BY-2 cells. FEBS Lett 391:175–180

Rider SD Jr, Henderson JT, Jerome RE, Edenberg HJ, Romero-Severson J, Ogas J (2003) Coordinate repression of regulators of embryonic identity by PICKLE during germination in Arabidopsis. Plant J 35:33–43

Rider SD Jr, Hemm MR, Hostetler HA, Li HC, Chapple C, Ogas J (2004) Metabolic profiling of the Arabidopsis pkl mutant reveals selective derepression of embryonic traits. Planta 219:489–499

Riechmann JL, Meyerowitz EM (1998) The AP2/EREBP family of plant transcription factors. Biol Chem 379:633–646

Riou-Khamlichi C, Huntley R, Jacqmard A, Murray JA (1999) Cytokinin activation of Arabidopsis cell division through a D-type cyclin. Science 283:1541–1544

Roudier F, Fedorova E, Lebris M, Lecomte P, Györgyey J, Vaubert D, Horvath G, Abad P, Kondorosi A, Kondorosi E (2003) The Medicago species A2-type cyclin is auxin regulated and involved in meristem formation but dispensable for endoreduplication-associated developmental programs. Plant Physiol 131:1091–1103

Rück A, Palme K, Venis MA, Napier R, Felle RH (1993) Patch-clamp analysis establishes a role for an auxin binding protein in the auxin stimulation of plasma membrane current in Zea mays protoplasts. Plant J 4:41–46

Satoh S, Kamada H, Harada H, Fujii T (1986) Auxin-controlled glycoprotein release into the medium of embryogenic carrot cells. Plant Physiol 81:931–933

Sauter M (1997) Differential expression of a CAK (cdc2-activating kinase)-like protein kinase, cyclins and cdc2 genes from rice during the cell cycle and in response to gibberellin. Plant J 11:181–190

Schaller GE, Bleecker AB (1995) Ethylene-binding sites generated in yeast expressing the Arabidopsis ETR1 gene. Science 270:1809–1811

Schmidt ED, Guzzo F, Toonen MA, de Vries SC (1997) A leucine-rich repeat containing receptor-like kinase marks somatic plant cells competent to form embryos. Development 124:2049–2062

Schoof H, Lenhard M, Haecker A, Mayer KF, Jurgens G, Laux T (2000) The stem cell population of Arabidopsis shoot meristems in maintained by a regulatory loop between the *CLAVATA* and *WUSCHEL* genes. Cell 100:635–644

Shimomura S, Watanabe S, Hiroaki I (1999) Characterization of auxin-binding protein 1 from tobacco: content, localization and auxin-binding activity. Planta 209:118–125

Somleva MN, Schmidt EDL, de Vries SC (2000) Embryogenic cells in Dactylis glomerata L. (Poaceae) explants identified by cell tracking and by *SERK* expression. Plant Cell Rep 19:718–726

Stals H, Inze D (2001) When plant cells decide to divide. Trends Plant Sci 6:359–364

Steffens B, Feckler C, Palme K, Christian M, Bottger M, Luthen H (2001) The auxin signal for protoplast swelling is perceived by extracellular ABP1. Plant J 27:591–599

Stone SL, Kwong LW, Yee KM, Pelletier J, Lepiniec L, Fischer RL, Goldberg RB, Harada JJ (2001) *LEAFY COTYLEDON2* encodes a B3 domain transcription factor that induces embryo development. Proc Natl Acad Sci USA 98:11806–11811

Swiatek A, Lenjou M, Van Bockstaele D, Inze D, Van Onckelen H (2002) Differential effect of jasmonic acid and abscisic acid on cell cycle progression in tobacco BY-2 cells. Plant Physiol 128:201–211

Takahashi Y, Kuroda H, Tanaka T, Machida Y, Takebe I, Nagata T (1989) Isolation of an auxin-regulated gene cDNA expressed during the transition from G0 to S phase in tobacco mesophyll protoplasts. Proc Natl Acad Sci USA 86:9279–9283

Thiel G, Blatt MR, Fricker MD, White IR, Millner P (1993) Modulation of K^+ channels in Vicia stomatal guard cells by peptide homologs to the auxin-binding protein C terminus. Proc Natl Acad Sci USA 90:11493–11497

Thomas C, Bronner R, Molinier J, Prinsen E, van Onckelen H, Hahne G (2002) Immunocytochemical localization of indole-3-acetic acid during induction of somatic embryogenesis in cultured sunflower embryos. Planta 215:577–583

Thomas C, Meyer D, Himber C, Steinmetz A (2004) Spatial expression of a sunflower *SERK* gene during induction of somatic embryogenesis and shoot organogenesis. Plant Physiol Biochem 42:35–42

Thomas TL (1993) Gene expression during plant embryogenesis and germination: An overview. Plant Cell 5:1401–1410

Tiwari SB, Wang XJ, Hagen G, Guilfoyle TJ (2001) AUX/IAA proteins are active repressors, and their stability and activity are modulated by auxin. Plant Cell 13:2809–2822

Ueguchi C, Sato S, Kato T, Tabata S (2001) The *AHK4* gene involved in the cytokinin-signaling pathway as a direct receptor molecule in Arabidopsis thaliana. Plant Cell Physiol 42:751–755

Ulmasov T, Hagen G, Guilfoyle TJ (1999) Activation and repression of transcription by auxin-response factors. Proc Natl Acad Sci USA 96:5844–5849

van der Fits L, Memelink J (2001) The jasmonate-inducible AP2/ERF-domain transcription factor ORCA3 activates gene expression via interaction with a jasmonate-responsive promoter element. Plant J 25:43–53

Wang H, Qi Q, Schorr P, Cutler AJ, Crosby WL, Fowke LC (1998) ICK1, a cyclin-dependent protein kinase inhibitor from Arabidopsis thaliana interacts with both Cdc2a and CycD3, and its expression is induced by abscisic acid. Plant J 15:501–510

Weigel D, Alvarez J, Smyth DR, Yanofsky MF, Meyerowitz EM (1992) LEAFY controls floral meristem identity in Arabidopsis. Cell 69:843–859

West M, Yee KM, Danao J, Zimmerman JL, Fischer RL, Goldberg RB, Harada JJ (1994) LEAFY COTYLEDON1 is an essential regulator of late embryogenesis and cotyledon identity in Arabidopsis. Plant Cell 6:1731–1745

Worley CK, Zenser N, Ramos J, Rouse D, Leyser O, Theologis A, Callis J (2000) Degradation of Aux/IAA proteins is essential for normal auxin signaling. Plant J 21:553–562

Yamamoto N, Kobayashi H, Togashi T, Mori Y, Kikuchi K, Kuriyama K, Tokuji Y (2005) Formation of embryogenic cell clumps from carrot epidermal cells is suppressed by 5-azacytidine, a DNA methylation inhibitor. J Plant Physiol 162:47–54

Yasuda H, Nakajima M, Ito T, Ohwada T, Masuda H (2001) Partial characterization of genes whose transcripts accumulate preferentially in cell clusters at the earliest stage of carrot somatic embryogenesis. Plant Mol Biol 45:705–712

Yazawa K, Takahata K, Kamada H (2004) Isolation of the gene encoding Carrot leafy cotyledon 1 and expression analysis during somatic and zygotic embryogenesis. Plant Physiol Biochem 42:215–223

Yeung EC (1995) Structural and developmental patterns in somatic embryogenesis. In: Thorpe TA (ed) In vitro Embryogenesis in Plants. Kluwer Academic Publishers, Dordrecht, p 205–248

Zhang K, Letham DS, John PC (1996) Cytokinin controls the cell cycle at mitosis by stimulating the tyrosine dephosphorylation and activation of p34cdc2-like H1 histone kinase. Planta 200:2–12

Zhao J, Morozova N, Williams L, Libs L, Avivi Y, Grafi G (2001) Two phases of chromatin decondensation during dedifferentiation of plant cells: distinction between competence for cell fate switch and a commitment for S phase. J Biol Chem 276:22772–22778

Zimmerman JL, Thomine S, Guern J, Barbier-Brygoo H (1994) An anion current at the plasma

Zuo J, Niu QW, Frugis G, Chua NH (2002) The *WUSCHEL* gene promotes vegetative-to-embryonic transition in Arabidopsis. Plant J 30:349–359

Plant Cell Monogr (2)
A. Mujib · J. Šamaj: Somatic Embryogenesis
DOI 10.1007/7089_036/Published online: 22 December 2005

Somatic Embryogenesis in Chestnut

E. Corredoira (✉) · A. Ballester · F. J. Vieitez · A. M. Vieitez

Instituto de Investigaciones Agrobiológicas de Galicia, CSIC, Avenida de Vigo, s/n, Apartado 122, 15080 Santiago de Compostela, Spain
elenac@iiag.cesga.es, aballester@iiag.cesga.es

Abstract Somatic embryogenesis is an important biotechnological tool that demonstrates significant benefits when applied to forest tree species; clonal propagation, cryostorage of valuable germoplasm and genetic transformation are among the most promising of its applications. In this chapter, the state of the art of somatic embryogenesis in chestnut (an important economical tree species of the genus *Castanea*) is assessed and discussed. The factors affecting the induction (type of explant, growth conditions, mineral media, plant growth regulators), maintenance and multiplication of the embryogenic cultures (through repetitive embryogenesis) and the maturation and conversion into plants of somatic embryos are described. The latest results achieved on the application of the process on both genetic transformation and cryopreservation of chestnut embryogenic lines are also mentioned.

1 Introduction

In vitro plant regeneration of forest trees (by either organogenesis or somatic embryogenesis) provides tools for cloning superior trees as well as engineering trees with similar efficiency that can be applied to other organisms (Merkle and Dean 2000). There is great interest in applying somatic embryogenesis, not only to mass propagation but also to the development of genetic transformation protocols in forest trees. However, there are several constraints when somatic embryogenesis is applied to these species: in many cases, successful induction only occurs from juvenile tissues (limiting its use for the propagation of mature elite trees), and the quality of the somatic embryos obtained and their conversion rate into plantlets are dependent upon the genotype of the original explant (Stasolla and Yeung 2003). The somatic embryogenesis process is considered to have great potential for sustained clonal propagation, especially when coupled with long-term cryostorage to preserve embryonal tissue juvenility (Park et al. 1998). As somatic embryogenesis is still difficult to achieve in material beyond the seedling stage, cryoconservation precludes genotypes from ageing during the whole selection stage of field-tested, clonally propagated seed progenies.

Chestnut is an important hardwood species of economical relevance that is found in natural stands, in small groves or grown as nut orchards and coppices throughout the range (Bounous 2002). The *Castanea* genus, belonging to the *Fagaceae* family, is native to the Northern Hemisphere, comprising 13 different species. Since the early 20th century, the populations of *Castanea sativa* Miller (European chestnut) and *C. dentata* Borkhausen (Marshall) (American chestnut) have been devastated by two diseases, ink disease and blight disease, the former caused by the fungi *Phytophthora cambivora* and *Ph. cinnamomi* and the latter by *Cryphonectria parasitica* (Bunrham 1988; Vieitez et al. 1996). To multiply disease-resistant trees or trees selected for specific traits, asexual propagation is required. Difficulties of conventional vegetative propagation by cuttings, air-layering, graft, stooling, etc., have been pointed out (Vieitez et al. 1986); these could be overcome, at least partially, with in vitro tissue culture techniques. Protocols for plant regeneration have been defined for both juvenile and mature tissues, mainly through the proliferation of axillary shoots (Sánchez et al. 1997a,b).

The first report on morphogenetic events associated with somatic embryogenesis in chestnut could probably be dated as early as 1978. Vieitez et al. (1978) showed the differentiation of isolated groups of unorganized sphere-like meristematic pockets in cotyledon explants. Although the work was not properly addressed for the induction of somatic embryos, the structures observed look like the initial stages of what today is known as an embryogenic process. Other authors have subsequently attempted to induce somatic embryogenesis in chestnut, especially in the two species most susceptible to fungal diseases, *C. sativa* and *C. dentata* (Table 1). In this review, the state of the art of chestnut somatic embryogenesis will be described, as well as its potential applications in clonal propagation, genetic transformation and cryopreservation.

2 Culture Initiation

To date, there have been few reports on somatic embryogenesis from members of the genus *Castanea*, in spite of their importance. Although considerable effort has been made in recent years, somatic embryogenesis in chestnut has mainly been successfully induced from immature zygotic embryos, as has been the case in many other forest tree species, both Gymnosperms and Angiosperms (Raemakers et al. 1999). The induction of somatic embryos in chestnut from leaf sections published by Corredoira et al. (2003a) opens up new possibilities for induction from mature, selected material.

2.1 Somatic Embryogenesis from Zygotic Embryos

Most of the chestnut embryogenic systems mentioned in Table 1 used both mature and immature zygotic embryos as initial explants. In a first attempt,

Table 1 Summary of somatic embryogenesis studies in *Castanea*

Initial explants	Induction medium[a]	Maintenance medium[a]	Maturation medium[a]	Response	Reference
Castanea sativa					
Cotyledon from immature zygotic embryos	1/2 MS NAA (5.4) or 2,4-D (5.4) + BA (2.2)	—	—	Somatic embryos	Piagnani and Eccher 1990
Ovaries/Ovules/ Immature zygotic embryos	P24 + 2,4-D (5) + BA (0.5)	P24 + BA (0.89)	P24 + BA (0.89)+ Agar 1.1% + cold storage (3 mo)	Somatic embryos Plantlet development	Ballester et al. 2001
Leaf sections	MS + BA (4.4) + NAA (5.4)	1/2 MS + BA (0.44) + NAA (0.54)	1/2 MS + Maltose 3% (4wk) + cold storage (2 mo)	Somatic embryos Plantlet development	Corredoira 2002; Corredoira et al. 2003a
Castanea sativa × *C. crenata*					
Cotyledon from immature zygotic embryos	MS + 2,4-D (0.45–45.3) ± BA, Z or Kin	—	—	Embryoids	González et al. 1985
Immature zygotic embryos	MS + 2,4-D (0.45) + Z (4.56) or 2,4-D (2.26–4.52) ± BA (4.49)	WPM + BA (0.04) or WPM PGR-free	—	Somatic embryos	Vieitez et al. 1990

Table 1 Continued

Initial explants	Induction medium[a]	Maintenance medium[a]	Maturation medium[a]	Response	Reference
Immature zygotic embryos	MS + 2,4-D (0.45) + Z (4.56) or 2,4-D (2.26–4.52) ± BA (4.49)	1/2 MS + Z (0.92) + IBA (0.25)	MS + Z (0.92) (4wk) + cold storage (10–14wk)	Somatic embryos Plantlet development	Vieitez 1995; Vieitez 1999
Castanea dentata					
Immature zygotic embryos	WPM + BA (1.1) + 2,4-D (18.1) or NAA (32.2)	Induction medium or WPM + BA (1.11) or WPM PGR-free	—	Somatic embryos	Merkle et al. 1991
Immature zygotic embryos	WPM + 2,4-D (13.5)	WPM + 2,4-D (13.5) + BA (1.11)	WPM + AC 0.5% (12wk) + cold storage (8–12wk)	Somatic embryos Plantlet development	Carraway and Merkle 1997
Immature zygotic embryos	WPM + 2,4-D (18.1) + BA (1.11)	WPM + 2,4-D (18.1) + BA (1.11)	B5 + Sac. 6% + BA (0.5) + NAA (0.5)	Somatic embryos Plantlet development	Xing et al. 1999

Table 1 Continued

Initial explants	Induction medium[a]	Maintenance medium[a]	Maturation medium[a]	Response	Reference
Immature zygotic embryos	—	WPM + 2,4-D (9.0)	WPM + Sac. 6% + AC 0.1% + Asparagine (25 mM) (4wk) + cold storage (4wk)	Plantlet development	Robichaud et al. 2004

[a] Quantities in brackets are expressed in μM unless otherwise stated.
— not mentioned.
Mineral media: B5 – Gamborg et al. (1968); MS – Murashige and Skoog (1962); P24 – Teasdale (1992); WPM – Woody Plant Medium (Lloyd and McCown, 1980).
Supplements: AC – activated charcoal; BA – N6 – benzyladenine; 2,4-D – 2,4-dichlorophenoxyacetic acid; IBA – indol-3-butyric acid; Kin – kinetin; NAA – 1-naphthaleneacetic acid; Z – zeatin.

González et al. (1985) observed the differentiation of embryoids in cotyledon sections excised from mature seeds of a *Castanea sativa* × *C. crenata* hybrid, and cultured on Murashige and Skoog (1962; MS) medium to which was added different concentrations of 2,4-dichorophenoxyacetic acid (2,4-D), either alone or combined with N_6-benzyladenine (BA), kinetin (kin) or zeatin (Z). Histological analyses of the embryogenic tissue showed the presence of globular to cotyledonary somatic embryos (bipolar structures), although the transfer of these structures to a medium without plant growth regulators (PGR) failed to bring about embryoid development into plantlets.

Although Piagnani and Eccher (1990) mentioned somatic embryo formation in one cultivar of *C. sativa*, the first clear report describing the induction of true somatic embryos in chestnut was published by Vieitez et al. (1990). In this study, samples consisting of zygotic embryos excised at different developmental stages were collected from mid-July to mid-October from two ink disease-resistant *Castanea sativa* × *C. crenata* trees. Embryogenic cultures were induced from immature seeds (15–20 mm long) collected 10–12 weeks post-anthesis, cultured on MS medium supplemented with either 0.45 μM 2,4-D plus 4.56 μM Z or 2,4-D (2.26–4.52 μM) with or without 4.49 μM BA for 2 months in darkness. They were then transferred to half-strength MS containing 0.44 μM BA with or without 0.27 μM NAA or 0.25 μM IBA and kept under a 16 h photoperiod (30 $\mu mol\, m^{-2}\, s^{-1}$) with 25 °C day/20 °C dark. After 2–3 months, embryogenic cultures consisting of friable yellowish embryogenic tissue or proembryogenic masses (PEMs) formed cotyledonary somatic embryos that were capable of regenerating plants. The overall embryogenic induction capacity was around 2% (Vieitez et al. 1990; Vieitez 1995).

Further experiments were carried out in our laboratory (unpublished results) to better define the induction of the embryogenic system. Plant material was sampled from *Castanea sativa* × *C. crenata* and *C. sativa* trees during the 2001–2003 seasons. Immature zygotic embryos were collected from the last week of August to the third week of September (approximately 10–13 weeks post-anthesis). After sterilization, zygotic embryos were dissected into cotyledon segments and embryonic axes, and were then cultured for 6 weeks on MS medium supplemented with 3% sucrose, 0.7% Bacto agar, 500 mg/l casein hydrolysate, 4.52 μM 2,4-D and 0.88 μM BA. After this period, the cultures were transferred to the same basal medium supplemented with BA and NAA at 0.44 and 0.54 μM, respectively. Four weeks later, the explants were transferred to PGR-free basal medium with subsequent monthly subculture to fresh medium. Depending on the genotype, the time required from the initiation of the experiment up to the appearance of the first somatic embryos ranged from 3 to 5 months. Somatic embryos formed on the surface of nodular friable masses induced on the embryonic axis, as well as on the cotyledon pieces, but the induction efficiency was twice as high in the former than in the latter. In the three years studied, the best response was obtained from material collected during the last week of August and the first week of September (10–11 weeks post-anthesis), and the induction rate was clearly influenced by both genotype and year of collection, ranging from 2.2% for the hybrid material collected in 2001, to 10% for *C. sativa* trees collected in 2003. Between 1 and 20 somatic embryos at different stages of development can be obtained from a single explant. Similar results were reported by Sauer and Wilhelm (Ballester et al. 2001) from immature zygotic embryos of *C. sativa* trees collected between 5 and 10 week post-anthesis.

The first report on the induction of somatic embryogenesis in American chestnut was by Merkle et al. (1991), who initiated the cultures from developing ovules and excised immature embryos collected during early and middle stages of fruit development (3–9 weeks post-pollination). Explants were cultured initially for 1 or 2 weeks on Woody Plant Medium (WPM; Lloyd and McCown 1980) containing 1.11 μM BA and either 18.1 μM 2,4-D or 32.2 μM NAA. The competence to initiate somatic embryos was very low, and appeared to depend on the developmental stage of explants, as only ovules collected 6 or 7 weeks post-anthesis produced embryogenic cultures. Ovules which were pulsed on NAA or 2,4-D supplemented medium produced somatic embryos, directly originated from the radicles of the zygotic embryos, and often continued development to the cotyledonary stage; however, explants maintained on auxin-supplemented medium initially generated a nodular growth that resembled proembryogenic masses (PEMs), which formed globular and heart-stage embryos, even while still exposed to auxin, but plantlets were not recorded (Merkle et al. 1991).

A more extensive study was made by Carraway and Merkle (1997) in which immature and mature zygotic embryos were used as explants sampled from 30 American chestnut trees. The effect of three auxins (2,4-D, NAA or 3-indoleacetic acid, IAA) and two cytokinins (BA or thidiazuron, TDZ) on the embryogenic capacity of seed tissues sampled at different developmental stages was investigated. Across all treatments, genotypes and explant types (12039 explants in total), the embryogenic response was 0.9%. According to these authors, the most efficient induction of embryogenic cultures was achieved from zygotic embryos less than 4 mm in length, and cotyledons smaller than 6 mm^2. Both IAA and 2,4-D induced embryogenic response; however, no embryogenic cultures were recorded on medium containing NAA or TDZ.

Following a similar procedure, Xing et al. (1999) also induced somatic embryogenesis from *C. dentata* developing ovules (4–7 weeks post-anthesis) cultured on an induction medium containing 18.8 μM 2,4-D and 1.11 μM BA. PEMs, identified within 5 weeks after plating, consisted of clusters of globular proembryos attached to the callus surface. An induction frequency of 1.6% was obtained, which did not differ greatly from the values reported by Carraway and Merkle (1997). In *C. sativa* and *C. dentata*, it seems that immature zygotic embryo tissues exhibited a lower competence for somatic embryogenesis induction than those of other related species, such as *Quercus robur* and *Q. suber* (Wilhelm 2000; Hernández et al. 2003).

As occurs in other tree species (Triggiano et al. 1999; Corredoira et al. 2002), the potential of zygotic embryo explants to form embryogenic cultures is influenced by their developmental stage, the developmental window of chestnut responsive material being very narrow. Less mature stages in zygotic embryos were responsive in *C. dentata* in comparison to *C. sativa* or hybrid materials. Another difference to be considered in the embryogenic induction protocols is the culture medium: MS supplemented with 4.52 μM 2,4-D was used for *C. sativa*, whereas WPM plus 13.6 to 18.8 μM 2,4-D was used for American chestnut.

2.2 Somatic Embryogenesis from Leaf Explants

To date, only one report on the induction of somatic embryogenesis from somatic tissues other than the zygotic embryos has been published in chestnut (Table 1). Somatic embryogenesis was initiated from leaf explants excised from stock shoot multiplication cultures of *C. sativa* maintained by sequential subculturing of shoot tips and nodal segments every 4–5 weeks (Corredoira 2002; Corredoira et al. 2003a). The 1–3 uppermost unfurled expanding leaves were excised from 4-week-old shoot cultures, and were cut transversally across the midvein. Proximal (basal) leaf halves were cultured (abaxial side down) on MS medium supplemented with 3% sucrose, 0.7% Bacto agar, 500 mg/l casein hydrolysate and different concentrations of NAA (5.37; 10.74; 20 μM) in combination with BA (2.22; 4.44; 8.87 μM). They were maintained in darkness at 25 °C for 6 weeks, and then transferred to the same medium with 0.54 μM NAA and 0.44 μM BA and also maintained in darkness for a further 30 days. After this period, leaves were transferred, at monthly intervals, to PGR-free basal medium and kept under a 16-h photoperiod (50–60 $\mu mol\,m^{-2}\,s^{-1}$) at 25 °C light/20 °C darkness. Generally, somatic embryos appeared in this medium on the surface of a callus 3–6 months after the culture initiation (Fig. 1a,b), a period that was longer than that observed for induction from zygotic embryo explants. The best results were obtained when leaf explants were initially cultured with 5.37 μM NAA and 4.44 μM BA, with an induction frequency of 1%, a lower value than those obtained from zygotic embryos, which could be expected in a more differentiated tissue, such as that of leaves.

The use of leaf explants excised from shoot cultures to initiate the embryogenic systems offers advantages over the zygotic embryo tissues, as clonal material could be a suitable source of explants for inducing somatic embryogenesis from selected, mature genotypes. In addition, when using leaves from in vitro cultures no sterilization procedure is required, and experiments can be programmed all year around. In contrast to what occurs when somatic

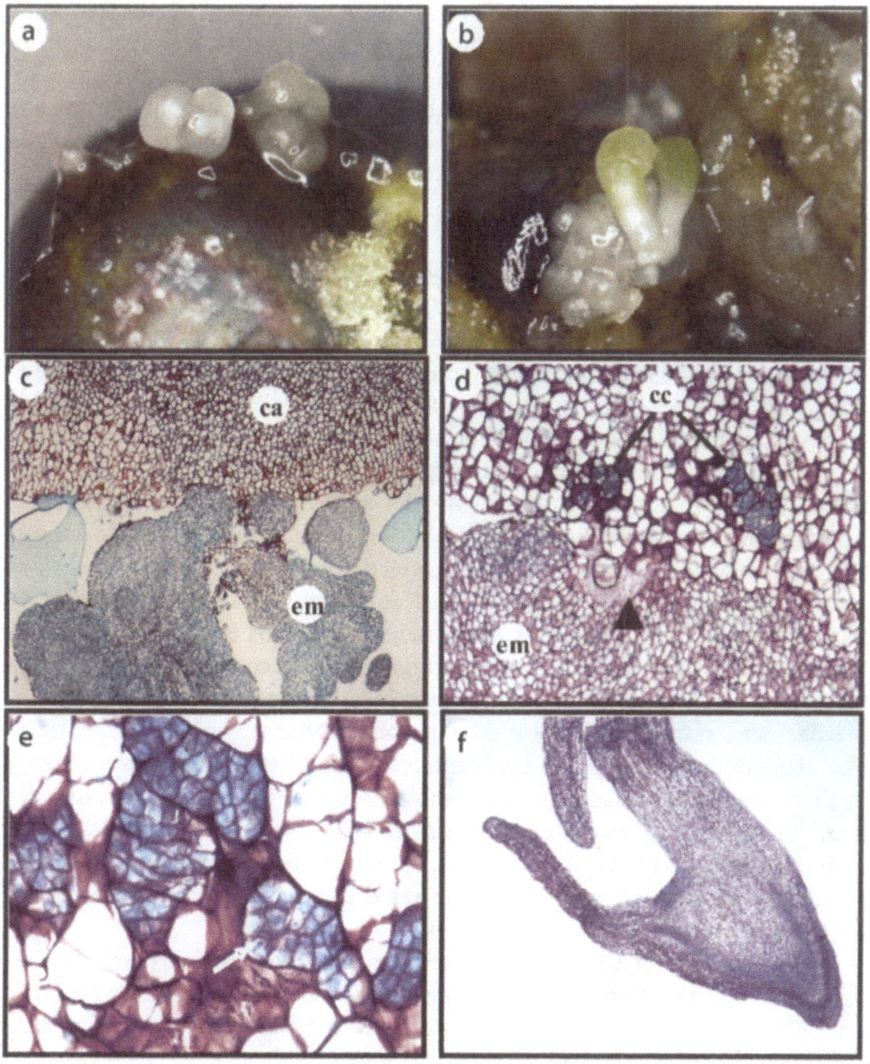

Fig. 1 Induction of *C. sativa* somatic embryos from leaf explants. (**a**),(**b**) Somatic embryos and nodular embryogenic masses emerging from callus formed on leaf tissues (6.2×). (**c**) Callus tissue consisting of large parenchymatic cells (ca), and nodular embryogenic masses (em) that arose from the callus (25×). (**d**) Embryogenic cell clumps (cc) differentiated in the callus tissue which is in contact with embryogenic masses (em). The disruption of callus tissue resulting in the separation of parenchymatic cells at the surface (*arrow head*) should be noted (62×). (**e**) Enlarged view of embryogenic cell clumps formed by densely cytoplasmic cells with presence of starch grains (*arrow*). Note the expanded vacuolated cells of the callus around the embryogenic clumps (247×). (**f**) Cotyledonary-stage somatic embryo showing shoot and root meristems and an independent vascular system (25×). (Safranin-fast *green* in (**c**); PAS-naphtol-blue *black* in (**d**)–(**f**))

embryogenesis is induced from zygotic embryos, 2,4-D was ineffective when applied to leaf sections. The combination of NAA and BA was also used for the induction of somatic embryos from leaf tissues in other Fagaceae, including oaks, where embryogenic cultures have been initiated from both juvenile (Cuenca et al. 1999) and adult (Hernández et al. 2003; Toribio et al. 2004) leaf explants.

2.3 Somatic Embryo Development in the Original Explants

When using immature zygotic embryos of *C. sativa* or the hybrid material, the initiation of globular embryogenic masses and/or somatic embryos occurred after transfer of explants to PGR-free medium. Nodular masses and somatic embryos appear as translucent white structures that seem to be directly differentiated from embryonic tissue explants (embryonic axes or cotyledon pieces). After isolation of somatic embryos, new embryos generally differentiated from the original explant. It was a common morphology for somatic embryos to have white or pale green cotyledons and a dense, yellowish root pole; fused embryos, embryos with their cotyledons fused together in a cup-like structure, and multiple or anomalous cotyledons were also produced. Vieitez et al. (1990) reported that nodular embryogenic tissue consisted of nodular masses of small parenchymatic cells, and exhibited areas of great meristematic activity, especially at its periphery, where preglobular and globular stage embryos were also apparent. No vascular tissue was differentiated in these nodular masses, which resembled the proembryogenic masses defined by Halperin (1966). The meristematic areas evolved to develop somatic embryos, which were typically bipolar structures with both shoot and root apices, a closed independent vascular system and no vascular connections with the subjacent embryogenic masses.

The generation of PEMs from American chestnut immature embryo explants was also mentioned by Carraway and Merkle (1997), who reported that after 6 weeks of culture initiation embryogenic cultures began as a mixture of both embryogenic and nonembryogenic callus. To produce cultures with embryogenic potential, 4–5 cycles of visual selection were needed. Approximately 5 months after the first embryogenic tissue was observed, culture lines producing PEMs were established. *C. dentata* embryogenic cultures proliferated as mixtures of embryogenic cell clusters and early cotyledonary stage somatic embryos, and most somatic embryos that differentiated in presence of 2,4-D grew in fused masses with multiple cotyledons; however, the removal of 2,4-D from the culture medium did not preclude the appearance of these anomalous embryos (Carraway and Merkle 1997).

When somatic embryos were originated from leaf tissues of *C. sativa*, the explants initially responded by enlargement followed by a small callus formation, which was mainly differentiated on the leaf cut surfaces. A greenish callus subsequently originated from the midvein, spreading to the rest of the explant. In some cases, translucent globular structures and somatic embryos at various developmental stages began to grow from this callus tissue at different times. The earliest could be seen after one week of transfer to PGR-free medium, although differentiation also occurred after 2–3 months' culture on this medium, which resulted in a 3- to 6-month period from culture initiation. The anatomical study (unpublished results) performed on cultured leaf explants showed that they yielded callus tissue comprising parenchymatic cells with vascular elements (Fig. 1c). Certain zones in the periphery of this callus exhibited a gradual disruption of tissue integrity, which gave rise to a friable callus area formed by expanded parenchymatic cells and large intercellular spaces that took on a disaggregating appearance (Fig. 1d). Within this zone, clumps of small densely cytoplasmic cells were differentiated, having a large centrally positioned nucleus with prominent nucleoli, and accumulation of starch grains (Fig. 1d,e). These characteristics correspond to those displayed by embryogenic cells, whereas the occurrence of embryogenic cell clumps undergoing a series of divisions with a common thick cell wall indicates a probable unicellular origin. Only a small number of these cell clumps continued to develop nodular embryogenic masses that emerged on the disaggregating callus surface, and they were generally formed of small vacuolated cells and zones of meristematic cells at the periphery (Fig. 1c,d); neither vascular elements nor starch grains were observed in these nodular masses. Somatic embryos at different developmental stages, including the cotyledonary stage (Fig. 1f), were differentiated from the meristematic areas of the nodular embryogenic masses, which were attached to the callus during initiation but became detached at later stages of development. Embryogenic masses seem to be of unicellular origin, although somatic embryos that originated later from these masses appear to be of either unicellular or multicellular origin.

It should be stressed that the generation of nodular embryogenic masses in leaf explants is an indirect process through the formation of an intermediate callus tissue, whereas the PEMs or embryogenic masses differentiate directly from immature zygotic embryo explants.

3 Culture Maintenance

In chestnut, the multiplication and maintenance of embryogenic capacity can be carried out via two methods: (1) secondary or repetitive embryogenesis from isolated somatic embryos in torpedo-cotyledonary stages which develop secondary embryos from the root-hypocotyl zone; and (2) subculture of both nodular embryogenic masses and PEMs. The embryogenic masses were produced from the surface of somatic embryos.

Medium-term maintenance of various embryogenic *Castanea sativa* × *C. crenata* culture lines on semi-solid medium has been thoroughly described by Vieitez (1995; 1999) and later updated by Vieitez and Merkle (2005). Essentially, embryogenic lines have been maintained by monthly subculture of PEMs on semi-solid half-strength MS containing 3 μM glutamine, 0.91 μM Z, 0.25 μM IBA and 3% sucrose under a 16 h photoperiod at 6–15 $\mu mol\,m^{-2}\,s^{-1}$. After more than 12 years of repeated subculture on this medium, the production of cotyledonary somatic embryos remains undiminished. The type and concentration of carbon source was also investigated for maintenance of hybrid embryogenic cultures, where sucrose at 3% was superior to fructose, glucose and maltose, maltose being the least effective (Vieitez 1999).

Corredoira et al. (2003a) also reported the proliferation of embryogenic cultures derived from European chestnut leaf explants, by both secondary embryogenesis and by subculture of nodular embryogenic masses originated from cotyledons of somatic embryos. Secondary embryos were induced by subculturing somatic embryos on proliferation medium consisting of MS mineral salts (half-strength macronutrients) and vitamins supplemented with 3% sucrose, 0.8% Sigma agar, 3 μM glutamine, and different concentrations of BA (0.44 and 4.4 μM) and NAA (0.54 and 5.4 μM). As in the hybrid material, low levels of an auxin and a cytokinin were necessary for secondary embryo proliferation. The best results, with a multiplication coefficient of 3.9 (this coefficient was defined as the product of the proportion of explants producing secondary embryos and the mean number of embryos per embryogenic explant), were achieved on medium supplemented with 0.44 μM BA and 0.54 μM NAA (Fig. 2a,b). In addition to secondary embryos, the subcultured primary embryos also began to develop nodular masses from their cotyledons as a form of repetitive embryogenesis. The frequency of nodular clumps producing somatic embryos (Fig. 2c) ranged from 31 to 50%, with the mean number of embryos per clump ranging from 4.2 to 11.3, the best results (4.6 multiplication coefficient) being obtained with the same PGR combination as for secondary embryogenesis.

The occurrence of both types of repetitive embryogenesis suggests that different cells from the same embryo respond differently to the same culture conditions. The embryonic cells in the hypocotyl-root zone of primary embryos of chestnut are probably embryogenically determined, and a single stimulus for cell division may be sufficient for the formation of secondary embryos. In the case of embryogenic masses originated from cotyledon cells (which are more differentiated), a number of mitotic divisions producing these masses seem to be necessary prior to somatic embryo development. Therefore, direct secondary embryogenesis and indirect proliferation through proembryogenic masses can be considered as two extremes of a continuum (Merkle 1995). A similar process for embryo proliferation was reported for the related species *Q. robur*, in which secondary embryos

Fig. 2 Maintenance of embryogenic cultures and plant recovery in European chestnut. (**a**),(**b**) Embryogenic cultures multiplied by secondary embryogenesis after 6 weeks of culture on proliferation medium ((**a**) 3.9×; (**b**) 4.9×). (**c**) Somatic embryos originated from a nodular embryogenic clump explant after 6 weeks of culture on proliferation medium (15.5×). (**d**),(**e**) Conversion into plantlets (**d**) and somatic embryo exhibiting only shoot development (**e**) after 8 weeks of culture on germination medium. (**f**) Somatic embryo derived trees 12 years after transplanting to soil. (**g**) GUS-positive somatic embryos transformed with *Agrobacterium tumefaciens* strain/plasmid combination EHA105/p35SGUSINT (4.8×)

developed both directly from primary embryos and indirectly from calli originated from cortical tissues (Zegzouti et al. 2001).

When embryo productivity of proliferating cultures reported in Vieitez (1995) and Corredoira et al. (2003a) is compared, it is higher in the former, although the important effect of the genotype as well as the different origin of the embryogenic systems (zygotic embryos vs. leaf explants) should be taken into consideration. We have observed that competence for repetitive embryogenesis in different embryogenic lines originated from zygotic embryos of *C. sativa* and hybrid material differs from line to line, highlighting the effect of the genotype on embryo proliferation, an effect that has been well documented in other species (Park et al. 1994; Corredoira et al. 2003b).

The culture of embryogenic masses in liquid medium has also been investigated. Vieitez (1995) established embryogenic cell suspension cultures by transferring proembryogenic masses to liquid medium consisting of MS (half-strength macronutrients) supplemented with 1.13 μM 2,4-D and 0.45 μM BA. Somatic embryos remained arrested at the globular stage, and their further development required the transfer of PEMs to solid maintenance medium. The suspension cultures were allowed to settle for 1 min, then the suspended fraction was discarded, and the settle fraction was resuspended and filtered through a 40 μm size; PEMs were collected and transferred to semi-solid maintenance medium where embryos at all stages of development were observed after 3–4 weeks of culture.

In *C. dentata*, production of secondary embryos was extremely slow and ceased after one or two cycles (Merkle et al. 1991). The maintenance and proliferation of embryogenic cultures has mainly been reported by subculture at monthly intervals of PEMs on semi-solid medium supplemented with 13.56 μM 2,4-D and 1.11 μM BA in the dark (Carraway and Merkle 1997). Suspension cultures were established by inoculating 0.5 g of PEMs in liquid medium with the aforementioned growth regulators, and these were maintained through transfer to fresh liquid medium at 3-week intervals. PEMs proliferated more rapidly in liquid than on semi-solid medium. Production of somatic embryos arrested at the early cotyledonary stage was achieved after removal of PGRs from suspension cultures. Further development of somatic embryos beyond the early cotyledonary stage was obtained when PEMs were transferred to semi-solid medium, where single embryos, clumps of fused somatic embryos and embryos that had multiple cotyledons were observed. In contrast, when PEMs were size-fractionated and transferred to semi-solid PGR-free medium the number of single somatic embryos increased. Addition of charcoal to the basal medium, enhanced the yield and growth of somatic embryos (Carraway and Merkle 1997).

Xing et al. (1999) multiplied American chestnut PEMs on semi-solid medium by subculturing on the initiation medium defined by Merkle et al. (1991) at 2-week intervals and maintaining them in continuous darkness. The development of somatic embryos from PEMs was achieved by transferring them to semi-solid medium supplemented with 0.5 μM BA and 0.5 μM NAA.

4 Embryo Maturation and Germination

The conversion of somatic embryos into plantlets is currently a limiting step for all chestnut embryogenic systems. It has been shown that in a number of species, low plant recovery rates are due to poor embryo quality and a lack of maturation and desiccation tolerance (Ettienne et al. 1993). In general, maturation of somatic embryos can be achieved through treatments with abscisic acid (ABA) and/or permeating osmotica (high concentrations of sugars, sugar alcohols, amino acids) or nonpermeating osmotica [polyethylene glycol (PEG) and dextran] which induce water stress in the culture medium (Lipavská and Konrádová 2004). However, in a number of species, including chestnut, the transfer of previously matured somatic embryos to a germination medium leads to a poor conversion rate, making it necessary to also apply pregermination treatments, among which we could include cold storage, partial desiccation or the application of gibberellic acid (GA_3), the aim of which is to break the dormancy imposed by ABA and/or osmotic stress.

4.1 Effect of Carbohydrates

Carbon source and concentration had a significant effect on the maturation and subsequent germination and conversion ability of *C. sativa* somatic embryos (Corredoira et al. 2003a). In this report, cotyledonary somatic embryos (4–6 mm) were isolated from embryogenic cultures and transferred to various maturation media consisting of PGR-free MS (half strength macronutrients) medium supplemented with sucrose (3 or 6%), maltose (3 or 6%), 3% sucrose + 6% sorbitol or 3% sucrose + 0.5% activated charcoal. After 4 weeks of culture on maturation medium, somatic embryos were transferred to basal medium with 3% sucrose and stored at 4 °C for 2 months, and then cultured for 8 weeks on germination medium (MS with half strength macronutrients and 0.44 μM BA). Plantlet conversion was achieved in embryos matured on media supplemented with 6% sucrose, and with 3% or 6% maltose, whereas mean shoot length, root length and leaf number of produced plants were not significantly affected by these maturation media, even though higher values were observed after maturation on medium with 6% maltose. Overall, the best results were obtained with 3% maltose-treated embryos, which converted to plants at 6%, in addition to 33% of somatic embryos that developed only shoots (Fig. 2d,e). These shoots were multiplied and rooted following the micropropagation procedure previously described for European chestnut

(Sánchez et al. 1997b). Maltose also promoted the somatic embryo maturation of various species (Tremblay and Tremblay 1991; Norgaard 1997), but its mechanism of action has yet to be elucidated (Lipavská and Konrádová 2004). Norgaard (1997) assumed that the beneficial effect of maltose in the maturation of *Abies normandiana* somatic embryos may be due to low hexose levels resulting from slow maltose hydrolysis, which limits cell carbon nutrition. Blanc et al. (2002) provided further information that supported the carbohydrate deficit hypothesis to explain the maltose effect.

Carbon source and concentration were also evaluated on the development and maturation of American chestnut somatic embryos. Carraway and Merkle (1997) reported that sugar type had a noticeable influence on number and morphology of cotyledonary stage somatic embryos produced per unit weight of PEMs. Very poor results were obtained with maltose, whereas sucrose promoted development of greater numbers of cotyledonary stage somatic embryo than did fructose, but fructose promoted development of single somatic embryos of normal appearance at higher levels than did sucrose. The contrasting results obtained with maltose, with respect to those achieved in *C. sativa* (Corredoira et al. 2003a), may be due to the genotype or the moment when maltose was applied (cotyledonary embryos in *C. sativa* vs. PEMs in *C. dentata*). The preference among carbohydrates has been shown to be species-specific or even cell line-specific (Lipavská and Konrádová 2004). Xing et al. (1999) improved embryo maturation following culture in Gamborg's B5 medium (Gamborg et al. 1968) supplemented with 0.5 μM BA and 0.5 μM NAA, and with sucrose concentration increased to 6%. Mature embryos then germinated in WPM containing 0.89 μM BA and 0.2% activated charcoal, giving rise to plant conversion, shoot regeneration and rooting rates of 3.3, 6.3 and 12.3%, respectively. The 6.3% of mature embryos developing only shoots could indirectly regenerate plantlets through a micropropagation procedure (Xing et al. 1997). By contrast, Robichaud et al. (2004) reported that sucrose level (3–7.5%) in the maturation medium had no effect on the germination frequency of American chestnut embryos, suggesting a possible influence of genotype in order to explain the differences obtained regarding previous studies (Carraway and Merkle 1997; Xing et al. 1999).

4.2 Effect of Cold Storage

As chestnut seeds require cold stratification to germinate, somatic embryos may also need the application of a cold period to break the epicotyl dormancy. In general, this treatment resulted in an overall enhancement of conversion in comparison to previous experiments without chilling. Thus, in hybrid material, plantlet conversion of cold-treated somatic embryos (10–14 weeks at 4 °C) was 18–19% (Vieitez 1995; 1999). The application of a 2-month cold treatment period was also essential to achieve plantlet conversion in *C. sativa* (Corredoira et al. 2003a). The best results, considering both the percentage of somatic embryos developing plants and the percentage of embryos developing only shoots, were obtained with the application of cold storage with or without partial desiccation, giving a total of 41.7 and 38.9% of mature embryos eventually producing plants, respectively. Partial desiccation did not appear to influence the conversion rate.

The effect of a chilling treatment was also investigated by Carraway and Merkle (1997), who concluded that cold storage (8–12 weeks at 4 °C) is necessary for the germination of American chestnut somatic embryos. However, Xing et al. (1999) did not apply this pretreament in their germination experiments.

4.3 Other Maturation Treatments

Activated charcoal had no positive effect on the germination and plantlet conversion of European chestnut somatic embryos (Corredoira et al. 2003a), whereas in American chestnut it was included in both maturation (Carraway and Merkle 1997; Robichaud et al. 2004) and germination media (Carraway and Merkle 1997; Xing et al. 1999; Robichaud et al. 2004).

The culture of isolated embryos of hybrid material on media supplemented with ABA (0.38–7.45 μM) failed to prevent secondary embryogenesis, and had no effect on their subsequent conversion on MS medium containing 0.92 μM Z and 150 μM Fe-Na-EDTA or on MS with GA_3 at various concentrations (Vieitez 1995). The application of ABA (0.37–37.8 μM) in combination with different gelling agents, as well as the effect of PEG_{8000} at 2–4% was also evaluated (Vieitez 1999); however, these treatments were very poor in supporting embryo maturation.

As in the case of *Castanea sativa* × *C. crenata* (Vieitez 1995; 1999), addition of ABA to the maturation medium did not increase plantlet conversion of American chestnut somatic embryos (Xing et al. 1999). In a further report, Robichaud et al. (2004) investigated the addition of ABA, PEG_{6000}, and amino acids (glutamine and asparagine) to the maturation medium prior to cold storage for 4 weeks. They found that some of these treatments increased the dry weight/fresh weight ratios and starch content, but did not increase germination ability; only the 25 μM asparagine treatment significantly enhanced the germination rate (14.17%) and the root length of the germinants.

We also noted that PGRs incorporated into the germination medium affected conversion ability, whereas the somatic embryo size (two classes of 2–5 mm and 6–8 mm) prior to culture on maturation medium did not significantly influence plantlet recovery. The best results (percentage of plantlet conversion and percentage of embryos forming only shoots) were obtained in treatments including 0.44 μM BA with or without auxin (0.54 μM

NAA or 0.49 μM IBA), although shoot length, root length and leaf number were enhanced in both PGR-free medium and BA plus IBA supplemented medium.

As has already been mentioned in culture initiation and culture maintenance sections, the genotype is also an important factor influencing germination and plantlet recovery of chestnut somatic embryos (Vieitez 1995; 1999; Xing et al. 1999; Robichaud et al. 2004). Thus, further efforts will be necessary to optimize maturation and germination protocols, in order for them to be applied to a wide range of genotypes.

5
Acclimatization and Growth in the Field

To date, there is scant information on acclimatization and transfer to soil of plantlets derived from chestnut somatic embryos. Although the results obtained so far indicate that somatic seedlings of *C. sativa* and their hybrids and *C. dentata* can be acclimatized and grown in the field, the number of field-grown plants is currently very low.

Vieitez (1995) transferred somatic plantlets to pots containing a 1 : 1 mixture of peat moss and quartz sand, and these were kept inside an acclimatization tunnel for hardening. Between 70–80% of embryo-derived plantlets (116 out 147 for E-431 line and 38 out 52 for E-HV line) survived and resumed growth within 4–8 weeks of transplantation. Surviving plants were moved to greenhouse conditions and allowed to grow for one year. Some 100 somatic plants were transferred to the field, and all of them survived in soil. After two years, their heights ranged from 70 to 110 cm. Surprisingly, many of these plants showed symptoms of precocious maturation, developing male catkins after 3 years, and beginning to regularly bear chestnuts the following year (Fig. 2f). European chestnut plants derived from seeds require around 10–15 years for flowering, although for *C. crenata* and *C. mollisima* this may be earlier, at 3–5 years (Paglietta and Bounous 1979). Precocity of somatic plants is an extremely valuable character which may be useful in breeding programmes.

In American chestnut, Xing et al. (1999) attained acclimatized plants in a growth chamber after transfer of germinated somatic embryos and plantlets micropropagated from shoot-producing embryos to potting mix. Of 20 plantlets acclimatized and grown in a greenhouse, the largest six were transferred to the field. These authors also observed that at the end of the second growing season, the four surviving plants averaged 27.3 cm in height in comparison to 61.7 cm achieved by normal seedlings (control). Similar results were recorded by Robichaud et al. (2004), who reported that 6 out of 23 somatic plants survived transfer to potting mix, acclimatization to greenhouse conditions, and transplanting to the field.

6 Applications of Chestnut Embryogenic Cultures

6.1 Genetic Transformation

Somatic embryogenesis is not only a promising method for clonal mass propagation, but it is also viewed as a valuable tool for genetic engineering. One of the most important goals in the genetic transformation of trees is to increase resistance to fungal pathogens by transferring genes encoding proteins that are involved in the defence mechanism, such as chitinases (Maynard et al. 1998). The definition of a transformation protocol using marker genes opens up the possibility of applying biotechnological tools to the genetic improvement of chestnut through the development of blight- and/or ink-resistant trees.

Genetic transformation was first attempted by Carraway et al. (1994) and Maynard et al. (1998), who used particle bombardment and *Agrobacterium tumefaciens*, respectively, to transform embryogenic cultures of American chestnut. However, only transgenic cell lines (and no transgenic somatic embryos) were produced. The development of a reliable and reproducible genetic transformation protocol for European chestnut in which embryogenic cultures initiated from leaf explants were used as the target material was reported by Corredoira et al. (2004a). In this study, a transformation efficiency of 25% was recorded when somatic embryos at the globular to early-torpedo stages were co-cultured for 4 days with *A. tumefaciens* strain EHA105 harboring the pUbiGUSINT plasmid containing marker genes. Transformation was confirmed by a histochemical β-glucuronidase (GUS) assay (Fig. 2g), PCR and Southern blot analyses for the *uidA* (GUS) and *npt*II (neomycin phosphotransferase II) genes, and germination and plant recovery was achieved from transformed somatic embryos.

6.2 Cryopreservation

Cryopreservation is currently the safest and most cost-effective method for the long-term conservation of species that are vegetatively propagated or which have seeds that are recalcitrant to storage. Chestnut embryogenic cultures are generally maintained by repetitive embryogenesis. To facilitate management of embryogenic lines and limit the risks of somaclonal variation and contamination, as well as to reduce labor and supply costs, cryopreservation may be a reliable alternative. The feasibility of long-term preservation of *C. sativa* germplasm via the cryopreservation of embryogenic cultures has recently been demonstrated by Corredoira et al. (2004b). In this work an embryogenesis resumption level of 68% was obtained by first preculturing

6–8 mg clumps of globular or heart-shaped somatic embryos on medium containing 0.3 M sucrose for three days, followed by 60 min application of PVS2 vitrification solution (Sakai et al. 1990) before direct immersion in liquid nitrogen. Successful cryostorage of embryogenic cultures of American chestnut has been achieved using the application of a cryoprotectant/slow-freezing method (Holliday and Merkle 2000), but the vitrification protocol used in European chestnut seems to be both simpler and less expensive.

The cryopreservation procedures developed for chestnut may be applied to the long-term storage of valuable embryogenic lines, such as those derived from selected genotypes or transformed material.

7 Conclusions and Future Prospects

Chestnut embryogenic cultures were initiated from immature zygotic embryos and leaf explants, although at low induction rates. The most important factors controlling somatic embryogenesis induction are the genotype, the developmental stage of the zygotic embryos, and the type of growth regulators used; an exogenous auxin (either 2,4-D or NAA alone or in combination with a cytokinin) was an essential pre-requisite to initiate chestnut embryogenic tissue. The long-term maintenance of the embryogenic capacity by repetitive embryogenesis makes the continuous supply of somatic embryos possible, as embryogenic cultures can be efficiently multiplied by both secondary embryogenesis and subculture of nodular embryogenic masses or PEMs.

In spite of the numerous maturation and germination treatments assayed, germination and conversion into plantlets is at present a limiting step in the embryogenic process. It should be stressed that cold storage significantly improved plantlet conversion. Although conversion rates are relatively low, an additional higher number of germinating embryos exhibiting only shoot development was also recorded. These shoots could be multiplied and rooted by using micropropagation techniques. Chestnut somatic seedlings can be acclimatized and grown in the field, where they display a normal appearance.

The recent publication (Corredoira et al. 2004a) describing the production of transgenic chestnut plants via somatic embryogenesis offers an additional alternative to the improvement of the species, specifically if plants with increased resistance to fungal diseases are produced. In addition, the combination of somatic embryogenesis and cryoconservation improves the ability to select superior genotypes, allowing the storage of cultures for several years while awaiting the results of field testing.

To optimize the scale-up of plant production, the following aspects of the embryogenic system need to be improved: (i) induction from mature material; (ii) enhancement of plantlet recovery by investigating embryo synchronization, maturation and germination; (iii) ascertainment of genetic fidelity of the regenerants. Most of the information gathered on embryo development in chestnut has been the result of empirical studies. Molecular biology approaches leading to the understanding of the different steps of the embryogenic process in forest trees are scarce, and to the best of our knowledge, no efforts have been addressed in this regard in chestnut. This is probably one of the most promising lines of research for the coming years.

Acknowledgements This research was partially supported by DGI (MEC) and Xunta de Galicia (Spain) through the projects AGL2004-00335, and PGIDIT03BTF40001PR and PGIDIT03RF40001PR, respectively.

References

Ballester A, Bourrain L, Corredoira E, Gonçalves JC, Lê C-L, Miranda-Fontaíña ME, San-José MC, Sauer U, Vieitez AM, Wilhelm E (2001) Improving chestnut micropropagation through axillary shoot development and somatic embryogenesis. For Snow Landsc Res 76:460–467

Blanc G, Lardet L, Martin A, Jacob JL, Carron MP (2002) Differential carbohydrate metabolism conducts morphogenesis in embryogenic callus of *Hevea brasiliensis* (Müll. Arg.). J Exp Bot 53:1453–1462

Bounous G (2002) Il Castagno. Coltura, ambiente ed utilizzazioni in Italia e nel mondo. Edagricole, Bologna, Italy

Bunrham CR (1988) The restoration of American chestnut. Am Sci 76:478–487

Carraway DT, Merkle SA (1997) Plantlet regeneration from somatic embryos of American chestnut. Can J For Res 27:1805–1812

Carraway DT, Wilde HD, Merkle SA (1994) Somatic embryogenesis and gene transfer in American chestnut. J Am Chestnut Found 8:29–33

Corredoira E (2002) Desarrollo de sistemas embriogénicos en olmo y castaño. Doctoral Thesis. University of Santiago de Compostela, Spain

Corredoira E, Vieitez AM, Ballester A (2002) Somatic embryogenesis in elm. Ann Bot 89:637–644

Corredoira E, Ballester A, Vieitez AM (2003a) Proliferation, maturation and germination of *Castanea sativa* Mill. somatic embryos originated from leaf explants. Ann Bot 92:129–136

Corredoira E, Vieitez AM, Ballester A (2003b) Proliferation and maintenance of embryogenic capacity in elm embryogenic cultures. In Vitro Cell Dev Biol-Plant 39:394–401

Corredoira E, Montenegro D, San-José MC, Vieitez AM, Ballester A (2004a) *Agrobacterium*-mediated transformation of European chestnut embryogenic cultures. Plant Cell Rep 23:311–318

Corredoira E, San-José MC, Ballester A, Vieitez AM (2004b) Cryopreservation of zygotic embryo axes and somatic embryos of European chestnut. CryoLetters 25:33–42

Cuenca B, San-José MC, Martínez MT, Ballester A, Vieitez AM (1999) Somatic embryogenesis from stem and leaf explants of *Quercus robur* L. Plant Cell Rep 18:538–543

Ettienne H, Montoro P, Michaux-Ferriere N, Carron MP (1993) Effects of desiccation, medium osmolarity and abscisic acid on the maturation of *Hevea brasiliensis* somatic embryos. J Exp Bot 44:1613–1619

Gamborg OL, Miller RA, Ojima K (1968) Nutrient requirements of suspension cultures of soybean root cells. Expt Cell Res 50:148–151
González ML, Vieitez AM, Vieitez E (1985) Somatic embryogenesis from chestnut cotyledon tissue cultured in vitro. Sci Hort 27:97–103
Halpering W (1966) Alternative morphogenetic events in cell suspensions. Am J Bot 53:443–453
Hernández I, Celestino C, Toribio B (2003) Vegetative propagation of *Quercus suber* L. by somatic embryogenesis. I. Factors affecting the induction in leaves from mature cork oak trees. Plant Cell Rep 21:759–764
Holliday C, Merkle S (2000) Preservation of American chestnut germplasm by cryostorage of embryogenic cultures. J Am Chestnut Found 14:46–52
Lipavská H, Konrádová H (2004) Somatic embryogenesis in conifers: the role of carbohydrate metabolism. In Vitro Cell Dev Biol-Plant 40:23–30
Lloyd G, McCown B (1980) Comercially feasible micropropagation of mountain laurel, *Kalmia latifolia*, by use of shoot-tip culture. Comb Proc Int Plant Prop Soc 30:421–427
Maynard C, Xing Z, Bickel S, Powell W (1998) Using genetic engineering to help save American chestnut: a progress report. J Am Chestnut Found 12:40–56
Merkle SA (1995) Strategies for dealing with limitations of somatic embryogenesis in hardwood trees. Plant Tiss Cult Biotechnol 1:112–121
Merkle SA, Dean JFD (2000) Forest tree biotechnology. Current Opin Biotech 2000:298–302
Merkle SA, Wiecko AT, Watson-Pauley BA (1991) Somatic embryogenesis in American chestnut. Can J For Res 21:1698–1701
Murashige T, Skoog F (1962) A revised medium for rapid growth and bioassays with tobacco tissue culture. Physiol Plant 15:473–497
Norgaard JV (1997) Somatic embryo maturation and plant regeneration in *Abies nordmanniana* Lk. Plant Sci 124:211–221
Paglieta R, Bounous G (1979) Il castagno da frutto. Edizioni Agricole, Bologna, Italy
Park YS, Pond SE, Bonga JM (1994) Somatic embryogenesis in white spruce (*Picea glauca*): genetic control in somatic embryos exposed to storage, maturation treatments, germination and cryopreservation. Theor Appl Genet 89:742–750
Park YS, Barrett JD, Bonga JM (1998) Application of somatic embryogenesis in high-value clonal forestry: deployment, genetic control, and stability of cryopreserved clones. In Vitro Cell Dev Biol-Plant 34:231–239
Piagnani C, Eccher T (1990) Somatic embryogenesis in chestnut. Acta Hortic 280:159–161
Raemakers K, Jacobsen E, Visser R (1999) Proliferative somatic embryogenesis in woody species. In: Jain SM, Gupta PK, Newton RJ (eds) somatic embryogenesis in woody plants, vol 4. Kluwer, Dordrecht, pp 29–59
Robichaud RL, Lessard VC, Merkle SA (2004) Treatments affecting maturation and germination of American chestnut somatic embryos. J Plant Physiol 161:957–969
Sakai A, Kobayashi S, Oiyama I (1990) Cryopreservation of nucellar cells of navel orange (*Citrus sinensis* Osb. var. *brasiliensis* Tanaka) by vitrification. Plant Cell Rep 9:30–33
Sánchez MC, Ballester A, Vieitez AM (1997a) Reinvigoration treatments for the micropropagation of mature chestnut trees. Ann Sci For 54:359–370
Sánchez MC, San-José MC, Ferro E, Ballester A, Vieitez AM (1997b) Improving micropropagation conditions for adult-phase shoots of chestnut. J Hort Sci 72:433–443
Stasolla C, Yeung EC (2003) Recent advances in conifer somatic embryogenesis: improving somatic embryo quality. Plant Cell Tiss Org Cult 74:15–35

Teasdale R (1992) Formulation of plant culture media and applications therefore. International publication N WO 92/07460, patent N Europe 92902531.0, Forbio Pty Ltd., Queensland, Australia

Toribio M, Fernández C, Celestino C, Martínez MT, San-José MC, Vieitez AM (2004) Somatic embryogenesis in mature *Quercus robur* trees. Plant Cell Tiss Org Cult 76:283–287

Tremblay L, Tremblay FM (1991) Carbohydrate requirements for the development of black spruce (*Picea mariana* (Mill.) B.S.P.) and red spruce (*P. rubens* Sarg.) somatic embryos. Plant Cell Tiss Org Cult 27:95–103

Trigiano RN, Buckley LG, Merkle SA (1999) Somatic embryogenesis in woody legumes. In: Jain MS, Gupta PK, Newton RJ (eds) Somatic embryogenesis in woody plants, vol 4. Kluwer, Dordrecht, pp 189–208

Vieitez AM, González ML, Vieitez E (1978) In vitro culture of cotyledon tissue of *Castanea sativa* Mill. Sci Hort 8:243–247

Vieitez AM, Vieitez ML, Vieitez E (1986) Chestnut (*Castanea* spp). In: Bajaj YPS (ed) Biotechnology in agriculture and forestry, vol 1. Springer, Berlin Heidelberg New York, pp 393–414

Vieitez E, Vieitez ML, Vieitez FJ (1996) El castaño. Edilesa, León, Spain

Vieitez FJ (1995) Somatic embryogenesis in chestnut. In: Jain MS, Gupta PK, Newton RJ (eds) Somatic embryogenesis in woody plants, vol 2. Kluwer, Dordrecht, pp 375–407

Vieitez FJ (1999) Mass balance of a long-term somatic embryo cultures of chestnut. In: Espinel S, Ritter E (eds) Proc Application of Biotechnology to Forest Genetics. BIOFOR-99 Vitoria-Gasteiz, Spain, pp 199–211

Vieitez FJ, Merkle SA (2005) *Castanea* spp. Chestnut. In: Litz RE (ed) Biotechnology of fruit and nut crops. CAB International, Wallingford, UK, pp 265–296

Vieitez FJ, San-José MC, Ballester A, Vieitez AM (1990) Somatic embryogenesis in cultured immature zygotic embryos in chestnut. J Plant Physiol 136:253–256

Wilhelm E (2000) Somatic embryogenesis in oak (*Quercus* spp.) In Vitro Cell Dev Biol-Plant 36:349–357

Xing Z, Satchwell MF, Powell WA, Maynard CA (1997) Micropropagation of American chestnut: increasing rooting rate and preventing shoot tip necrosis. In Vitro Cell Dev Biol-Plant 33:43–48

Xing Z, Powell WA, Maynard CA (1999) Development and germination of American chestnut somatic embryos. Plant Cell Tiss Org Cult 57:47–55

Zegzouti R, Arnould M-F, Favre J-M (2001) Histological investigation of the multiplication step in secondary somatic embryogenesis of *Quercus robur* L. Ann For Sci 58:681–690

Plant Cell Monogr (2)
A. Mujib · J. Šamaj: Somatic Embryogenesis
DOI 10.1007/7089_033/Published online: 10 January 2006

Somatic Embryogenesis in *Cryptomeria japonica* D. Don: Gene for Phytosulfokine (PSK) Precursor

T. Igasaki (✉) · N. Akashi · K. Shinohara

Department of Molecular and Cell Biology, Forestry and Forest Products Research Institute (FFPRI), P.O.Box 16, 305-8687 Tsukuba, Japan
iga@ffpri.affrc.go.jp

Abstract Genetic transformation requires a reproducible system for the regeneration of plants via somatic embryogenesis or organogenesis. We established a reproducible system of plant regeneration based on somatic embryogenesis in *Cryptomeria japonica* D. Don. The developmental stage of the zygotic embryos was critical in the induction of embryogenic tissue. Embryogenic tissues that proliferated in liquid medium included small and loosely packed cells and elongating or elongated cells. Phytosulfokine, which has been identified as a plant growth factor, had a dramatic stimulatory effect on the formation of somatic embryos of *C. japonica* in the presence of polyethylene glycol. Induced somatic embryos germinated with synchronous sprouting of cotyledons, hypocotyls and roots, and most of the seedlings grew normally. This system of somatic embryogenesis in *C. japonica* should allow the genetic engineering of transgenic *C. japonica* with allergen-free pollen grains.

1 Introduction

Genetic engineering has the potential to allow the selective improvement of individual traits in forest trees without the loss of any of the desired traits of the parental lines. Using such techniques, we can overcome the difficulties associated with the breeding of long-lived perennials, where the production of progeny takes a long time. The genetic transformation of conifers by both microprojectile bombardment (Ellis et al. 1993; Charest et al. 1996; Klimaszewski et al. 1997; Walter et al. 1998) and by *Agrobacterium* (Shin et al. 1994; Tzfira et al. 1996; Levée et al. 1997; Wenck et al. 1999) has been reported. However, many difficulties have been encountered in attempts to regenerate transgenic woody plants and, in many cases, appropriate regeneration systems have not yet been established.

Sugi, *Cryptomeria japonica* D. Don (Taxodiaceae), is one of the most commercially important conifers in Japan. However, sugi pollinosis is one of the most serious allergic diseases in Japan. We are interested in the genetic engineering of transgenic *C. japonica* that produces allergen-free pollen grains. Recently, we established a simple and reliable procedure for introducing DNA into mature zygotic embryos of three species of Japanese conifer, including

C. japonica (Mohri et al. 2000), and a system for the reproducible regeneration of plants via somatic embryogenesis (Igasaki et al. 2003b). However, appropriate techniques were necessary to enhance the efficiency of plant regeneration. We found that the addition of synthesized phytosulfokine (PSK) to both the medium used for proliferation and that used for embryogenesis has a dramatic stimulatory effect on the formation of somatic embryos (Igasaki et al. 2003a). In this paper, we introduce a simple, reliable and highly efficient procedure for somatic embryogenesis and regeneration of *C. japonica.*

1.1 Induction and Maintenance of Embryogenic Cells

Fifteen specimens of *C. japonica* that had been planted in an experimental field of the Forestry and Forest Products Research Institute (FFPRI) were used as sources of material. Flowering of *C. japonica* was induced by treatment with 290 μM gibberellin A_3 (GA_3) once at the end of July and once at the beginning of August (Nagao et al. 1989). In the following year of GA_3 treatment, seeds were collected after open pollination and sterilized with sodium hypochlorite and ethanol. After seed coats had been removed, the megagametophytes that contained intact immature zygotic embryos or isolated immature zygotic embryos were used for the induction of embryogenic cells.

To determine the optimal developmental stage of immature zygotic embryos of *C. japonica* for the induction of embryogenic cells, we collected immature seeds at approximately weekly intervals from the middle of June to the beginning of August (Fig. 1). We examined more than 800 explants at each stage. We counted the number of induced tissues (Fig. 2a) that had characteristics similar to embryogenic tissues of loblolly pine (Gupta and Durzan 1987). Immature zygotic embryos that were collected from the end of June to the beginning of July yielded a higher frequency (5%) more inducion of embryogenic tissue than in the other samples (Fig. 1). These embryos corresponded to the early embryos before the formation of cotyledons (Yokoyama 1975). The average lengths of immature zygotic embryos at these stages ranged from 0.5 to 1.0 mm. However, the frequency of induction varied slightly among sampling years and among mother trees.

We also determined the optimal medium for the induction of embryogenic tissues. We tested media derived from three basal media, namely MSG (Becwar et al. 1988), GP (Gupta and Pullman 1991) and EMM (Smith 1996), supplementing them with 2,4-dichlorophenoxyacetic acid (2,4-D) and N^6-benzyladenine (BA) at various concentrations (Table 1). We found that solid medium (SMSG medium) which contained MSG basal salts and vitamins, 0.01% (w/v) myo-inositol, 0.15% (w/v) glutamine, 3.2 μM 2,4-D, 1.8 μM BA and 3% (w/v) sucrose, supplemented with 0.4% (w/v) gellan gum, gave the highest frequency of induction.

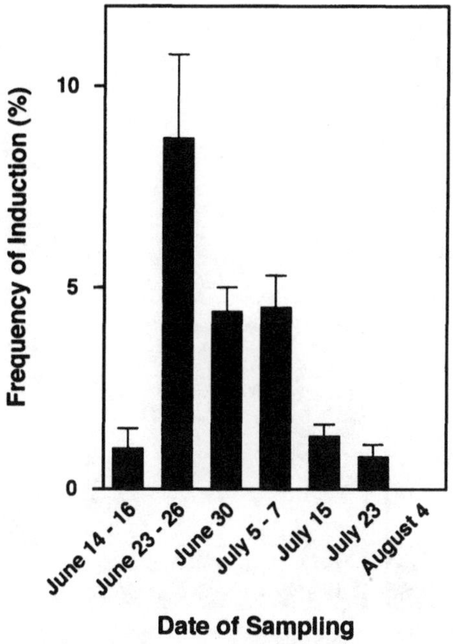

Fig. 1 Frequency of induction of embryogenic tissues from immature zygotic embryos of *C. japonica* that were harvested from the middle of June to the beginning of August. Average frequencies were determined from an analysis of more than 800 immature zygotic embryos from four mother trees at each stage for four years (1998–2001). Values are means ± S.E. of results

Table 1 Frequency of induction of embryogenic tissues on media with various concentrations of 2,4-D and BA

Basal medium	2,4-D (μM)	BA (μM)	Frequency of induction (%)[a]
MSG[b]	1.8	1.8	8.7 ± 1.1
	3.2	1.8	10.6 ± 1.0
	3.2	3.2	3.1 ± 0.3
	10.0	1.8	5.1 ± 1.8
	10.0	3.2	4.3 ± 1.2
GP[c]	3.2	3.2	1.9 ± 0.5
	10.0	3.2	1.7 ± 0.4
EMM[d]	3.2	3.2	0.5 ± 0.2
	10.0	3.2	1.1 ± 0.2

[a] Values are means ± S.E. of results ($n = 48$ to 96);
[b] Becwar et al. (1988);
[c] Gupta and Pullman (1991);
[d] Smith (1996)

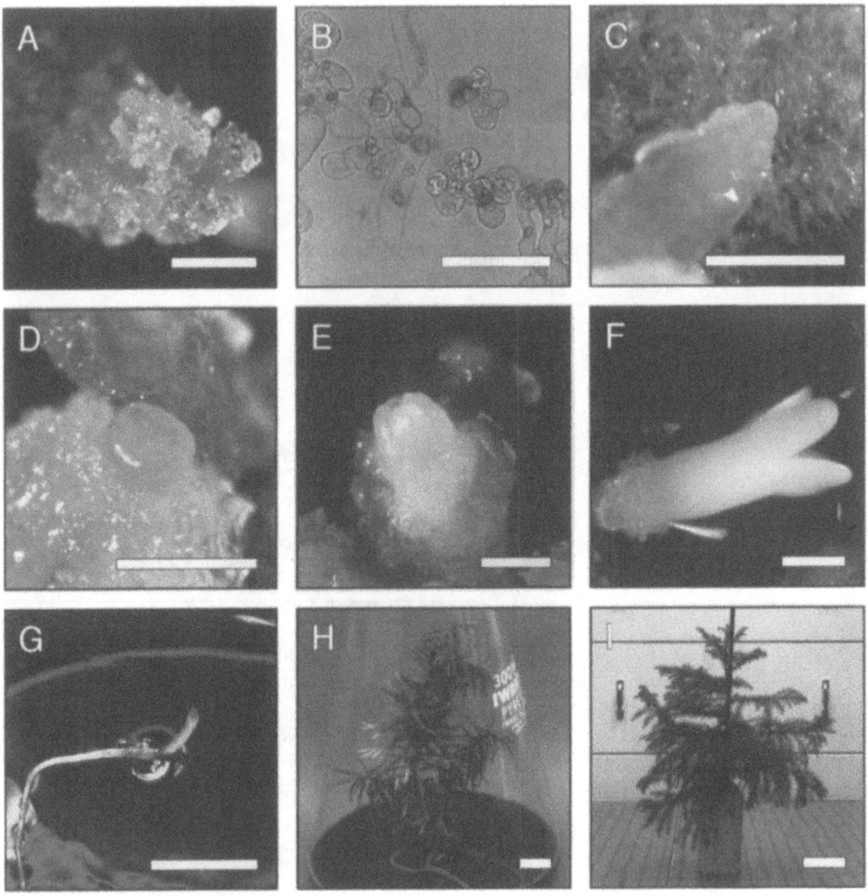

Fig. 2 Somatic embryogenesis of *C. japonica*. **A** Induced embryogenic tissue. **B** Cells proliferating in liquid medium. **C** to **F** Somatic embryos at various stages of development. **G** Germination of somatic embryos. **H** and **I** Growth of seedlings in vitro. *Bars*: 250 μm (**B**), 1 mm (**A**)and (**C**) to (**F**), 1 cm (**G**) or 10 cm (**H**)

1.2 Proliferation of Embryogenic Cells

For proliferation, we transferred the embryogenic tissues to liquid medium (LMSG: SMSG medium without solidification; 10 ml in 50-ml flasks) and cultured them on a rotary shaker operated at 110 rpm, in darkness, at 25 °C. The fresh weight of cells in LMSG medium increased 8- to 10-fold during culture for two weeks, and proliferated cells were subcultured at two-week intervals in the same medium. When we examined subcultured cells by light microscopy, we observed small loosely packed cells and some elongating or elongated cells (Fig. 2b) but no typical embryogenic cell clusters with large, dense embryonic regions and long suspensor cells (Gupta and Durzan 1987). The cells in these suspension cultures were able to form mature somatic embryos, and the ratio of globular cells to elongating or elongated cells was approximately 2 to 1 in these cultures (Fig. 2b). Moreover, cell lines that yielded cultures of rather typical cell clusters (Gupta and Durzan 1987; Maruyama et al. 2000) never produced mature somatic embryos. Our results are partially consistent with previous findings for *Picea abies* by Bellarosa et al. (1992), who found that both small loosely packed cells and embryogenic cell clusters can produce mature somatic embryos. Embryogenic tissues subcultured on SMSG medium and cell lines subcultured in LMSG medium have maintained their ability to differentiate mature somatic embryos for approximately two years.

2 Somatic Embryogenesis

Using ten cell lines that were induced from different immature embryos (Igasaki et al. 2003b), we determined the optimal medium for the development of somatic embryos by testing several media derived from basal media, namely MSG (Becwar et al. 1988) and EMM (Smith 1996). After proliferation, cells were collected on a cell strainer with 100-μm pores (Falcon 2360; Becton Dickinson Labware, NJ, USA) and rinsed twice with a liquid medium. Approximately 1×10^5 to 2×10^5 cells in 1 ml of liquid medium were plated on filter paper disks (Advantec no. 2, 70 mm in diameter; Toyo Roshi Kaisha, Ltd, Tokyo, Japan) on 90 × 20 mm petri dishes that contained liquid medium supplemented with 0.2% (w/v) activated charcoal and solidified with 0.3% (w/v) gellan gum (50 ml per petri dish). Petri dishes were sealed with Parafilm "M" (American National Can Co., Chicago, IL, USA) and incubated in darkness at 24 °C/16 °C (day/night; 12 h/12 h) for four to eight weeks.

We found that SEMM medium derived from EMM (Smith 1996), which contained 1431 mg/l KNO_3, 310 mg/l $NaNO_3$, 25 mg/l $CaCl_2 \cdot 2H_2O$, 0.2 mg/l $CoCl_2 \cdot 6H_2O$, 400 mg/l $MgSO_4 \cdot 7H_2O$, 27.3 mg/l $MnSO_4 \cdot H_2O$, 25 mg/l $ZnSO_4 \cdot 7H_2O$, 2.4 mg/l $CuSO_4 \cdot 5H_2O$, 30 mg/l $FeSO_4 \cdot 7H_2O$, 40 mg/l Na_2-EDTA, 225 mg/l $NH_4H_2PO_4$, 1 mg/l KI, 8 mg/l H_3BO_3, 0.2 mg/l $Na_2MoO_4 \cdot 2H_2O$, 0.5 mg/l pyridoxine-HCl, 5 mg/l thiamine-HCl, 5 mg/l nicotinic acid, 1000 mg/l myo-inositol, 7300 mg/l, glutamine, 2100 mg/l asparagine, 700 mg/l argnine, 79 mg/l citrulline, 76 mg/l ornithine, 55 mg/l lysine, 40 mg/l alanine, 35 mg/l proline, 5% (w/v) polyethylene glycol 4000 (PEG), 3% (w/v) maltose and 100 μM abscisic acid, allowed highly efficient formation of mature somatic embryos (Igasaki et al. 2003b). The presence of amino acids, activated charcoal and PEG in the SEMM medium was essential for the formation of mature somatic embryos. The addition of 5% (w/v) PEG stimulated the formation of embryos, but at concentrations above 5% (w/v)

PEG was no more effective than it was at 5% (w/v). This result is somewhat inconsistent with reported results for Sawara cypress (Maruyama et al. 2002).

We also examined other culture conditions, such as numbers of cells plated on the medium, temperature conditions, and petri dish seals. The number of embryos formed increased when there were up to 10^5 cells per petri dish, but decreased at more than 10^6 cells per petri dish. Similar numbers of embryos formed at 26 °C/18 °C (day/night: 12 h/12 h), 24 °C/16 °C (day/night: 12 h/12 h) and 22 °C, but no embryos appeared at 26 °C. By contrast, non-embryogenic tissues proliferated at 26 °C/18 °C but not at 24 °C/16 °C or at 22 °C. Parafilm "M", with its low air permeability, was a much better petri dish seal than surgical tape (21N ; Nichiban Co. Ltd., Tokyo, Japan), which had higher air permeability.

Under optimal conditions, as identified and described above, somatic embryos at early to mature stages were observed (Figs. 2c to 2f), and mature embryos (Fig. 2f) were obtained after about four weeks. The potential for development of somatic embryos varied among the cell lines in the suspension cultures, and embryos did not appear in all of the petri dishes (Igasaki et al. 2003b).

2.1 Germination and Plant Regeneration

Somatic embryos were collected from the SEMM medium and transferred to Smith's germination medium (Smith 1996) supplemented with 0.2% (w/v) activated charcoal and 10 μM GA_3. Cultures were kept in darkness at 24 °C/16 °C (day/night; 12 h/12 h). After germination, the plantlets were transferred to the same medium without GA_3 and maintained at 25 °C under cool white fluorescent light (30 $\mu mol\,m^{-2}\,s^{-1}$, 16-h photoperiod) for regeneration of plantlets. Upon germination, somatic embryos sprouted cotyledons, hypocotyls and roots synchronously (Fig. 2g). The presence of GA_3 in the germination medium did not affect the frequency of germination of somatic embryos, but GA_3 had a positive effect on the elongation of hypocotyls (Fig. 2g) and on the survival of seedlings. The frequency of germination differed among the various cell lines (Igasaki et al. 2003b). Most of the germinated seedlings developed normally (Figs. 2h and 2i).

2.2 Effects of PSK on the Maintenance of Embryogenic Cells

PSK, a small sulfated peptide (Fig. 4a), acts as an extracellular ligand in the initial steps of cellular dedifferentiation, proliferation and redifferentiation. PSK has been found in both monocotyledonous and dicotyledonous plants, for example *Asparagus officinalis*, *Oryza sativa*, *Daucus carota* and *Arabidopsis thaliana* (Matsubayashi and Sakagami 1996; Matsubayashi et al.

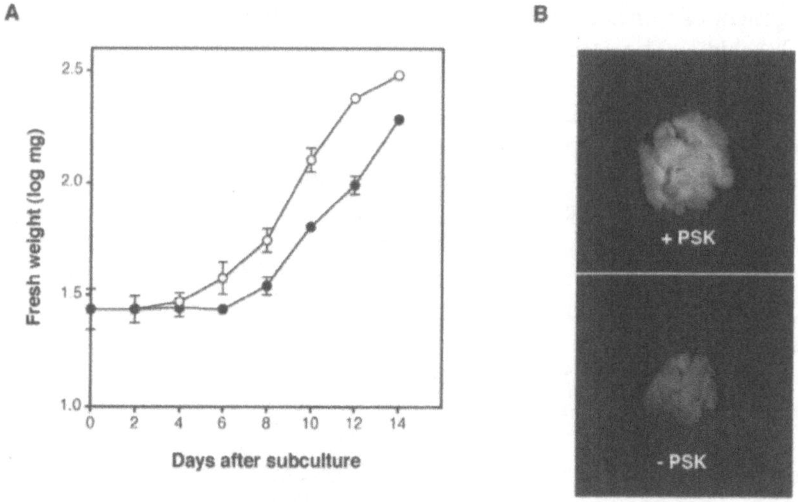

Fig. 3 Effects of PSK on the proliferation and maintenance of embryogenic cells of *C. japonica*. **a** Growth proflles of embryogenic cells with 32 nM PSK (*open circles*) or without PSK (*closed circle*) during culture for two weeks. The fresh weight of embryogenic cells collected from 10 ml of suspension culture was determined. Values are means ± SE of results from five replicates. The *symbols without bars* indicate that the SEs lie within the symbols. **b** Two week-cultured embryogenic cells in LEMM medium with 32 nM PSK (*upper*) or without PSK (*lower*). Cell line L-6, which had been subcultured for more than two years in the presence of 32 nM PSK, was used

1996; 1997; Yang et al. 1999; 2000; Hanai et al. 2000; Yang et al. 2001). PSK has also been shown to stimulate somatic embryogenesis in carrot (Kobayashi et al. 1999). Therefore, we examined the effects of PSK on somatic embryogenesis in *C. japonica* (Igasaki et al. 2003a).

We used ten lines of embryogenic cells whose ability to produce somatic embryos had been confirmed (Igasaki et al. 2003b). In many cases, embryogenic cells that had been maintained in LMSG medium lost the capacity to proliferate and to regenerate and they often turned brown during repeated subculture. However, the addition of PSK at 32 nM to the medium maintained both the capacity to proliferate and regenerate and the freshness (bright yellow color) of embryogenic cells for more than five years (Figs. 3 a and 3b). This observation suggests that PSK might play an important role in the maintenance of the capacity for cell division and the juvenility of embryogenic cells.

2.3
Effects of PSK on Somatic Embryogenesis

PEG has a stimulatory effect on the formation of somatic embryos of *Picea glauca*, *Chamaecyparis pisifera* and *C. japonica* (Attree et al. 1995; Maruyama et al. 2002; Igasaki et al. 2003b). We examined the effects of PSK on the development of somatic embryos of *C. japonica* in the presence and absence of PEG (Table 2). The addition of either PSK or PEG could increase the efficiency of formation of somatic embryos. Moreover, the addition of both PSK and PEG had a dramatic stimulatory effect on the formation of somatic embryos. Furthermore, while cell line L-5 never produced somatic embryos in the absence of PSK, it formed the embryos in the presence of PSK (Table 2). It seems likely that embryogenic cells of *C. japonica* produce a smaller amount of active PSK compared to those of *P. grauca* and *C. pisifera*, from which one can easily induce somatic embryogenesis through the addition of only PEG.

We observed and obtained embryos at early to mature stages after about four weeks (Figs. 2c to 2e). The time required for the generation of somatic embryos was unaffected by PSK (Igasaki et al. 2003b). The optimal concentration of PSK for the formation of somatic embryos was 32 nM. PSK produced an obvious effect even at 1 nM, but at levels above 32 nM PSK was no more effective than it was at 32 nM. Our results are almost consistent with the previous findings of *D. carota* by Kobayashi et al. (1999), who found that 100 nM PSK was most effective for somatic embryogenesis in *D. carota*, but they did not test at 32 nM. Approximately 80% of the induced somatic embryos germinated, with synchronous sprouting of cotyledons, hypocotyls and roots, and the germinated seedlings grew normally (Figs. 2g to 2i). Thus, PSK clearly had a positive effect on the development of somatic embryos, and our results suggest that the PSK signaling pathway, previously identified by Matsubayashi et al. (2002) in angiosperms, is also operative in *C. japonica*.

Table 2 Effects of PSK on the frequency of formation of embryos, in the presence and absence of PEG, in ten lines of embryogenic cells

Cell line	Number of embryos per petri dish			
	+ PEG		– PEG	
	+ PSK	– PSK	+ PSK	– PSK
L-1	26.3 ± 4.7	15.4 ± 2.8	4.6 ± 1.0	3.7 ± 1.5
L-2	2.3 ± 0.3	1.1 ± 0.5	0.3 ± 0.2	0.1 ± 0.1
L-3	4.2 ± 1.9	0.8 ± 0.4	0.3 ± 0.1	0.5 ± 0.2
L-4	32.3 ± 6.4	16.3 ± 5.1	0.7 ± 0.3	0
L-5	8.3 ± 3.9	0	4.6 ± 1.5	0
L-6	11.3 ± 2.0	1.5 ± 0.5	1.2 ± 0.3	0
L-7	0.8 ± 0.1	0.4 ± 0.1	0.3 ± 0.1	0
L-8	1.3 ± 0.2	0.8 ± 0.3	0.7 ± 0.2	0
L-9	6.7 ± 1.4	1.9 ± 0.4	2.9 ± 0.6	0
L-10	5.5 ± 1.9	1.1 ± 0.4	1.9 ± 0.9	0.1 ± 0.1

Values are means ± SE of results ($n = 3$ to 6).
See text for full details

3 Gene for PSK Precursor in *C. japonica*

To examine the presence and expression of a gene for the precursor to PSK in *C. japonica*, we surveyed a database of *C. japonica* expressed sequence tags (EST) using the amino acid sequence of PSK (YIYTQ). We found one EST clone, CC4124 (accession no. AB105536) that encoded PSK within a putative open reading frame (ORF), and determined the complete sequence of its coding region (Igasaki et al. 2003a). The ORF that we identified is 306 bp long and encodes 102 amino acids (Fig. 4b). Application of the rules proposed by von Heijne (1986) allowed us to predict that the ORF encodes an amino-terminal hydrophobic signal sequence of 28 amino acids. The predicted polypeptide includes the sequence YIYTQ at amino acid positions 93 through 97 and a conserved Asp residue at position 92 (Figs. 4b and 4c). These three features are conserved in other precursors to PSK in angiosperms (Yang et al. 1999; 2000). Thus, a gene for the precursor to PSK is present and expressed in *C. japonica*, supporting the hypothesis that a PSK signaling pathway exists in this conifer.

4 Conclusion

We established a simple and reliable procedure for somatic embryogenesis and regeneration of *C. japonica* with high efficiency. To our knowledge, this is the first report of a reproducible system for the regeneration of *C. japonica*. PSK had positive effects on both the proliferation and maintenance of embryogenic cells and on the formation of somatic embryos of *C. japonica* (Fig. 3, Table 2). We also found evidence that suggests that a PSK signaling pathway is present in a gymnosperm, as it is in angiosperms. Furthermore, the gene for a precursor to PSK was found in the genome of *C. japonica* (Fig. 4). Our findings allowed us to establish a simple and reliable procedure for somatic embryogenesis and the regeneration of *C. japonica*. In our system, embryogenic cells can be induced from various genotypes of *C. japonica*, and somatic embryos can be easily produced in any season by the addition of PSK. Our system also allowed us to repeat the induction of somatic embryos via embryogenic cells from newly induced somatic embryos. Such a system for the reproducible regeneration of plants from embryogenic callus is essential for the genetic transformation of *C. japonica*.

In previous studies, we established a simple and reliable procedure for the regeneration of transgenic Japanese broad-leaved trees (Mohri et al. 1996; Mohri et al. 1997; Mohri et al. 1999; Igasaki et al. 2000; Igasaki et al., 2002). However, to our knowledge, no studies of the transformation of Japanese coniferous species have been reported. Recently, we established an effective

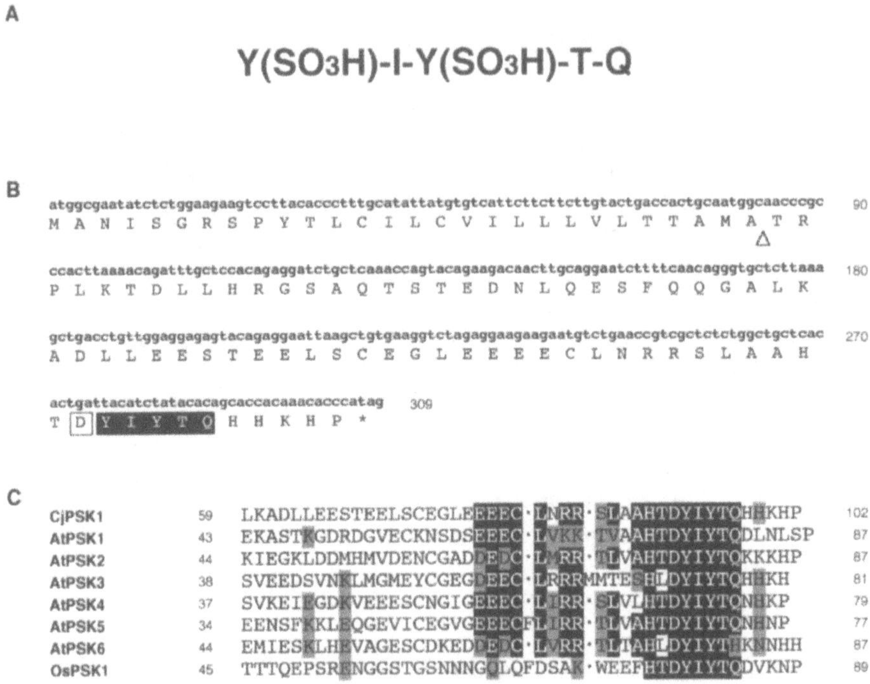

Fig. 4 Presence of a gene for the precursor to PSK in *C. japonica*. **a** Chemical structure of PSK. **b** Nucleotide sequence of the cDNA for the putative precursor to PSK of *C. japonica* (taken from the database, as indicated in the text) and the deduced amino acid sequence (CjPSK1). Amino acids in *black* and *white boxes* are those of the PSK peptide and the conserved aspartate residue, respectively. A putative processing site is indicated by the *open triangle*. **c** Alignment of the deduced amino acid sequence of CjPSK1 with other precursors to PSK. The amino acid sequence of CjPSK1 is compared with the deduced amino acid sequences of peptide precursors to PSK from *Arabidopsis thaliana* [AtPSK1, AGI (Arabidopsis Genome Initiative; http://www.arabidopsis.org) code At1g13590; AtPSK2, AGI code At2g22860; AtPSK3, AGI code At3g44735; AtPSK4, AGI code At3g49780; AtPSK5, AGI code At5g65870; AtPSK6, AGI code At4g37720], and *Oryza sativa* (OsPSK1, accession number AB020505). Amino acids in *black boxes* and in *gray boxes* are identical and similar, respectively, in at least six of the ten precursors to PSK. *Dots* indicate gaps introduced to maximize the extent of homology among sequences. The Arabic numerals in the sequences represent the positions of amino acid residues from the beginning of the signal peptides

procedure for the introduction of DNA into mature zygotic embryos of three species of Japanese conifer, including *C. japonica* (Mohri et al. 2000). In addition, we have also isolated genes for various allergens from *C. japonica* (Sone et al. 1994; Namba et al. 1994; Komiyama et al. 1994; Futamura et al. 2002; Kawamoto et al. 2002). Therefore, in the near future, the present system for the regeneration of *C. japonica* should permit the genetic engineering of transgenic *C. japonica* with allergen-free pollen grains.

Acknowledgements The authors express their gratitude to Dr. Yoshikatsu Matsubayashi of Nagoya University for the generous gift of the synthesized PSK used in this study. The authors are also grateful to Dr. Tokuko Ujino-Ihara of the FFPRI for useful information about *C. japonica* EST. This work was supported by a Grant-in-Aid from the Ministry of Agriculture, Forestry and Fisheries of Japan and, in part, by the Program for the Promotion of Basic Research Activities for Innovative Biosciences (PROBRAIN).

References

Attree SM, Pomeroy MK, Fowke LC (1995) Development of white spruce (*Picea glauca* (Moench.) Voss) somatic embryos during culture with abscisic acid and osmoticum, and their tolerance to drying and frozen storage. J Exp Bot 46:433–439

Becwar MR, Wann SR, Johnson MA, Verhagen SA, Feirer RP, Nagmani R (1988) Development and characterization of in vitro embryogenic systems in conifers. In: Ahuja MR (ed). Somatic Cell Genetics of Woody Plants. Kluwer Academic, Dordrecht, The Netherlands, pp 1–18

Bellarosa R, Mo LH, von Arnold S (1992) The influence of auxin and cytokinin on proliferation and morphology of somatic embryos of *Pices abies* (L.) Karst. Ann Bot 70:199–206

Charest PJ, Devantier Y, Lachance D (1996) Stable genetic transformation of *Picea mariana* (black spruce) via particle bombardment. In Vitro Cell Dev Biol Plant 32:91–99

Ellis DD, McCabe DE, Mclnnis S, Ramachandran R, Russell DR, Wallace KM, Martinell BJ, Robert DR, Raffa KF, McCown BH (1993) Stable transformation of *Picea glauca* by particle acceleration. Bio/Technology 11:84–89

Futamura N, Mukai Y, Sakaguchi M, Yasueda H, Inouye S, Midoro-Horiuchi T, Goldblum RM, Shinohara K (2002) Isolation and characterization of cDNAs that encode homologs of a pathogenesis-related protein allergen from *Cryptomeria japonica*. Biosci Biotechnol Biochem 66:2495–2500

Gupta PK, Durzan DJ (1987) Biotechnology of somatic polyembryogenesis and plantlet regeneration in loblolly pine. Bio/Technology 5:147–151

Gupta PK, Pullman GS (1991) Method for reproducing coniferous plants by somatic embryogenesis using abscisic acid and somatic potential variation. U.S. Patent no. 5,036,007

Hanai H, Matsuno T, Yamamoto M, Matsubayashi Y, Kobayashi T, Kamada H ,Sakagami Y (2000) A secreted peptide growth factor, phytosulfokine, acting as a stimulatory factor of carrot somatic embryo formation. Plant Cell Physiol 41:27–32

Igasaki T, Akashi N, Ujino-Ihara T, Matsubayashi Y, Sakagami Y, Shinohara K (2003a) Phytosulfokine stimulates somatic embryogenesis in *Cryptomeria japonica*. Plant Cell Physiol 44:1412–1416

Igasaki T, Ishida Y, Mohri T, Ichikawa H, Shinohara K (2002) Transformation of *Populus alba* and direct selection of transformants with the herbicide bialaphos. Bull FFPRI 1:235–240

Igasaki T, Mohri T, Ichikawa H, Shinohara K (2000) Agrobacterium tumefaciens-mediated transformation of *Robinia pseudoacacia*. Plant Cell Rep 19:448–453

Igasaki T, Sato T, Akashi N, Mohri T, Maruyama E, Kinoshita I, Walter C, Shinohara K (2003b) Somatic embryogenesis and plant regeneration from immature zygotic embryos of *Cryptomeria japonica* D. Don. Plant Cell Rep 22:239–243

Kawamoto S, Fujimura T, Nishida M, Tanaka T, Aki T, Masubuchi M, Hayashi T, Suzuki O, Shigeta S, Ono K (2002) Molecular cloning and characterization of a new Japanese cedar pollen allergen homologous to plant isoflavone reductase family. Clin Exp Allergy 32:1064–1070

Klimaszewska K, Devantier Y, Lachance D, Lelu MA, Charest PJ (1997) *Larix laricina* (tamarack): somatic embryogenesis and genetic transformation. Can J For Res 27:538–550

Kobayashi T, Eun C-H, Hanai H, Matsubayashi Y, Sakagami Y, Kamada H (1999) Phytosulphokine-a, a peptidyl plant growth factor, stimulates somatic embryogenesis in carrot. J Exp Bot 50:1123–1128

Komiyama N, Sone T, Shimizu K, Morikubo K, Kino K (1994) cDNA cloning and expression of Cry j II, the second major allergen of Japanese cedar pollen. Biochem Biophys Res Commun 201:1021–1028

Levée V, Lelu M-A, Jouanin L, Cornu D, Pilate G (1997) Agrobacterium tumefaciens-mediated transformation of hybrid larch (*Larix kaempferi* x L. decidua) and transgenic plant regeneration. Plant Cell Rep 16:680–685

Maruyama E, Hosoi Y, Ishii K (2002) Somatic embryogenesis in Sawara Cypress (*Chamaecyparis* pisifera Sieb. et Zucc.) for stable and efficient plant regeneration, propagation and protoplast culture. J For Res 7:23–34

Maruyama E, Tanaka T, Hosoi Y, Ishii K, Morohoshi N (2000) Embryogenic cell culture, protoplast regeneration, cryopreservation, biolistic gene transfer and plant regeneration in Japanese cedar (*Cryptomeria japonica* D. Don). Plant Biotech 17:281–296

Matsubayashi Y, Hanai H, Hara O, Sakagami Y (1996) Active fragments and analogs of the plant growth factor, phytosulfokine: structure-activity relationships. Biochem Biophys Res Commun 225:209–214

Matsubayashi Y, Ogawa M, Morita A, Sakagami Y (2002) An LRR receptor kinase involved in perception of a peptide plant hormone, phytosulfokine. Science 296:1470–1472

Matsubayashi Y, Sakagami Y (1996) Phytosulfokine, sulfated peptides that induce the proliferation of single mesophyll cells of *Asparagus officinalis* L. Proc Natl Acad Sci USA 93:7623–7627

Matsubayashi Y, Takagi L, Sakagami Y (1997) Phytosulfokine-α, a sulfated pentapeptide, stimulates the proliferation of rice cells by means of specific high- and low-affinity binding sites. Proc Natl Acad Sci USA 94:13357–13362

Mohri T, Igasaki T, Futamura N, Shinohara K (1999) Morphological changes in transgenic poplar by expression of the rice homeobox gene OSH1. Plant Cell Rep 18:816–819

Mohri T, Igasaki T, Sato T, Shinohara K (2000) Expression of genes for β-glucuronidase and luciferase in three species of Japanese conifer (*Pinus thunbergii*, *P. densiflora* and *Cryptomeria japonica*) after transfer of DNA by microprojectile bombardment. Plant Biotech 17:49–54

Mohri T, Mukai Y, Shinohara K (1997) Agrobacterium tumefaciens-mediated transformation of Japanese white birch (*Betula platyphylla* var. japonica). Plant Sci 123:53–60

Mohri T, Yamamoto N, Shinohara K (1996) Agrobacterium-mediated transformation of lombardy poplar (*Populus nigra* L. var. italica Koehne) using stem segments. J For Res 1:13–16

Nagao A, Sasaki S, Pharis RP (1989) *Cryptomeria japonica*, In: Halevy AH (ed), CRC Handbook of Flowering, vol VI. CRC, Boca Raton, FL, pp 247–269

Namba M, Kurose M, Torigoe K, Hino K, Taniguchi Y, Fukuda S, Usui M , Kurimoto M (1994) Molecular cloning of the second major allergen, Cry j II, from Japanese cedar pollen. FEBS Lett 353:124–128

Shin D-I, Podila GK, Huang YH, Karnosky DF (1994) Transgenic larch expressing genes for herbicide and insect resistance. Can J For Res 24:2059–2067

Smith DR (1996) Growth medium. U.S. Patent no. 5,565,355

Sone T, Komiyama N, Shimizu K, Kusakabe T, Morikubo K, Kino K (1994) Cloning and sequencing of cDNA coding for Cry j I, a major allergen of Japanese cedar pollen. Biochem Biophys Res Commun 199:619–625

Tzfira T, Yarnitzky O, Vainstein A, Altman A (1996) Agrobacterium rhizogenes-mediated DNA transfer in *Pinus halepensis* Mill. Plant Cell Rep 16:26–31

von Heijne G (1986) A new method for predicting signal sequence cleavage sites. Nucleic Acids Res 14:4683–4690

Walter C, Grace LJ, Wagner A, White DWR, Walden AR, Donaldson SS, Hinson H, Gardner RC, Smith DR (1998) Stable transformation and regeneration of transgenic plants of *Pinus radiata* D. Don. Plant Cell Rep 17:460–468

Wenck AR, Quinn M, Whetten RW, Pullman G, Sederoff R (1999) High-efficiency Agrobacterium-mediated transformation of Norway spruce (*Picea abies*) and loblolly pine (*Pinus taeda*). Plant Mol Biol 39:407–416

Yang H, Matsubayashi Y, Hanai H, Nakamura K, Sakagami Y (2000) Molecular cloning and characterization of OsPSK, a gene encoding a precursor for phytosulfokine-α, required for rice cell proliferation. Plant Mol Biol 635:635–647

Yang H, Matsubayashi Y, Nakamura K, Sakagami Y (1999) *Oryza sativa* PSK gene encodes a precursor of phytosulfokine-α, a sulfated peptide growth factor found in plants. Proc Natl Acad Sci USA 96:13560–13565

Yang H, Matsubayashi Y, Nakamura K, Sakagami Y (2001) Diversity of Arabidopsis genes encoding precursors for phytosulfokine, a peptide growth factor. Plant Physiol 127:842–851

Yokoyama T (1975) Embryogenesis and cone growth in *Cryptomeria japonica*. Bull Gov For Exp Sta 277:1–20

Plant Cell Monogr (2)
A. Mujib · J. Šamaj: Somatic Embryogenesis
DOI 10.1007/7089_029/Published online: 30 November 2005

Protein Markers for Somatic Embryogenesis

Magdalena I. Tchorbadjieva

Department of Biochemistry, Faculty of Biology, Sofia University, 8 Dragan Zankov str., 1164 Sofia, Bulgaria
magd@biofac.uni-sofia.bg

Abstract The capacity for somatic embryogenesis is a remarkable property of plant cells. Somatic embryogenesis is the process by which somatic cells develop into plants through characteristic morphological changes, thus rendering it a good model system for studying early plant development. Most of the important crops and grasses are recalcitrant for in vitro culturing, which hampers the development of reliable regeneration techniques. Better understanding of the fundamental processes that trigger and control somatic embryogenesis will lead to more rational regeneration protocols. The characterization and functional analysis of protein markers for somatic embryogenesis offer the possibility of determining the embryogenic potential of plant cells in culture long before any morphological changes have taken place, and of gaining further information on the molecular basis of induction and differentiation of plant cells.

The present review aims to summarize recent work that employs a variety of experimental approaches for the identification and use of protein markers for somatic embryogenesis in different species. The role of extracellular proteins as markers for somatic embryogenesis is especially emphasized.

1 Introduction

Somatic embryogenesis is a remarkable biological phenomenon. It is an ideal system for investigating the entire process of differentiation in plants, as well as of the mechanisms of expression of totipotency in plant cells. The three steps of embryogenesis from somatic cells, which comprise (a) induction of cell division, (b) induction of embryogenic potential, and (c) expression of the embryogenic program, include reprogramming of the gene expression pattern of the cells. The molecular basis of this unique developmental pathway, particularly the transition of somatic cells into embryogenic ones, is still the least understood (for a review, see Fehér et al. 2003). Markers for somatic embryogenesis help to establish embryogenic potential in plant cells for obtaining reasonable regeneration frequencies and provide information on the molecular mechanisms of plant cell differentiation. Different experimental approaches have been applied to isolate and characterize markers for somatic embryogenesis. In most cases, the comparative analysis of the total protein patterns from embryogenic and nonembryogenic cells resulted in a large number of specific proteins, making it difficult to use them as markers (Hahne et al. 1988; Hilbert et al. 1992). The observation that the extracellular proteins are indispensable for differentiation and morphogenesis of plant cells, as well as their limited number when compared to the whole protein pattern, makes them appropriate candidates as markers for somatic embryogenesis. Indeed, many extracellular protein markers for embryogenic potential have been described (Sterk et al. 1991; De Jong et al. 1992; Kreuger and Van Holst 1993; Egertsdotter and Von Arnold 1995; Domon et al. 2000). Monoclonal antibodies against marker proteins have been useful in elucidating the complex structure of the plant cell surface, as well as for marking cells destined to develop somatic embryos (Toonen et al. 1996; Knox 1997; McCabe et al. 1997).

Differential cDNA screening has been widely applied to identify and characterize embryogenic markers (Schmidt et al. 1997; McCabe et al. 1997; Chugh and Khurana 2002). Differential display has been successfully used to isolate low-abundant genes (Alexandrova and Conger 2002; Yamazaki and Saito 2002; Charbit et al. 2004).

In this review, data are presented on the identification and use of early markers for somatic embryogenesis in different species by applying various experimental approaches.

2 Comparative Analysis of Proteins

2.1 Comparison of Protein Patterns after One- and/or Two-Dimensional Gel Electrophoresis

Biochemical aspects of the induction phase of somatic embryogenesis have so far been investigated at the protein level in many species. The first studies on carrot were reported by Sung and Okimoto (1981) who evidenced two 77- and 43-kD embryo-specific proteins. Similar studies performed on rice revealed the presence of several polypeptides in the range of 40 to 44 kD, which were more abundant in embryogenic calli than in nonembryogenic calli (Chen and Luthe 1987). The detection of embryogenesis-related proteins from total protein extracts has been reported for *Cichorium intybus* (Hilbert et al. 1992), *Dactylis glomerata* L. (Hahne et al. 1988), and *Cupressus sempervirens* (Sallandrouze et al. 1999). The analysis of total protein extracts from embryogenic versus nonembryogenic primary explants of the same origin allowed Pedroso et al. (1995a) to detect two polypeptides E_1 and E_2, specifically related to the process of proembryo induction and globular embryo development of *Camelia japonica*. Fellers et al. (1997) identified two proteins with 43 kD/pI 7.6 and 27 kD/pI 8.2 that can be used as markers for embryogenic potential in wheat callus. Blanco et al. (1997) found a marker protein for the regeneration potential of sugarcane embryogenic callus. Hvoslef-Eide and Corke (1997) detected proteins specific for embryogenic cultures of birch. An investigation of total protein expression using two-dimensional gel electrophoresis during the ontogeny of carrot somatic embryogenesis enabled Dodeman and Ducreux (1996b) to identify markers of the induction phase and different developmental stages.

2.2 Comparison of Isoenzyme Patterns

The development of cells into embryogenic cell clusters and afterward into somatic embryos is accompanied by specific changes in protein pattern: new proteins are synthesized, others decrease and disappear. Changes in isozyme patterns have proved to be an efficient tool for analyzing the different stages in somatic embryogenesis. Isozyme expression is part of the controlled functional program involved both in acquisition of embryogenic potency and in the subsequent differentiation of the embryo. It has been shown earlier that isozyme responses vary with tissue organization during development and differentiation. Coppens and Dewitte (1990) found the esterase system to be very sensitive for the detection of embryogenesis in barley callus before somatic embryos are formed. In carrot, Chibbar et al. (1988) were able to detect two esterase isoenzyme systems differentially expressed in embryogenic and nonembryogenic cells. Esterase and peroxidase were found to be appropriate to discriminate between embryogenic and nonembryogenic callus in sweet potato (Cavalcante et al. 1994). Bapat et al. (1992) found several enzyme isoforms that discriminate between wheat embryogenic calli with regeneration potential and nonembryogenic calli that remain unorganized. A comparative analysis of ten somatic embryogenesis stages of carrot using a seven-enzyme system did not evidence any somatic embryogenesis-specific isozyme (Dodeman and Ducreux 1996a). Still other data indicate the potential of some enzymes to function as stage-specific markers for somatic embryogenesis. According to Bagnoli et al. (1998), the antioxidant enzymes superoxide dismutase and catalase could be convenient markers for defining the developmental stages in *Aesculus hippocastanum* somatic and zygotic embryos. The same role was postulated for peroxidase, whose isoenzyme patterns were shown to reflect the embryogenic potential of *Medicago sativa* (Hrubcová et al. 1994). The analysis of the electrophoretic patterns of specific enzymes proved to be an effective approach to the characterization of the main steps of *Vitis rupestris* somatic embryogenesis (Martinelli et al. 1993).

3 Antibodies Against Marker Proteins

3.1 Monoclonal Antibodies

Somatic embryogenesis involves a set of molecular events, both differential gene expression and various signal transduction pathways, for activating and/or repressing numerous sets of genes (Chugh and Khurana 2002). Studies on gene expression have revealed that embryo-specific genes are low-abundant genes and difficult to isolate. Differential hybridoma screening for the selection of monoclonal antibodies against marker proteins for somatic embryogenesis is more sensitive than two-dimensional gel electrophoresis, giving a chance of detecting low-abundant proteins. Antibodies are produced which may be used to monitor marker protein expression in different tissues and species. Smith et al. (1988) described a monoclonal antibody designated 21D7 that reacted with a nuclear protein associated with cell division in carrot somatic embryogenesis. Fukuda et al. (1994) proved that the 21D7 protein could be a candidate as an early marker of totipotency when cells start to divide and a competent cell becomes an embryogenic one. Kiyosue et al. (1990) generated a monoclonal antibody 1D11 against a 31-kD glycoprotein expressed in embryogenic cells but not in somatic embryos or nonembryogenic cells, and proposed that it should be a useful marker of embryogenic competence. Altherr et al. (1993) selected a monoclonal antibody 7C5 directed against a putative non-histone protein in *Pisum sativum* L. The acidic 50-kD protein was detected in other species, both dicots and monocots, and could serve as a marker for embryogenic potential. Monoclonal antibodies have been selected against germins (Lane et al. 1993). These proteins are associated with the cell wall and are one of the best-characterized markers for somatic embryogenesis in cereals.

The surface of plant cells includes the outer side of the plasma membrane, cell wall, middle lamella, and intercellular spaces. The monoclonal antibodies prepared against different components of the plant cell wall and extracellular proteins from the culture medium are useful molecular probes for studying the complex organization and dynamics of interaction between single components of the cell wall as a part of the plant extracellular matrix (Knox 1997, 1999; Smallwood et al. 1995, 1996; Willats et al. 2000). Arabinogalactan proteins (AGPs) are a class of proteoglycans implicated in diverse processes of plant growth and development, including somatic embryogenesis (for a review, see Showalter 2001). Presumably, AGPs are involved in molecular interactions and cellular signaling at the cell surface. Several antibodies have been prepared against diverse AGPs and were used to mark specific cell types (for reviews, see Knox 1997; Willats et al. 2000). A JIM4 antibody recognizing AGP epitopes in the protoderm of proembryogenic masses (PEMs) and the culture medium of *Daucus carota* suspension cultures has been described (Stacey et al. 1990). Immunofluorescence using monoclonal antibody JIM4 has shown that the extracellular matrix surface network that covers the surface of embryogenic cells in friable maize callus is equipped with JIM4 epitope, while nonembryogenic callus cells are devoid of this epitope. Thus, JIM4 antibody can serve as an early marker of embryogenic competence in maize callus cultures (Samaj et al. 1999). The epitope of monoclonal antibody JIM13 is localized in epidermal cells (Knox et al. 1991), and Filonova et al. (2000) used it to distinguish PEMs from somatic embryos in *Picea abies*. JIM16 antibody recognized AGPs localized in the cell wall of peripheral cells of globular embryos and the culture medium and can be used as a marker for somatic embryogenesis in *Cichorium* (Chapman et al. 2000). ZUM18 recognizes AGPs with stimulatory effect on somatic embryogenesis in carrot (Kreuger and Van Holst 1995). Tchorbadjieva et al. (1998) isolated a monoclonal antibody 1D1, which recognizes two extracellular proteins from *D. glomerata* L. suspension cultures. The monoclonal antibodies against a range of polysaccharides and proteoglycan epitopes have been very useful in providing markers of developmental state and developmental potential. They have also helped to provide insight into aspects of cell-derived developmental signals (McCabe et al. 1997; Pennell 1998).

3.2 Phage Display Antibodies

Antibody technology has advanced in line with the development of molecular biological techniques. With the advent of phage display antibody technology there has been an extension of cell-based methods of generating monoclonal antibodies to gene-based methods (Winter et al. 1994). Phage antibody production is rapid and requires only very small amounts of antigen compared to hybridoma technology (Willats et al. 2000). A phage display monoclonal antibody PAM1 with specificity for de-esterified blocks of pectic homogalacturonan (HG) has been described (Willats et al. 1999a). In an intact cluster of suspension-cultured cells of *Arabidopsis thaliana* the PAM1 epitope is restricted to regions of cell-to-cell adhesion at the cell wall surface. A phage display antibody against the pectic component rhamnogalacturonan (RG) II has been isolated (Williams et al. 1996). Using a phage display subtraction method, Shinohara et al. (2000) were able to isolate monoclonal antibodies recognizing vascular development-specific cell wall components from *Zinnia* differentiating cells.

In conclusion, using both techniques, generation of monoclonal antibodies and phage display antibodies against components of the plant cell surface will provide further useful probes for studying the cell wall complexity and its structure–function relationships during somatic embryogenesis.

4 cDNA Differential Screening and Differential Display

4.1 cDNA Differential Screening

Many genes with altered expression during somatic embryogenesis have been identified; however, most of these are in late developmental stages (for reviews, see Chugh and Khurana 2002; Fehér et al. 2003). In the present review, only those experiments that aimed to isolate genes activated in the early stages of induction of somatic embryogenesis, with emphasis on their use as markers, will be described. Several different genes that are induced during somatic embryogenesis and are putative molecular markers have been isolated, typically by differential screening of cDNA libraries. These include genes encoding late embryogenesis abundant (LEA) proteins. The ECP31 transcripts were preferentially localized in the peripheral cells of embryogenic cells, and the authors suppose that ECP31 protein participates in the induction and/or maintenance of embryogenic competence (Kiyosue et al. 1992). *Emb-1* accumulates in the stage of maturation of somatic embryos (Wurtele et al. 1993). A cDNA clone for germin-like proteins (*PcGER1*) has been isolated whose transcripts are abundant in all embryogenic lines and absent from nonembryogenic lines of pine (Neutelings et al. 1998). They are localized in the walls of preglobular embryos and are markers for this early developmental stage. The approaches to identify genes activated during the early phases of chicory embryogenesis resulted in the identification of cDNAs of a β-1,3-glucanase (Helleboid et al. 1998).

The processes that govern the property of embryogenic competence in plant cells remain largely unknown (Mordhorst et al. 1997; Fehér et al. 2003). At present, there is only one gene known to play a role in the acquisition of embryogenic competence in plant cells. This is the somatic embryogenesis receptor kinase (*SERK*) gene (Schmidt et al. 1997). In carrot, *SERK* expression was shown to be characteristic of embryogenic cell cultures and somatic embryos whose expression ceased after the globular stage. Cell tracking experiments showed that *SERK*-expressing single cells could develop into somatic embryos; thus, *SERK* is considered to mark cells competent to form embryos in cell culture. The *Arabidopsis* homologue of the carrot *SERK* cDNA has also been cloned, and it was shown that the *AtSERK1* gene is highly expressed during embryogenic cell formation in culture and during early embryogenesis (Hecht et al. 2001). It was also established that the *AtSERK1* product is sufficient to confer embryogenic competence in culture. A carrot *SERK* homologue was shown to exist in embryos of *D. glomerata* L., and this gene can be used as a convenient marker to monitor embryogenic cell formation in monocots (Somleva et al. 2000). A *SERK* gene from *Medicago truncatula* (*MtSERK1*) has been isolated, orthologous to *AtSERK1*, which in legumes may have a broader role in morphogenesis in cultured tissue rather than being specific for somatic embryogenesis (Nolan et al. 2003).

4.2 Differential Display

Genes involved in early stages of somatic embryogenesis have very low expression (Heck et al. 1995). Therefore, an alternative cloning method was developed in place of differential screening or subtractive hybridization. The differential display (DD) was first reported by Liang and Pardee (1992). In the last ten years, DD has been actively applied for the isolation of various genes from plants (review, Yamazaki and Saito 2002). It also turned out to be very effective in the isolation of genes involved in very early stages of somatic embryogenesis (Yoshida et al. 1994; Momiyama et al. 1995; Linkiewicz et al. 2004). Alexandrova and Conger (2002) identified two somatic embryogenesis-related genes DGE1 and DGE2 that were expressed in embryogenic but not in nonembryogenic leaf cultures from *D. glomerata* L. with possible nuclear regulatory functions. Charbit et al. (2004) isolated five cDNAs that could be used to distinguish between calli prior to induction, thus enabling an early diagnosis of callus embryogenic potential. Transcripts unique to embryogenic cell clusters in *Coffea arabica* (Rojas-Herrera et al. 2002), in cell clusters at the earliest stages of carrot somatic embryogenesis (Yasuda et al. 2001), and in embryogenic calli of *Lycium barbarum* (Kairong et al. 1999) have been detected.

5 Extracellular Proteins as Markers for Somatic Embryogenesis

The molecular basis of the unique developmental pathway of somatic embryogenesis, particularly the transition of somatic cells into embryogenic ones, is still the least understood (for review, see Fehér et al. 2003). Somatic embryogenesis in cell suspension cultures provides an alternative way to address this problem. The growth medium of plant cell cultures may be regarded as a large extension of the intercellular space; soluble secreted molecules that inhabit the apoplast *in planta* will accumulate in the medium when cells are grown in suspension. Thus, the complex array of molecules mainly derived from cell walls reflects the growth and development of cultured cells (Mordhorst et al. 1997). This opens up the possibility of studying the role of these molecules in early plant development, as well as searching for early markers for somatic embryogenesis among the secreted molecules. Suspension cultures secrete into the medium glycoproteins that play an important role in somatic embryogenesis by their ability to stimulate (De Vries et al. 1988; Kreuger and Van Holst 1993; Toonen et al. 1997a; Egertsdotter and

Von Arnold 1998; Domon et al. 2000) or inhibit (Gavish et al. 1992; Maës et al. 1997) somatic embryo development.

Comparison of extracellular protein patterns after one-dimensional or two-dimensional (2-D) gel electrophoresis showed that some proteins specifically appeared in embryogenic but not in nonembryogenic cell lines (De Vries et al. 1988; Nielsen and Hansen 1992; Tchorbadjieva et al. 1992; Kreuger and Van Holst 1993). Besides, it has been shown that suspension cultures of *Digitalis lanata* (Reinbothe et al. 1992) and *Dactylis glomerata* L. (Tchorbadjieva et al. 2004) differentiating into somatic embryos secreted proteins into the growth medium in a stage-specific manner. Analysis of extracellular proteins with the aid of 2-D protein gels was used to distinguish between different stages of somatic embryogenesis, and to identify putative candidates of proteins as markers for somatic embryogenesis (Tchorbadjieva et al. 2004). Some of these proteins were identified as an acidic esterase (Tchorbadjieva and Odjakova 2001), acidic lipid transfer protein-like proteins (Tchorbadjieva 2001), and an acidic endochitinase (Tchorbadjieva and Pantchev, 2006). All of these extracellular proteins were detected in a very early stage of somatic embryogenesis in *D. glomerata* L. embryogenic suspension cultures only, and could be used as early markers of embryogenic potential. Esqueda et al. (1998) identified two 34- and 36-kD polypeptides present in embryogenic cell suspension and involved in embryogenic development of sugarcane. An extracellular protein (46 kD, pI 6.1) was found that correlated with the embryogenic capacity of *Hordeum vulgare* L. cell cultures (Stirn et al. 1995). Domon et al. (1995) identified three glycoproteins secreted from embryogenic cell cultures of pine as germin-like proteins, one of the best-characterized markers of cereal embryo development (Lane et al. 1993). It was shown that during somatic embryogenesis of *Cichorium*, the change of the protein pattern in the medium is associated with the induction and initiation of somatic embryogenesis (Hilbert et al. 1992; Helleboid et al. 1995). Mo et al. (1996) observed that the morphology of somatic embryos of *Picea abies*, and especially that of the embryogenic regions, correlated with the presence of specific extracellular proteins that could be used to distinguish between normally developing embryos and embryos blocked in their development.

A first characterization of embryogenic suspension cultures, with respect to secreted esterases at defined stages of *D. glomerata* L. somatic embryogenesis, identified a unique acidic esterase that could discriminate on a biochemical level between *D. glomerata* L. embryogenic suspension cultures that regenerate whole plants and nonembryogenic suspension cultures (Tchorbadjieva and Odjakova 2001). Extracellular proteins secreted by distinct cell structures from embryogenic and nonembryogenic suspension cultures originating from the same genotype were submitted to isoelectric focusing (IEF) and stained for esterase activity (Fig. 1a). A new esterase A1 (pI 3.8) appeared in the phase when PEMs form from microclusters (Fig. 1a, lane 2). This isoenzyme persisted throughout the next phases until mature embryos developed

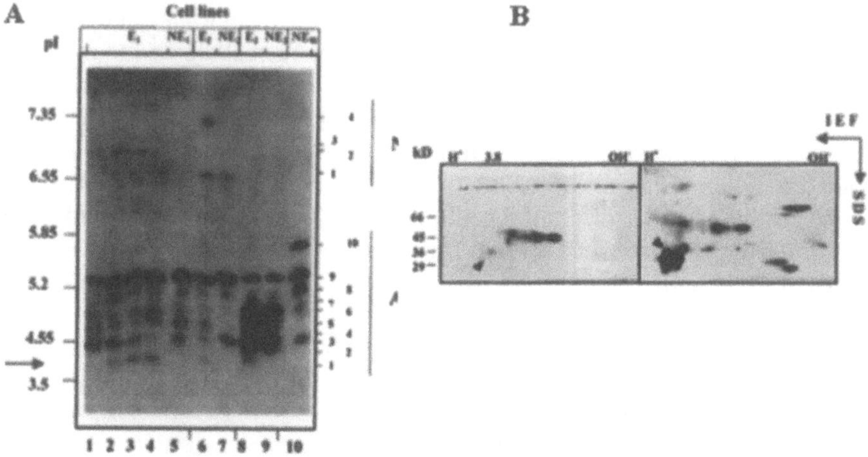

Fig. 1 Isoenzyme pattern of esterase activity of *Dactylis glomerata* L. suspension cultures after isoelectric focusing (**a**) and renaturation of esterases after two-dimensional gel electrophoresis (**b**). **a** Extracellular proteins harvested from the medium of: single cells → microclusters (*lane 1*); microclusters → PEMs (*lane 2*); PEMs → embryos (*lane 3*); embryos (*lane 4*) of E_1 embryogenic suspension culture; microclusters → PEMs (*lanes 6* and *8*) of E_2 and E_3 embryogenic suspension cultures; microclusters from NE_1, NE_2, NE_3, NE_W (*lanes 5, 7, 9, 10*, respectively) nonembryogenic suspension cultures. *Numbers on the right* refer to the position of the various isoforms of esterase activities of the A and N groups. Equal amounts of protein (7 μg) were loaded on each lane. The acidic esterase A1 (pI 3.8) is marked with an *arrow*. **b** Renaturation of extracellular esterases secreted by PEMs from E_1 embryogenic suspension culture in two-dimensional gel; *left panel*, slab gel stained for esterase activity only; *right panel*, the same gel subsequently silver-stained for protein. Molecular weight markers are as shown. The 36-kD esterase A1 is marked with an *arrow*

(Fig. 1a, lanes 3, 4). Among all esterase isoforms, only the presence of A1 was common to all embryogenic suspension cultures (Fig. 1a, lanes 4, 6, 8). In the nonembryogenic control lines (NE_1, NE_2, NE_3, NE_W) this enzyme was virtually absent (Fig. 1a, lanes 7, 9, 10). After 2-D SDS-PAGE electrophoresis and a successful renaturation, A1 occurred as a single polypeptide with an apparent molecular mass of 36 kD and pI 3.8 (Fig. 1b). Silver staining of the same gel showed it to be a moderately abundant protein (Fig. 1b). This unique esterase would allow for the identification of embryogenic potential at early stages of development before morphological changes have taken place.

One of the secreted proteins shown to play a key role in carrot somatic embryogenesis was identified as a 10-kDa lipid transfer protein designated EP2 (Sterk et al. 1991). It was found to be secreted only by embryogenic cells and somatic embryos as well as zygotic embryos. Studies revealed that expression was restricted to peripheral cells of proembryogenic masses (PEMs) and to protoderm cells of somatic embryos.

Nonspecific lipid transfer proteins (ns-LTPs) represent a protein family that is ubiquitous in plants (Kader 1996). These proteins are characterized by their ability to transfer phospholipids between membranes and to bind fatty acids in vitro. Several in vivo functions have been attributed to ns-LTPs, including transport of cuticular compounds (Sterk et al. 1991) and inhibition of the growth of bacterial and fungal pathogens (Molina et al. 1993). Cutin is only present in embryogenic regions and on embryos as a homogeneous and continuous layer. One of the roles of a lipophilic substance like cutin in the cell wall of embryogenic cells is the physiological isolation of embryogenic competent cells from their neighbors as a prerequisite for organized development (Pedroso and Pais 1995b). The other role refers to the formation of a protective layer around the young embryo, which serves as protection against water loss, or the action of hydrolytic cell wall-degrading enzymes that are abundant in the conditioned medium. Expression of LTP gene is a well-known early marker of somatic embryogenesis induction in different systems (Sterk et al. 1991; Poulsen et al. 1996; Schmidt et al. 1997; Sabala et al. 2000). It is a marker for embryo differentiation as it is linked to the formation of the protoderm layer in developing somatic and zygotic embryos (Thoma et al. 1994). Furthermore, the *D. carota* EP2 is already expressed in precursor cell clusters from which somatic embryos develop. Taken together, a correct expression of *ltp* genes is required for normal embryo development. Five acidic LTP-like proteins have been found in the cell wall and the conditioned medium of microcluster cells from embryogenic suspension cultures of *D. glomerata* L. that could discriminate between embryogenic and nonembryogenic suspension cultures (Tchorbadjieva 2001).

One of the secreted proteins shown to have a positive effect on somatic embryogenesis in carrot was identified as a 32-kDa acidic endochitinase classified as a chitinase IV (De Jong et al. 1992). The endochitinase was able to rescue somatic embryogenesis in the mutant carrot cell line *ts*11. Chitinases (EC 3.2.1.14) catalyze the hydrolysis of β-1,4 linkages in chitin, a polymer of *N*-acetyl-D-glucosamine. Chitinases are expressed in many plant species in response to pathogen attack or to other environmental stresses (for a review, see Kasprzewska 2003). In the search for a plant-derived substrate for chitinase, Van Hengel et al. (2001) showed that AGPs from embryogenic suspension cultures contain *N*-acetyl-D-glucosamine and have cleavage sites for endochitinase. Pretreatment of AGPs with EP3 endochitinase resulted in optimal somatic embryo-forming activity. In addition to their putative role in plant defense responses, chitinases may also function in the development of somatic embryos, perhaps by releasing endogenous factors acting as signal molecules (Van Hengel et al. 2002). Chitinases released into the culture medium of *D. carota* (De Jong et al. 1992), as well as *Picea abies* (Mo et al. 1996) and *Pinus caribaea* (Domon et al. 2000) embryogenic cell lines, have been reported to influence somatic embryo development. In *D. glomerata* L. suspension cultures a 32-kD acidic endochitinase has been found to be expressed constitutively in embryogenic suspension cultures and during all stages of somatic embryogenesis (Tchorbadjieva and Pantchev 2006),

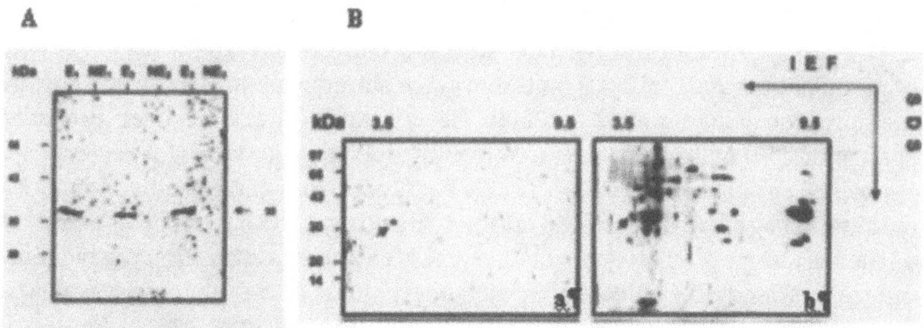

Fig. 2 Detection of a chitinase-like protein in culture media of *Dactylis glomerata* L. suspension cultures. **a** Immunoblot with extracellular proteins from embryogenic (E_1, E_2, E_3) and nonembryogenic (NE_1, NE_2, NE_3) suspension cultures with anti-32-kDa chitinase serum (De Jong et al. 1995). **b** Immunoreactivity of the extracellular proteins secreted by PEMs from E_3 embryogenic suspension culture with anti-32-kD serum from carrot after 2-D gel electrophoresis; *panel a*, immunoblot; *panel b*, silver-stained duplicate gel. The 32-kD acidic chitinase-like protein (pI 3.6) is shown with an *arrow*. Molecular mass markers are shown on the *left*

and could serve possibly as a marker for embryogenic potential (Fig. 2a). Two-dimensional gel electrophoresis and immunoblotting with anti-chitinase antiserum showed that the band of 32 kDa obtained after 1-D separation of E_3 extracts resolved in a unique spot located in the acidic part of the electrophoretogram (Fig. 2b, panel a). We assume that it could possibly serve as a marker for the embryogenic potential of *D. glomerata* L. suspension cultures. This is in agreement with the results of Mo et al. (1996), who found a correlation of chitinase secretion in a *Picea abies* in vitro culture with the ability of PEMs to form normal somatic embryos. Domon et al. (2000) reported the identification of a 48-kDa chitinase-like protein, ionically bound to the surfaces of preglobular somatic embryos of Caribbean pine. Two chitinase isoforms were shown to accumulate in the medium of embryo cultures to a much higher level compared to that in the medium of a nonembryogenic *Cichorium* variety (Helleboid et al. 2000). Wiweger et al. (2003) revealed that *Chia 4-Pa* chitinase genes were expressed in a subpopulation of proliferating cells and at the base of the somatic embryo in *Picea abies*, and that the protein promotes PEM-to-somatic embryo transition. Egertsdotter and Von Arnold (1998) observed a stimulating effect of a chitinase-4 related chitinase on early embryo development in Norway spruce suspension cultures.

Arabinogalactan proteins (AGPs) are proteoglycans commonly found in the cell wall, cell matrix, and cell membrane of plants. Different hypotheses propose that AGPs may be involved in cell proliferation, cell expansion, and regulation of somatic embryo development (for a review, see Showalter 2001). Promotive and inhibitory to somatic embryogenesis effects of certain exogenously added AGPs were reported for carrot cultures (Kreuger and Van Holst 1993; Toonen et al. 1997a) and Norway spruce cultures (Egertsdotter and Von Arnold 1998). In *Cichorium*, immunofluorescence studies localized AGPs to the outer cell wall of globular somatic embryos, and they were abundantly present in the culture medium, too (Chapman et al. 2000). Several antibodies have been prepared against diverse AGPs and were used to mark specific cell types (for reviews, see Knox 1997; Willats et al. 2000). An AGP epitope from carrot cell-conditioned medium recognized by the JIM8 antibody was originally described as a marker of the very early transitional stage of cultured carrot cells after embryogenic induction (Pennell et al. 1992). Subsequently it was shown that most embryos develop from cells lacking the JIM8 epitope (Toonen et al. 1996). Finally, it was found that the JIM8 epitope marks a specific cell type that, upon cell division, asymmetrically transferred the JIM8 epitope to a $JIM8^-$ embryogenic and $JIM8^+$ apoptotic cell type. It was further demonstrated that the JIM8 epitope represents a soluble signal produced by $JIM8^+$ cells to stimulate embryo development of $JIM8^-$ cells (McCabe et al. 1997). We isolated a monoclonal antibody MAb 3G2 against a cell wall protein designated EP48 secreted by the earliest morphological structures (microclusters) in *D. glomerata* L. embryogenic suspension cultures (Tchorbadjieva et al., 2005) (Fig. 3a). Screening of

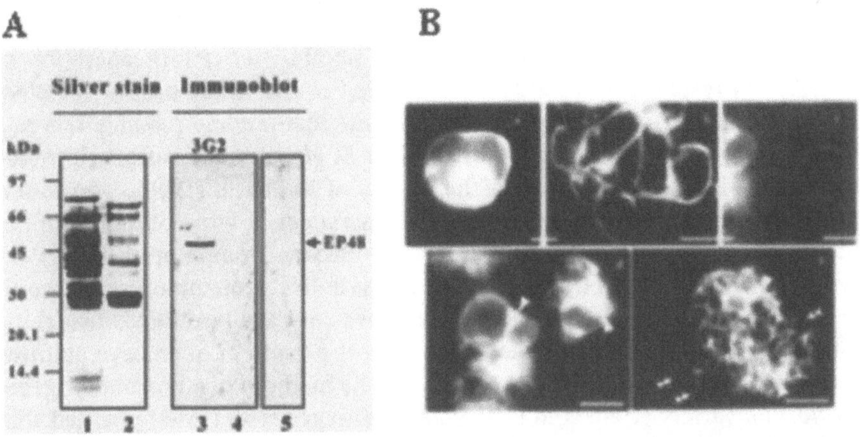

Fig. 3 Immunoblot analysis of extracellular proteins with monoclonal antibody MAb 3G2 (**a**) and indirect immunofluorescent localization of EP48 on intact *D. glomerata* L. suspension cells during somatic embryogenesis (**b**). **a** Immunoblot of extracellular proteins from embryogenic (*lanes 1* and *3*) and nonembryogenic (*lanes 2* and *4*) microcluster cells after SDS-PAGE and transfer to PVDF membrane. MAb 3G2 recognized a single protein (M_r 48 000) (*arrow*). The control with preimmune serum (*lane 5*) was negative. Molecular mass markers are indicated on the *left* in kD. **b** MAb 3G2 labeled the cell wall of small, isodiametric single cells (*a*) as well as elongated, banana-shaped single cells (*b*); many single cells (*c*) remain unstained. The fluorescence due to the antibody binding is most intense at the regions of cell adhesion of microcluster cells (*d*) and PEMs (*e*) (*single arrowheads*), while regions of cell wall without neighbors are unlabeled in PEMs (*double arrowheads*). Bars = 10 μm (a); 30 μm (b–e)

the extracellular proteins from microclusters of three embryogenic (E_1, E_2, and E_3) and nonembryogenic (NE_1, NE_2, and NE_3) suspension cultures on immunoblots showed that EP48 was found exclusively in the embryogenic cell lines. Immunofluorescence localized EP48 on the cell surface of some single cells, microclusters, and PEMs. Interestingly, in microclusters immunofluorescence was located at sites of cell–cell contact but could also be found on cell surface regions that were not in direct contact with neighboring cells, while in PEMs the distribution of EP48 was uneven, and was less intense or even absent from the regions of the surface of PEMs where cells had no neighbors (Fig. 3b). Possibly, during development of PEMs a local change in the cell wall of some cells occurred leading to the loss of MAb 3G2 epitope. Whether the monoclonal antibody marks cells destined for embryogenesis remains to be elucidated, but based on its localization and pattern of accumulation we conclude that it can be useful to monitor the embryogenic potential of *D. glomerata* L. suspension cultures.

It is now widely recognized that the extracellular proteins are indispensable for differentiation and morphogenesis, taking part in signal transduction, cell–cell recognition, cell expansion, and adhesion.

6
Conclusion

In the preceding section, protein markers for somatic embryogenesis and the different experimental approaches for their identification and use have been discussed. The protein markers are useful probes for defining embryogenic potential and for marking different phases in plant development. To gain a better insight into the mechanisms of somatic embryogenesis, a combination of more advanced methods such as the phage display subtraction method, differential display, and proteome analysis is indispensable. Immunomagnetic sorting and cell tracking could be successfully applied to determine the fate of embryogenic cells. All this will greatly accelerate the functional analysis of protein markers, and will contribute to the improvement of crop species together with the establishment of efficient propagation technologies.

References

Alexandrova K, Conger B (2002) Isolation of two somatic embryogenesis-related genes from orchardgrass (*Dactylis glomerata*). Plant Sci 162:301–307

Altherr S, Stirn S, Jacobsen H-J (1993) Immunobiochemical analysis of a nuclear protein marker for regeneration potential in higher plants. J Plant Physiol 141:415–422

Bagnoli F, Capuana M, Racchi M (1998) Developmental changes of catalase and superoxide dismutase isoenzymes in zygotic and somatic embryos of horse chestnut. Aust J Plant Physiol 25:909–913

Blanco M, Nieves N, Sánchez M, Borroto C, Castillo R, González J, Escalona M, Báez E, Hernández Z (1997) Protein changes associated with plant regeneration in embryogenic calli of sugarcane (*Saccharum* sp.). Plant Cell Tissue Organ Cult (3)51:153–158

Cavalcante AJM, Sihachakr D, Allot M, Tizroutine S, Mussio I, Servaes A, Ducreux G (1994) Isozyme modification and plant regeneration through somatic embryogenesis in sweet potato (*Ipomoea batatas* L. Lam). Plant Cell Rep 13(8):437–441

Chapman A, Blervacq AS, Vasseur J, Hilbert JL (2000) Arabinogalactan proteins in *Cichorium* somatic embryogenesis: effect of beta-glucosyl Yariv reagent and epitope localization during embryo development. Planta 211(3):305–314

Charbit E, Legavre T, Lardet L, Bourgeois E, Ferrière N, Carron M (2004) Identification of differentially expressed cDNA sequences and histological characteristics of *Hevea brasiliensis* calli in relation to their embryogenic and regenerative capacities. Plant Cell Rep 22(8):539–548

Chen L, Luthe D (1987) Analysis of proteins from embryogenic and nonembryogenic rice (*Oryza sativa* L.) calli. Plant Sci 48:181–188

Chibbar R, Shylu J, Georges F, Mallard C, Constabel F (1988) Esterase isozymes as markers of somatic embryogenesis in cultured carrot cells. J Plant Physiol 133:367–370

Choi JH, Sung ZR (1984) Two-dimensional gel analysis of carrot somatic embryonic proteins. Plant Mol Biol Rep 2:19–25

Chugh A, Khurana P (2002) Gene expression during somatic embryogenesis: recent advances. Curr Sci 83(6):715–730

De Jong A, Cordewener J, LoSchiavo F, Terzi M, Vandekerckhove J, Van Kammen A, De Vries S (1992) A carrot somatic embryo is rescued by chitinase. Plant Cell 4:425–433

De Vries S, Booij H, Janssens R, Vogels R, Saris L, LoSchiavo F, Terzi M, Van Kammen A (1988) Carrot somatic embryogenesis depends on the phytohormone-controlled presence of correctly glycosylated extracellular proteins. Genes Dev 2:62–476

Dodeman VL, Ducreux G (1996a) Isozyme patterns in zygotic and somatic embryogenesis of carrot. Plant Cell Rep 16:101–105

Dodeman VL, Ducreux G (1996b) Total protein pattern expression during induction and development of carrot somatic embryos. Plant Sci 120:57–69

Domon JM, Dumas B, Laine E, Meyer Y, David A, David H (1995) Three glycosylated polypeptides secreted by several embryogenic cell cultures of pine show highly specific serological affinity to antibodies directed against the wheat germin apoprotein monomer. Plant Physiol 108:141–148

Domon JM, Neutelings G, Roger D, David A, David H (2000) A basic chitinase-like protein secreted by embryogenic tissues of *Pinus caribaea* acts on arabinogalactan proteins extracted from the same cell line. J Plant Physiol 156:33–39

Egertsdotter U, Von Arnold S (1995) Importance of arabinogalactan proteins for the development of somatic embryos of Norway spruce (*Picea abies*). Physiol Plant 93:334–345

Egertsdotter U, Von Arnold S (1998) Development of somatic embryos in Norway spruce. J Exp Bot 49(319):155–162

Esqueda M, Oropeza M, De Garcia E (1998) Extracellular proteins secreted in the culture medium of embryogenic and nonembryogenic cell suspensions of sugarcane (*Saccharum* sp. L.) cv. 78-1. Acta Cient Venez 49(3):160–165

Fehér A, Pasternak T, Dudits D (2003) Transition of somatic plant cells to an embryogenic state. Plant Cell Tissue Organ Cult 74:201–228

Fellers JP, Guenzi AC, Porter DR (1997) Marker proteins associated with somatic embryogenesis of wheat callus cultures. J Plant Physiol 151:201–208

Filonova L, Bozhkov P, Von Arnold S (2000) Developmental pathway of somatic embryogenesis in *Picea abies* as revealed by time-lapse tracking. J Exp Bot 51:249–264

Fukuda F, Ito M, Sugiyama M, Komamine A (1994) Mechanisms of the proliferation and differentiation of plant cells in cell culture systems. Int J Dev Biol 38:287–299

Gavish H, Vard A, Fluhr R (1992) Suppression of somatic embryogenesis in *Citrus* cell cultures by extracellular proteins. Planta 186:511–517

Giroux R, Pauls KP (1996) Characterization of embryogenesis-related proteins in alfalfa. Physiol Plantarum 96:585–592

Hahne G, Mayer JE, Lörz H (1988) Embryogenic and callus-specific proteins in somatic embryogenesis of the grass, *Dactylis glomerata* L. Plant Sci 55:267–279

Hecht V, Vielle-Calzada J, Hartog M, Schmidt E, Boutilier K, Grossniklaus U, De Vries S (2001) The *Arabidopsis* SOMATIC EMBRYOGENESIS RECEPTOR KINASE 1 gene is expressed in developing ovules and embryos and enhances embryogenic competence in culture. Plant Physiol 127:803–816

Heck G, Perry S, Nichols K, Fernandez D (1995) AGL15, a MADS domain protein expressed in developing embryos. Plant Cell 7:1271–1282

Helleboid S, Bauw G, Belingheri L, Vasseur J, Hilbert J (1998) Extracellular β-1,3-glucanases are induced during early somatic embryogenesis in *Cichorium*. Planta 205:56–63

Helleboid S, Coullerot J, Hilbert J, Vasseur J (1995) Effects of α-difluoromethylarginine on embryogenesis, polyamine content, and protein patterns in a *Cichorium* hybrid. Planta 196:571–576

Helleboid S, Hendriks T, Bauw G, Inze D, Vasseur J, Hilbert J-L (2000) Three major somatic embryogenesis-related proteins in *Cichorium* identified as PR proteins. J Exp Bot 51(348):1189-1200

Hilbert J, Dubois T, Vasseur J (1992) Detection of embryogenesis-related proteins during somatic embryo formation in *Cichorium*. Plant Physiol Biochem 30:733–741

Hrubcová M, Cvikrova M, Eder J (1994) Peroxidase activities and content of phenolic acids in embryogenic and nonembryogenic alfalfa cell suspension cultures. Biol Plant 39:175–182

Hvoslef-Eide A, Corke F (1997) Embryogenesis-specific protein changes in birch suspension cultures. Plant Cell Tissue Organ Cult 51(1):35–41

Kader JC (1996) Lipid-transfer proteins in plants. Annu Rev Plant Physiol Plant Mol Biol 47:627–654

Kairong C, Gengsheng X, Lin Q, Xinmin L, Yafu W (1999) The analysis of differential gene expression in early somatic embryogenesis on *Lycium barbarum*. Plant Cell Tissue Organ Cult 59(3):169–174

Kasprzewska A (2003) Plant chitinases—regulation and function. Cell Mol Biol Lett 8:809–824

Kiyosue T, Dong J, Satoh S, Kamada H, Harada H (1990) Detection of an embryogenic cell antigen in carrot. Plant Cell Physiol 31(7):947–950

Kiyosue T, Yamaguchi-Shinozaki K, Shinozaki K, Higashi K, Satoh S, Kamada H, Harada H (1992) Isolation and characterization of a cDNA that encodes ECP31, an embryogenic-cell protein from carrot. Plant Mol Biol 19:239–249

Knox JP (1997) The use of antibodies to study the architecture and developmental regulation of plant cell walls. Int Rev Cytol 171:79–120

Knox JP (1999) Intriguing, complex and everywhere: getting to grips with arabinogalactan proteins. Trends Plant Sci 4:123–125

Knox JP, Linstead P, Peart J, Cooper C, Roberts K (1991) Developmentally regulated epitopes of cell surface arabinogalactan proteins and their relation to root tissue pattern formation. Plant J 1:317–326

Kreuger M, Van Holst G-J (1993) Arabinogalactan proteins are essential in somatic embryogenesis of *Daucus carota* L. Planta 189:243–248

Kreuger M, Van Holst G-J (1995) Arabinogalactan-protein epitopes in somatic embryogenesis of *Daucus carota* L. Planta 197:135-141

Lane BG, Dunwell JM, Ray J, Schmitt MR, Cuming AC (1993) Germin, a protein marker of early plant development, is an oxalate oxidase. J Biol Chem 268:12239–12242

Liang P, Pardee A (1992) Differential display of eukaryotic messenger RNA by means of the polymerase chain reaction. Science 257:967–971

Linkiewicz A, Filipecki M, Tomczak A, Grabowska A, Malepszy S (2004) The cloning of sequences differentially transcribed during the induction of somatic embryogenesis in cucumber (*Cucumis sativus* L.). Cell Mol Biol Lett 9(4B):795–804

Maës O, Coutos-Thevenot P, Jouenne T, Boulay M, Guern J (1997) Influence of extracellular proteins, proteases, and protease inhibitors on grapevine somatic embryogenesis. Plant Cell Tissue Organ Cult 50:97–105

Martinelli L, Scienza A, Villa P, De Ponti P, Gianazza E (1993) Enzyme markers for somatic embryogenesis in Vitis. J Plant Physiol 141:476–481

McCabe P, Valentine T, Forsberg L, Pennell R (1997) Soluble signals from cells identified at the cell wall establish a developmental pathway in carrot. Plant Cell 9:2225–2241

Mo L, Egertsdotter U, Von Arnold S (1996) Secretion of specific extracellular proteins by somatic embryos of *Picea abies* is dependent on embryo morphology. Ann Bot 77:143–152

Molina A, Segura A, Garcia-Olmedo F (1993) Lipid transfer proteins (nsLTPs) from barley and maize leaves are potent inhibitors of bacterial and fungal plant pathogens. FEBS Lett 316(2):119–122

Momiyama T, Afele J, Takeo S, Toshiaki K, Yutaka T, Fumio T, Kenji T, Shigeo N (1995) Differential display identifies developmentally regulated genes during somatic embryogenesis in eggplant (*Solanum melongena* L.). Biochem Biophys Res Commun 213(2):376–382

Mordhorst AP, Toonen MAJ, De Vries SC (1997) Plant Embryogenesis. Crit Rev Plant Sci 16(6):535–576

Neutelings G, Domon JM, Membré N, Bernier F, Meyer Y, David A, David H (1998) Characterization of a germin-like protein gene expressed in somatic and zygotic embryos of pine (*Pinus caribaea* Morelet). Plant Mol Biol 38(6):1179–1190

Nielsen K, Hansen I (1992) Appearance of extracellular proteins associated with somatic embryogenesis in suspension cultures of barley (*Hordeum vulgare* L.). J Plant Physiol 139:489–497

Nolan K, Irwanto R, Rose R (2003) Auxin up-regulates *MtSERK1* expression in both *Medicago truncatula* root-forming and embryogenic cultures. Plant Physiol 133:218–230

Pedroso MC, Hilbert J-L, Vasseur J, Pais S (1995a) Polypeptides associated with the induction of direct embryogenesis in *Camellia japonica* leaves. I. Identification of embryo-specific polypeptides. J Exp Bot 46(291):1579–1584

Pedroso MC, Pais S (1995b) Factors controlling somatic embryogenesis. Plant Cell Tissue Organ Cult 43:147–154

Pennell R (1998) Cell walls: structures and signals. Curr Opin Plant Biol 1:504–510

Pennell R, Janniche L, Scofield G, Booij H, De Vries S, Roberts K (1992) Identification of a transitional cell state in the developmental pathway to carrot somatic embryogenesis. J Cell Biol 119:1371–1380

Poulsen G, Frugis G, Albrechtsen M, Mariotti D (1996) Synthesis of extracellular proteins in embryogenic and nonembryogenic cell cultures of alfalfa. Plant Cell Tissue Organ Cult 44:257–260

Reinbothe C, Tewes A, Reinbothe S (1992) Altered gene expression during somatic embryogenesis in *Nicotiana plumbaginifolia* and *Digitalis lanata*. Plant Sci 82:47–58

Rojas-Herrera R, Quiroz-Figueroa F, Monforte-González M, Sánchez-Teyer L, Loyola-Vargas V (2002) Differential gene expression during somatic embryogenesis in *Coffea arabica* L., revealed by RT-PCR differential display. Appl Biochem Biotechnol 21(1):43–50

Sabala I, Elfstrand M, Farbos I, Clapham D, Von Arnold S (2000) Tissue-specific expression of Pa18, a putative lipid transfer protein gene, during embryo development in Norway spruce (*Picea abies*). Plant Mol Biol 42:461–478

Sallandrouze A, Faurobert M, Maataoui M, Espagnac H (1999) Two-dimensional electrophoretic analysis of proteins associated with somatic embryogenesis development in *Cupressus sempervirens* L. Electrophoresis 20:1109–1119

Samaj J, Baluska F, Bobak M, Volkman D (1999) Extracellular matrix surface network of embryogenic units of friable maize callus contains AGPs recognized by mAb JIM4. Plant Cell Rep 18:369–374

Schmidt E, Guzzo F, Toonen M, De Vries S (1997) A leucine-rich repeat containing receptor-like kinase marks somatic plant cells competent to form embryos. Development 124:2049–2062

Shinohara N, Demura T, Fukuda H (2000) Isolation of a vascular cell wall-specific monoclonal antibody recognizing a cell polarity by using a phage display subtraction method. Proc Natl Acad Sci USA 97(6):2585–2590

Showalter A (2001) Arabinogalactan proteins: structure, expression and function. Cell Mol Life Sci 58:1399–1417

Smallwood M, Martin H, Knox JP (1995) An epitope of rice threonine-and hydroxyproline-rich glycoprotein is common to cell wall and hydrophobic plasma-membrane glycoproteins. Planta 196:510–522

Smallwood M, Yates EE, Willats WGT, Martin H, Knox JP (1996) Immunological comparison of membrane-associated and secreted arabinogalactan proteins in rice and carrot. Planta 198:452–459

Smith J, Krauss M, Borkird C, Sung Z (1988) A nuclear protein associated with cell divisions in plants. Planta 174:462–472

Somleva M, Schmidt E, De Vries S (2000) Embryogenic cells in *Dactylis glomerata* L. (Poaceae) explant identified by cell tracking and by *SERK* expression. Plant Cell Rep 19:718–726

Stacey N, Roberts K, Knox JP (1990) Patterns of expression of the JIM4 arabinogalactan-protein epitope in cell cultures and during somatic embryogenesis in *Daucus carota* L. Planta 180:285–292

Sterk P, Booij H, Schellekens G, Van Kammen A, De Vries S (1991) Cell-specific expression of the carrot EP2 lipid transfer protein gene. Plant Cell 3:907–921

Stirn S, Mordhorst A, Fuchs S, Lörz H (1995) Molecular and biochemical markers for embryogenic potential and regenerative capacity of barley (*Hordeum vulgare* L.) cell cultures. Plant Sci 106:195–206

Sung Z, Okimoto R (1981) Embryonic proteins in somatic embryos of carrot. Proc Natl Acad Sci USA 78:3683–3687

Tchorbadjieva M (2001) Early markers for somatic embryogenesis in suspension cultures of *Dactylis glomerata* L. PhD thesis, Sofia University

Tchorbadjieva M, Kalmukova R, Pantchev I, Kyurkchiev S (2005) Monoclonal antibody against a cell wall marker protein for embyogenic potential of *Dactylis glomerata* L. suspension cultures. Planta (in press)

Tchorbadjieva M, Odjakova M (2001) An acidic esterase as a biochemical marker for somatic embryogenesis in orchardgrass (*Dactylis glomerata* L.) suspension cultures. Plant Cell Rep 20(1):28–33

Tchorbadjieva M, Odjakova M, Kyurkchiev S (1998) Generation of monoclonal antibodies against extracellular proteins from *Dactylis glomerata* L. suspension cultures. C R Acad Bulg Sci 51(1–2):97–100

Tchorbadjieva M, Pantchev I (2006) Secretion of a chitinase-like protein in embryogenic suspension cultures of *Dactylis glomerata* L. Biol Plant 50:142–145

Tchorbadjieva M, Pantchev I, Harizanova N (2004) Two-dimensional protein pattern analysis of extracellular proteins secreted by embryogenic and nonembryogenic suspension cultures of *Dactylis glomerata* L. Biotechnol Biotechnol Equip 18(2):20–27

Tchorbadjieva M, Somleva M, Odjakova M, Panchev I, Nikolaev N (1992) Glycoprotein release into the medium of embryogenic and nonembryogenic suspension cultures of the grass, *Dactylis glomerata* L. C R Acad Bulg Sci 45(7):103–106

Thoma S, Hecht U, Kippers A, Botella J, De Vries S, Somerville C (1994) Tissue-specific expression of a gene encoding a cell wall-localized lipid transfer protein from *Arabidopsis*. Plant Physiol 105:35–45

Toonen M, Schmidt E, Hendriks T, Verhoeven H, Van Kammen A, De Vries S (1996) Expression of the JIM8 cell wall epitope in carrot somatic embryogenesis. Planta 200:167–173

Toonen M, Schmidt E, Van Kammen A, De Vries S (1997a) Promotive and inhibitory effects of diverse arabinogalactan proteins on *Daucus carota* L. somatic embryogenesis. Planta 203:188–195

Toonen M, Verhees J, Schmidt E, Van Kammen A, De Vries S (1997b) *AtLTP1* luciferase expression during carrot somatic embryogenesis. Plant J 12:1213–1221

Van Hengel A, Tadesse Z, Immerzeel P, Schols H, Van Kammen A, De Vries S (2001) *N*-Acetylglucosamine- and glucosamine-containing arabinogalactan proteins control somatic embryogenesis. Plant Physiol 125:1880–1890

Van Hengel A, Van Kammen A, De Vries S (2002) A relationship between seed development, arabinogalactan proteins (AGP), and the AGP-mediated promotion of somatic embryogenesis. Physiol Plant 114:637–644

Willats W, Gilmartin P, Mikkelsen J, Knox JP (1999a) Cell wall antibodies without immunization: generation and use of de-esterified homogalacturonan block-specific antibodies from a naïve phage display library. Plant J 18(1):57–65

Willats W, Steele-King C, Marcus S, Knox JP (1999b) Side chains of pectic polysaccharides are regulated in relation to cell proliferation and cell differentiation. Plant J 20(6):619–628

Willats W, Steele-King C, McCartney L, Orfila C, Marcus S, Knox JP (2000) Making and using antibody probes to study plant cell walls. Plant Physiol Biochem 38(1/2):27–36

Williams M, Freshour G, Darvill A, Albersheim P, Hahn M (1996) An antibody Fab selected from a recombinant phage display library detects deesterified pectic polysaccharide rhamnogalacturonan II in plant cells. Plant Cell 8:673–685

Winter G, Griffiths A, Hawkins R, Hoogenboom H (1994) Making antibodies by phage display technology. Annu Rev Immunol 12:433–455

Wiweger M, Farbos I, Ingouff M, Lagercrantz U, Von Arnold S (2003) Expression of *Chia 4-Pa* chitinase genes during somatic and zygotic embryo development in Norway spruce (*Picea abies*): similarities and differences between gymnosperm and angiosperm class IV chitinases. J Exp Bot 54(393):2691–2699

Wurtele ES, Wang H, Durgerian S, Nikolau BJ, Ulrich TH (1993) Characterization of a gene that is expressed early in somatic embryogenesis of *Daucus carota*. Plant Physiol 102:303–312

Yamazaki M, Saito K (2002) Differential display analysis of gene expression in plants. Cell Mol Life Sci 59:1246–1255

Yasuda H, Nakajima M, Ito T, Ohwada T, Masuda H (2001) Partial characterization of genes whose transcripts accumulate preferentially in cell clusters at the earliest stages of carrot somatic embryogenesis. Plant Mol Biol 45(6):705–712

Yoshida K, Naito S, Takeda G (1994) cDNA cloning of regeneration-specific genes in rice by differential screening of randomly amplified cDNAs using RAPD primers. Plant Cell Physiol 35(7):1003–1009

Plant Cell Monogr (2)
A. Mujib · J. Šamaj: Somatic Embryogenesis
DOI 10.1007/7089_022/Published online: 20 October 2005

Cytological, Physiological and Biochemical Aspects of Somatic Embryo Formation in Flax

Anna Pret'ová (✉) · Jozef Šamaj · Bohuš Obert

Institute of Plant Genetics and Biotechnology, Slovak Academy of Sciences,
Akademická 2, P.O.Box 39A, 950 07 Nitra, Slovakia
anna.pretova@savba.sk

Abstract The cytological, physiological and some biochemical aspects of somatic embryo formation in flax are discussed. From the review it is obvious that the manifestation of the embryogenic potential in flax is rather low and very often the somatic embryos develop with no correctly formed shoot apices. [The terms somatic embryos and embryo-like structures (ELS) in this review are used as authors used them in their papers. Somatic embryos are fully developed cotyledonary-stage embryos, while ELS represent a wider range of globular and bipolar structures, and particularly structures with root or shoot poles that are not well defined, but that always have a vascular system between them.]

1 Introduction

Flax (*Linum usitatissimum* L.) is an ancient cultivated species that still has an important impact on the world economy. Traditionally cultivated for its main product—fibre and seed oil—this species has gained new interest in the emerging market of functional food owing to its high content in fatty acids, mainly α-linolenic acid, and lignan oligomers. Besides, flax fibre is a valuable component of modern composite materials also used in the automobile industry. Flax is also considered to be a very important diversification crop on land set aside.

2 Direct and Indirect Somatic Embryo Formation in Flax (*Linum usitatissimum* L.)

Besides the values mentioned in the "Introduction", flax has been the focus of a great deal of both applied and basic research efforts in plant cell and biotechnology studies in recent years. Interestingly, this species has a long history of research and applications, particularly in plant tissue culture (reviewed by Millam et al. 2005). A lot was done in embryo culture work undertaken on this species early in the evolution of plant tissue culture and the results have had a significant impact on modern methodologies of embryo culture studies, including somatic embryogenesis.

The process of somatic embryogenesis in flax was first derived from immature zygotic embryos (cultivar Glenelg) excised in the late heart or the early torpedo stage (Pret'ová and Williams 1986). In this case, the cells of flax hypocotyl epidermis of immature zygotic embryos can be considered as already proembryogenically predetermined cells at the moment of setting them into culture, and therefore there was no need to apply auxins to the culture medium. Only 6-benzylaminopurine (BAP), glutamine and yeast extract were sufficient external stimuli to promote somatic embryo formation. It was found that BAP was responsible for mitotic division, that yeast extract inhibited the growth of the embryo axis of the cultured zygotic embryo and that glutamine supported embryo proliferation and embryo development. From the flax embryo culture studies it is obvious that glutamine is a very significant factor in early flax (globular stage) embryo development (Pret'ová 1983, 1986).

Somatic embryogenesis via direct formation was also achieved when 2-mm-long hypocotyl segments from 6-day-old flax seedlings (cultivar Szegedi 30) were cultivated in liquid Murashige and Skoog medium (Murashige and Skoog 1962) supplemented with 2 mg l^{-1} 2,4-dichlorophenoxyacetic acid (2,4-D) for 2 weeks. After first subculture to a hormone-free medium, embryo-like structures (ELS) appeared on the cut ends of the segments. More structures were formed on the "shoot" end of the segments and fewer on the "root" end. Such a gradient was observable on each cultivated segment. The structures formed were liberated from the primary tissues after 3 weeks when they reached the heart stage, and freely floated in the medium (Pret'ová and Obert 2005b). Approximately one third of these somatic embryos reached the cotyledonary stage with well-formed shoot apices and were capable of germinating. The rest of them failed to form shoot apices; instead, secondary embryogenic structures were formed or sometimes the ELS possessed malformed (coalesced) cotyledons. All ELS were light green (Pret'ová and Obert 2005b). Secondary somatic embryo formation in flax was also reported by Tejavathi et al. (2000).

It is striking to notice that hypocotyls have been used as primary explants in the majority of experiments in connection with flax organogenesis and somatic embryogenesis. These experiments have shown that hypocotyls are the most responsive explants of flax (Kaul and Williams 1987; Millam and Davidson 1993; Bretagne et al. 1994; Cunha and Ferreira 1996; Dedičová et al. 2000; Mundhara and Rashid 2002). Even the already mentioned direct somatic embryo formation from immature zygotic flax embryos (Pret'ová and Williams 1986) occurred on the hypocotyl.

Variable effects of 2,4-D pretreatment on primary hypocotyl explants were obtained (oilseed flax cultivar Szegedi 30) in experiments using Monnier (Mo) medium (Monnier 1978). After a short 2,4-D pretreatment (5 mg l^{-1} for 24 h) the oilseed cultivar Szegedi 30 produced white granular callus on the cut ends of the hypocotyl segments. The callus produced light-green parts with dark-green globular ELS which later developed into maturer stages during long-term cultivation on Mo medium supplemented with exactly the same amount of 2,4-D. The highest production of ELS was obtained after a short 2,4-D pre-treatment (5 mg l^{-1}) with subsequent cultivation on medium supplemented with zeatin (2 mg l^{-1}). Experiments clearly showed a correlation between auxin concentration and duration of treatments (Dedičová et al. 2000).

The globular and heart-shaped ELS contained an epidermis on their surfaces, two or three subepidermal layers of parenchyma cells and meristematic tissues in their inner parts. The bipolar structures had elongated hypocotyls and two small polarly located regions interconnected with vascular tissue which showed a deviation on the top of the structure typical for entering of the vascular strands to cotyledons. Often, the organization of the shoot and root pole was abnormal. Occasionally, root meristem did not develop a normal root cap, and shoot meristem was not located on top of the structure and normal leaf primordia failed to develop. Sometimes, the ELS formed poorly defined and/or fused cotyledons, and apparently no shoot apices. Such structures were described as horn-shaped embryos in soybean (Lazzeri et al. 1987), or more recently as *shoot meristemless mutants (stm)* in *Arabidopsis* (Mordhorst et al. 1998). As a consequence of the structural abnormalities, these arrested ELS were unable to produce mature embryos and complete plants (Dedičová et al. 2000). Sometimes, additional postembryogenic shoot apices were formed on the top of these ELS.

It is known that initiation of shoot meristem is strictly dependent on auxin transport (Ling and Binding 1992). The correct auxin transport is essential for the final initiation of cotyledons and right bilateral symmetry of early embryos. The formation of cotyledons was a critical stage in the further development of globular zygotic embryos of flax cultured in vitro (Pret'ová 1986, 1990). It can be assumed that strong disturbance of polar auxin transport occurred in the system of somatic embryo formation from the hypocotyl segments, likely due to the addition of exogenous auxin. Root pole formation seemed to be less affected by disturbed auxin transport since only few abnormalities occurred there. With supplement of auxin (2,4-D) some more abnormalities appeared. A relatively high concentration of 2,4-D in the induction medium inhibited further development of somatic embryos and caused abnormal development of apical meristem in carrot (Halperin and Wetherell 1964). On the other hand, we have found a "cumulative" effect of 2,4-D on the cells in respect of expressing their totipotency (Pret'ová and Obert 2005b). Nearly all callus cells derived from flax hypocotyl segments turned embryogenic upon 2 weeks of 2,4-D treatment (Pret'ová and Obert 2005a, b); however, the resulting ELS showed weakly differentiated apical meristems. Additionally, meristemless embryos resembling *stm* mutants of *Arabidopis* (Mordhorst et al. 1998) were formed which did not germinate; *stm* mutation prevents both embryogenic and postembryogenic meristem formation (Barton and Poethig 1993). It seems likely that there is a checkpoint during transition phase between globular and heart stage of embryo development. Not all zygotic flax embryos excised in the globular stage were able to pass this checkpoint in the culture (Pret'ová 1986). Interestingly, numerous *Arabidopsis* embryogenic mutants are blocked exactly at this stage (Mayer et al. 1998).

A wide range of bipolar structures which can be classified as ELS have been formed during experiments focused on flax somatic embryogenesis (Fig. 1).

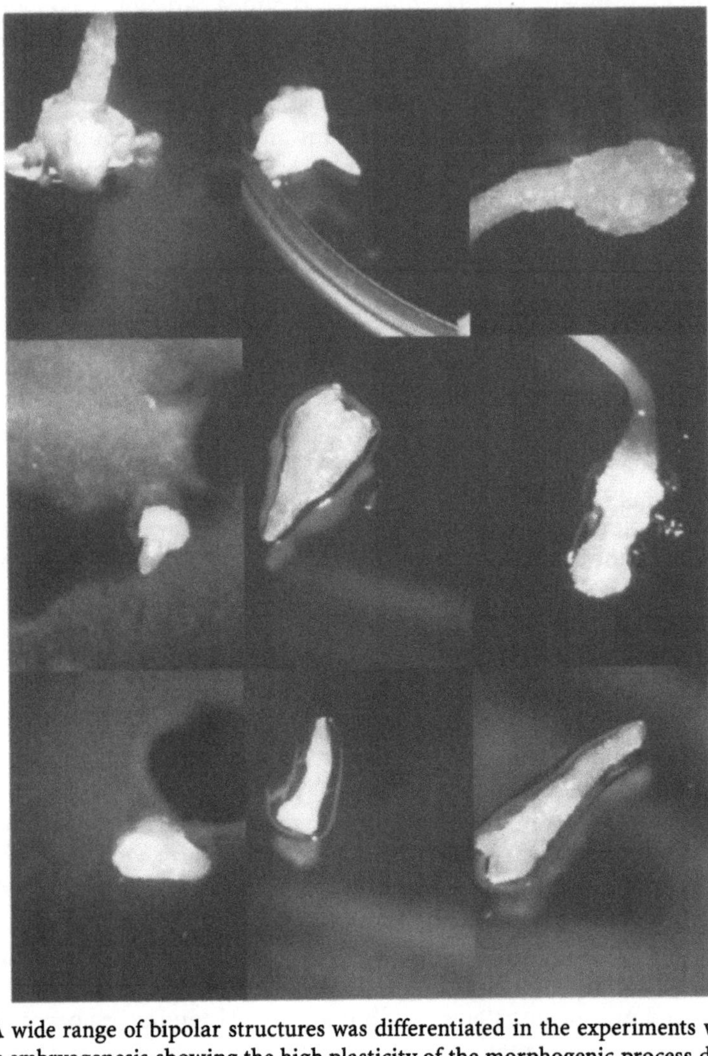

Fig. 1 A wide range of bipolar structures was differentiated in the experiments with flax somatic embryogenesis showing the high plasticity of the morphogenic process during in vitro culture

This indicates that in vitro conditions provide more morphogenic plasticity than is available to the cell restricted within the complete plant body, showing a high extent of cell, tissue and organ differentiation (Pret'ová and Obert 2005a).

Hypocotyl tissue represents a complex explant composed of several cell types. Certain cells, e.g. epidermis, subepidermis and cortical cells, can respond to inductive culture conditions by the reactivation of cell division, while the cells in the vascular tissue do not divide. Generally, the first sign of epidermal reaction can be observed after 1 day in culture when the nuclei of the epidermal cells increase in size and are positioned in the middle of the cell. Similar early structural events can also take place in the subepidermal layer and cortex. Competent cells do not increase in size and undergo several divisions leading to compact meristemoidal structures that can develop further. During the process of the somatic embryo formation the competent cells are regularly covered by an almost continuous layer of extracellular material ruptured by the emergence of shoots and ELS (Šamaj et al. 1997; Dedičová et al. 2000). Generally, organogenesis (shoot and root formation) in vitro and somatic embryo formation have several common features: the occurrence of the extracellular matrix is one of them. Organogenesis is considered to be a morphogenic pathway by which a cell is unable fully to express its totipotency. Somatic embryogenesis as well as organogenesis require a certain degree of cell dedifferentiation, reinitiation of cell division and morphogenic control over cell expansion under appropriate inductive and permissive conditions (Šamaj et al. 1997). In a wider sense, somatic embryogenesis can be considered as an extreme case of adaptation that is based on the phenotypic plasticity of an individual somatic cell. Phenotypic plasticity allows individuals to adapt or acclimate themselves to a wide range of environments, including in vitro conditions (Dedičová et al. 2000).

Indirect formation of somatic embryos in flax can also be induced from nearly mature (28-day-old) zygotic embryos. In such embryos the accumulation of reserve materials is in progress, and lipids together with proteins (aleurone grains) are synthesized (Pret'ová 1978; 1990). The flax zygotic embryos are unable to express their totipotency at this developmental stage. A series of cell divisions promoted by auxin have to take place, resulting in dedifferentiation and callus formation. This dedifferentiation is needed in order to allow cells to set up a new developmental programme. A certain degree of dedifferentiation is also required in the case of direct somatic embryo development. Even if no visible callus is formed, the transcriptional profiles are altered (Pret'ová and Obert 2005a). The cells of nearly mature zygotic flax embryos can be considered as induced proembryogenically determined cells.

During the flax somatic embryogenesis it was very important that the cohesion between callus cells was very tight. Callus with loosely attached cells did not form somatic embryos because of disturbed cell–cell contacts. The more cohesive the cell clusters at the moment of application of external stimuli, the greater was the yield of somatic embryos (Pretova and Obert, unpublished data). These results support the idea of active intercellular communication during the acquisition of embryogenic potential.

The different morphogenic response in flax may also be correlated with changes in oxidative status during early and later tissue cultivation. Reactive oxygen species (ROS) can accumulate in response to biotic and abiotic stress. They can have a detrimental effect on the metabolism, growth and development, through their ability to initiate and maintain reaction cascades. However, ROS may also have a positive role in plant growth and development since they can serve as signal molecules stimulating defence responses (Jabs et al. 1997). The balance between essential and damaging oxidative reactions is influenced by the physiological and developmental status of tissues and exogenous factors such as stress, disease, wounding and the application of plant growth regulators (Benson 2000a, b). Moreover, it was also demonstrated that toxic aldehydes originating from lipid peroxidation are produced in plant tissue during culture initiation and routine subculture. Some ROS are also produced by dedifferentiated tissues (Benson and Withers 1987). Free radicals may also be implicated in plant recalcitrance for regeneration. Flax can still be considered as a recalcitrant plant from the point of view of its regeneration potential via somatic embryogenesis.

Obert et al. (2005) designed their experiments on the basis of the aforementioned knowledge. Several flax cultivars (Atalante, Flanders, Jitka, Szegedi 30 and Super) were screened for organogenesis (shoot and root formation) and ELS production. A nondestructive assay for hydroxyl radicals utilizing dimethyl sulfoxide as a radical trap was used to determine OH radical formation during culture and mòrphogenesis. It was found that morphogenic response in flax can be moderated by oxidative stress. Significant differences were found in the level of hydroxyl radicals in relation to the type of induced morphogenic pathway. A lower level of hydroxyl radicals was observed when shoots were regenerated and the highest level was detected during ELS induction (Obert et al. 2005). The higher the ELS induction, the higher the level of hydroxyl radicals detected, which also depended on the genotype used. This indicates that hydroxyl radicals and peroxidation reactions are involved in the early stages of ELS development in flax. Oilseed cultivars Atalante and Flanders showed a higher response than fibre cultivars Super and Jitka (Obert et al. 2005). Externally added hydrogen peroxide also increased the number of induced ELS from flax hypocotyl explants (Takáč and Pret'ová, 2004).

The results obtained from these experiments strongly indicate that somatic embryogenesis is a stress response and that it is a way in which the plant cell realizes its survival strategy under completely changed and unusual conditions using its unique feature—totipotency (Pret'ová and Obert 2003). The cells are exposed to suboptimal nutrient and hormone supply which generate a significant degree of stress during tissue culture. The oxidative stress responses may be linked to auxin signalling and cell cycle regulation through mitogen-activated protein kinase phosphorylation cascades (reviewed by Hirt 2000).

In addition to hypocotyl cultures, somatic embryogenesis in flax was investigated using a protoplast regeneration system by Ling and Binding (1992). Further very complex studies on flax somatic embryogenesis were performad by Cunha and Ferreira (1996, 1999, 2001, 2003). Initial experiments focused on the examination of free sterol content and its variation during the process of somatic embryo formation (Cunha and Ferreira 1997). The content and composition of free sterols and lipids was analysed in nonmorphogenic cells, competent callus cells, somatic embryos and shoots. The induction of somatic embryos and shoot organogenesis was associated with an increase of total sterols in the competent callus, and an increased ratio of stigmasterol to β-sitosterol in derived embryos. They found a lower content of total lipids in embryogenic competent callus cells, suggesting their utilization in the process of emergence and maturation of somatic embryos. Cell division activity increased during somatic embryo formation. The intensive cell division also means new membrane formation. Lipids, predominantly phospholipids such as phytosterols, are the main components of plasma membrane.

The influence of the carbon source, total inorganic nitrogen as well as interactions between calcium and zeatin during embryo formation from hypocotyls was studied later (Cunha and Ferreira 1999). Subsequently, Cunha and Ferreira (2001) looked at the composition and distribution of *n*-alkanes in developing somatic embryos of flax. The highest content of *n*-alkanes was found in primary hypocotyl explants and at the early stage of competent callus development. The content of *n*-alakanes was significantly lower in somatic embryos and competent callus compared with that in other tissues. These results suggest that utilization of *n*-alkanes should occur in developing embryos. Further, biochemical studies were extended by determination of free and esterified fatty acid content during somatic embryo development (Cunha and Ferreira 2003). Both free and esterified fatty acids, representing fractions of total lipids, increased with the dedifferentiation and early callus formation. With the progress of development the content of total lipids dropped. The specific ratio between the long-chain and the short-chain fatty acids was considered as a potential indicator for partial autotrophy of flax somatic embryos. These results suggest an active membrane formation in mitochondria (18 : 2) and in plastids (18 : 3). The higher proportion of 18 : 3 in relation to 18 : 2 in differentiated green tissues could reflect the photosynthetic potential of somatic embryos grown in vitro. Importantly in this respect, flax belongs to the group of plants with green embryos called *Chloroembryophyta* (Yakovlev and Zhukova 1973). Investigations on the composition of pigments in flax zygotic embryos in situ and in vitro were described by Pret'ová (1977, 1978, 1990). Flax somatic embryos formed from zygotic embryos in vitro are dark green (Pret'ová and Williams 1986). On the other hand, ELS derived from hypocotyl segments either via direct or indirect regeneration are mostly light green (Dedičová et al. 2000). In this context, the results of Cunha and Ferreira (2003) may reflect a change in the nutrition mode of more developed stages of somatic flax embryos from heterotrophic to autotrophic or mixotrophic.

The genetic basis of embryogenic competence is not well defined. Eosinophil cationic protein (ECP) genes were found to be involved in carrot somatic embryogenesis (Kiyosue et al. 1992; Yang et al. 1996, 1997). Using carrot ECP genes as probes, the presence of homologues genes (ECP 31 and ECP 63) in flax genomic DNA was tested by Southern analysis. In flax, two transcripts of ECP 63 were found in the 6-day-old seedlings, and one transcript of ECP 31 was found in seeds (Hajduch et al. 1997).

Some studies were devoted to the chitinase activity in an attempt to find a reliable marker indicating the embryogenic potential of flax cultures. Recently, it was found that flax protoplasts derived from hypocotyl segments entrapped in either agarose or calcium alginate secrete basic chitinases (Roger et al. 1998). The authors hypothesized that flax chitinases associated with cell walls of ELS could generate Nod-like oligosaccharides which might represent signals for differentiation processes in flax. Interestingly, Petrovská et al. (2004) also reported on chitinase activity (acidic chitinase of approximately 25 kDa) in flax suspension culture able to form ELS.

3 Conclusions and Future Prospects

In conclusion, the embryogenic potential in flax is rather low and very often the somatic embryos are formed together with shoots (organogenesis) and it is hard to distinguish between them. Both somatic embryogenesis and organogenesis are considered to be a result of either fully or partially expressed totipotency of plant cells (Šamaj et al. 1997). Possibly, a dysfunction in expression of the embryogenic potential occurs in flax in vitro cultures. Because of the lack of discriminating morphological and/or histological features, great attention should be given to molecular markers. Functional genomics will also provide valuable information since ELS described in papers dealing with flax somatic embryogenesis resemble *stm* mutants described for *Arabidopsis thaliana* L. (Mordhorst et al. 2002). The corresponding gene remains to be cloned and its product characterized in more detail. Nevertheless, these recent experiments showed that it is entirely feasible to employ genetic tools such as different embryo mutants in order to help to answer the question of embryogenic competence. This approach is now possible in *Arabidopsis* where a collection of mutants is available. Since shoot apical meristem development is difficult in flax, characterization of mutants allelic to *stm* may provide a clue and show where flax somatic embryos "go wrong".

Acknowledgements This work was supported by grants from the Slovak Grant Agency APVT (APVT-51-002302 and APVT-51-028602) and VEGA (2/5079/5), Bratislava, Slovakia.

References

Barton MK, Poethig RS (1993) Formation of the shoot apical meristem in *Arabidopsis thaliana*: an analysis of development in the wild-type and in the shoot meristemless mutant. Development 119:823–831

Benson EE (2000a) In vitro plant recalcitrance: an introduction. In Vitro Cell Dev Biol Plant 36:141–148

Benson EE (2000b) Do free radicals have a role in plant tissue culture recalcitrance? In Vitro Cell Dev Biol Plant 36:163–170

Benson EE, Whiters LA (1987) Gas chromatographic analysis of volatile hydrocarbon production by cryopreserved plant tissue cultures: a non destructive method for assessing stability. Cryo Lett 8:35–46

Bretagne B, Chupeau MC, Chupeau Y, Fouilloux G (1994) Improved flax regeneration from hypocotyls using thidiazuron as a cytokinin source. Plant Cell Rep 14:120–124

Cunha AC, Ferreira MF (1996) Somatic embryogenesis, organogenesis and callus growth kinetics of flax. Plant Cell Tissue Org Cult 47:1–8

Cunha AC, Ferreira MF (1997) Differences in free sterols content and composition associated with somatic embryogenesis, shoot organogenesis and calli growth of flax. Plant Sci 124:97–105

Cunha AC, Fernandes-Ferreira M (1999) Influence of medium parameters on somatic embryogenesis from hypocotyl explants of flax (*Linum usitatissimum* L.)—effect of carbon source, total inorganic nitrogen and balance between ionic forms and interaction between calcium and zeatin. J Plant Physiol 155:591–597

Cunha AC, Fernandes-Ferreira M (2001) Ontogenic variations in n-alkanes during somatic embryogenesis of flax (*Linum usitatissimum* L.). Plant Sci 160:1137–1143

Cunha AC, Fernandes-Ferreira M (2003) Ontogenic variations in free and esterified fatty acids during somatic embryogenesis of flax (*Linum usitatissimum* L.). Plant Sci 164:863–872

Dedičová B, Hricová A, Šamaj J, Obert B, Bobák M, Preťová A (2000) Shoots and embryo-like structures regenerated from cultured flax (*Linum usitatissimum* L.) hypocotyl segments. J Plant Physiol 157:327–334

Hajduch M, Preťová A, Kamada H (1997) Carrot ECP genes in flax. In: Molecular biology for agriculture. Second symposium in the series Recent Advances in Plant Biotechnology. Ceske Budejovice, Czech Republic, 25–29 August, p 75

Halperin W, Wetherell DP (1964) Adventive embryony in tissue cultures of the wild carrot *Daucus carota*. Am J Bot 51:274–283

Hirt H (2000) Connecting oxidative stress, auxin, and cell cycle regulation through a plant mitogen-activated protein kinase pathway. Proc Natl Acad Sci USA 97:2405–2407

Jabs T, Tschope M, Colling C, Hahlbrock K, Scheel D (1997) Elicitor-stimulated ion fluxes and O_2^- from the oxidative burst are essential components in triggering defense gene activation and phytoalexin synthesis in parsley. Proc Natl Acad Sci USA 94:4800–4805

Kaul V, Williams EG (1987) Multiple shoot induction in vitro from the hypocotyl of germinating embryos of flax (*Linum usitatissimum* L.). J Plant Physiol 131:441–448

Kiyosue T, Yamaguchishinozaki K, Higashi K, Satoh S, Kamada H, Harada H (1992) Isolation and characterization of a cDNA that encodes ECP 31, an embryogenic-cell protein from carrot. Plant Mol Biol 19:239–249

Lazzeri PA, Hildebrand DF, Collins GB (1987) Soybean somatic embryogenesis—effect of hormones and culture manipulations. Plant Cell Tissue Org Cult 10:197–208

Ling HQ, Binding H (1992) Improvement of plant regeneration from *Linum* protoplasts by the induction of somatic embryogenesis. J Plant Physiol 139:422–426

Mayer KF, Schoof H, Haecker A, Lenhard M, Jurgens G, Laux T (1998) Role of WUSCHEL in regulating stem cell fate in the Arabidopsis shoot meristem. Cell 95:805–815

Millam S, Davidson D (1993) Evidence for an interactive response of flax (*Linum usitatissimum*) hypocotyl tissue cultures to auxin and carbohydrates. J Plant Physiol 141:353–356

Millam S, Obert B, Pret'ová A (2005) Plant cell and biotechnology studies in *Linum usitatissimum* L.—a review. Plant Cell Tissue Org Cult 82:93–103

Monnier M (1978) Culture of zygotic embryos. In: Thorpe TA (ed) Proceedings of the 4th international congress on plant tissue cultures, Calgary, Canada, pp 79–98

Mordhorst AP, Voerman KJ, Hartog MV, Meijer EA, van Went J, Koornneef M, de Vries SC (1998) Somatic embryogenesis in *Arabidopsis thaliana* is facilitated by mutations in genes repressing meristematic cell divisions. Genetics 149:549–563

Mordhorst AP, Hartog MV, El Tamer MK, Laux T, de Vries SC (2002) Somatic embryogenesis from *Arabidopsis* shoot apical meristem mutants. Planta 214:829–836

Mundhara R, Rashid A (2002) Stimulation of shoot-bud regeneration on hypocotyl of *Linum* seedlings, on a transient withdrawal of calcium: effect of calcium, cytokinin and thidiazuron. Plant Sci 162:211–214

Murashige T, Skoog F (1962) A revised medium for rapid growth and bioassays with tobacco tissue cultures. Physiol Plant 15:473–497

Obert B, Benson EE, Millam S, Pret'ová A, Bremner D (2005) Moderation of morphogenetic and oxidative stress responses in flax in vitro cultures by hydroxynonenal and desferrioxamine. J Plant Physiol 162:537–547

Petrovská B, Moravčiková J, Obert B, Salaj T, Pret'ová A, Salaj J (2004) Suspension cultures of flax—extracellular proteins and morphogenic response. Book of abstracts and programme, 5th IVCHB symposium, Debrecen, Hungary, 12–17 September, p 110

Pret'ová A (1977) Pigments in young embryos of *Linum usitatissimum* L. Photosynthetica 11:217–219

Pret'ová A (1978) Flax embryogenesis. Changes in the contents of pigments and fats in vitro and in situ. Biologia 33:29–34

Pret'ová A (1983) Cultivation of globular and heart stage flax embryos. Fertilization and embryogenesis in ovulated plants. Proceedings 7th international embryological symposium 1982, High Tatras, Czechoslovakia, pp 307–309

Pret'ová A (1986) Influence of kinetin on the growth of zygotic flax embryos in vitro. Plant Cell Rep 3:210–211

Pret'ová A (1990) Embryo culture studies in *Linum*. In: Bajaj YPS (ed) Biotechnology in agriculture and forestry, vol 10. Legumes and oil seed crops. I. Springer, Berlin Heidelberg New York, pp 515–538

Pret'ová A, Obert B (2003) Flax (*Linum usitatisimum* L.)—-a plant system for study of embryogenesis. Acta Biol Cracov Bot 45:15–18

Pret'ová A, Obert B (2005a) Some aspects of embryo development in vitro. Acta Hortic (in press)

Pret'ová A, Obert B (2005b) Hypocotyl segments are a good source for somatic embryo production in flax. J Plant Physiol (in press)

Pret'ová A, Williams EG (1986) Direct somatic embryogenesis from immature zygotic embryos of flax (*Linum usitatisimum* L.). J Plant Physiol 126:155–161

Roger D, Gallusci P, Meyer Y, David A, David H (1998) Basic chitinases are correlated with the morphogenic response of flax cells. Physiol Plant 103:271–279

Šamaj J, Bobák M, Ovečka M, Blehová A, Pret'ová A (1997) Structural features of plant morphogenesis in vitro. Veda, Bratislava, p 122

Takáč T, Pret'ová A (2004) The effect of hydrogen peroxide on flax embryo-like structure regeneration. In: Proceedings of the X days of plant physiology, Comenius University, Bratislava, Slovak Republic, 5–9 September, p 34

Tejavathi DH, Sita GL, Sunita AT (2000) Somatic embryogenesis in flax. Plant Cell Tissue Org Cult 63:155–159

Yakovlev MS, Zhukova GYA (1973) Angiospermous plants with green and colourless embryos. Nauka, Leningrad (in Russian)

Yang HP, Saitou T, Komeda Y, Harada H, Kamada H (1996) Late embryogenesis abundant protein in *Arabidopsis thaliana* homologous to carrot ECP31. Physiol Plant 98:661–666

Yang HP, Saitou T, Komeda Y, Harada H, Kamada H (1997) *Arabidopsis thaliana* ECP 63 encoding a LEA protein is located in chromosome 4. Gene 184:83–88

Plant Cell Monogr (2)
A. Mujib · J. Šamaj: Somatic Embryogenesis
DOI 10.1007/7089_037/Published online: 9 December 2005

Somatic Embryogenesis in Rose: Gene Expression and Genetic Transformation

S. S. Korban

Department of Natural Resources & Environmental Sciences, 310 ERML, University of Illinois, 1201 W. Gregory, Urbana, IL 61801, USA
korban@uiuc.edu

Abstract Induction of somatic embryogenesis in roses involves several critical steps requiring specific tissue culture media compositions and particular manipulations of explants. However, it is important to note that although there are various reports on successful induction of somatic embryogenesis in rose, these are often limited to particular genotypes. Therefore, to date, there is no single protocol for inducing somatic embryogenesis that can be used for multiple rose genotypes. Nevertheless, advances have been made in studying regulation of gene expression during somatic embryogenesis. Moreover, successful genetic transformation of rose has been achieved using embryogenic cultures. Transgenic rose lines with desirable traits have now been obtained. Further opportunities for exploiting somatic embryogenesis for genetic manipulation and improvement of roses will become available with all these current achievements and future efforts.

1 Introduction

Somatic cells of plant tissues have the capacity to undergo cellular dedifferentiation into a mass of unorganized cells, or callus, as well as the ability to generate differentiated cells. It is this latter ability to produce morphologically and developmentally normal organs from somatic plant cells that presents an intriguing and unique phenomenon in plants. In recent years, this observed phenomenon, referred to as totipotency of plant cells, has become critical for successful asexual propagation of plants. Moreover, it serves as a limiting step in the ever-expanding area of transgenic plant development. Therefore, this fascinating phenomenon is worthy of investigation to expand our fundamental knowledge of cellular behavior by elucidating the regulatory and morphogenetic events in plant cell growth and development.

Induction of in vitro embryogenesis from somatic plant tissues is an alternative developmental process that occurs in response to high concentrations of auxin or better yet to a functional analog of auxin, namely 2,4-D, added to the culture medium. This unique ability of vegetative plant cells to undergo cellular differentiation into somatic embryos has provided a valuable model system for fundamental studies on embryogenesis as the developmental process of somatic embryogenesis is considerably similar to that of zygotic embryogenesis (Zimmernman 1993). Cell competence for embryogenesis is acquired in the presence of auxin in the medium as cells form proembryogenic masses (PEMs). Upon removal of auxin from the culture medium, these PEMs then undergo differentiation from the globular stage to the heart/torpedo stage, and into plantlets. It is believed that in the presence of auxin, the PEMs synthesize all gene products necessary to complete the globular stage of embryogenesis, but new gene products are needed for the transition to the heart stage which can only be synthesized when the exogenous auxin is removed from the medium (Zimmerman 1993). It is likely that there are other gene products that are synthesized in PEMs in the presence of auxin that prohibit globular embryos from further development into the heart stage. Therefore, these developmental switches are most likely regulated at the transcriptional level, and it is generally believed that somatic embryogenesis is mediated by a signal transduction pathway that is triggered by exogenous auxin.

Successful development of regeneration systems for a number of rose species has already been reported. Embryogenic callus has been initiated from in vitro-derived leaf or stem segments of *Rosa hybrida* cv. Carl Red and *R. canina* (Visessuwan et al. 1997), *R. hybrida* cv. Carefree Beauty, and R. chinensis minima cv. Baby Katie (Hsia and Korban 1996). Embryogenic callus has also been induced in leaves of *R. hybrida* cvs. Domingo and Vicky Brown (De Wit et al. 1990), petioles and roots of *R. hybrida* cvs. Trumpeter and Glad Tidings (Marchant et al. 1996), root explants of both *R. hybrida* cv. Moneyway (van der Salm et al. 1996) and R. Heritage × Alista Stella Gray (Sarasan et al. 2001), petals of *R. hybrida* cv. Arizona (Murali 1996), and immature seeds of R. rugosa (Kunitake et al. 1993). This has also been achieved using immature leaf or stem segments of *R. hybrida* cv. Landora (Rout et al. 1991), in vivo mature leaves of *R. hybrida* cv. Soraya (Kintzios et al. 1999), anther filaments of *R. hybrida* cv. Royalty (Noriega and Söndahl 1991), as well as anthers, petals, receptacles, and leaves of *R. hybrida* cv. Meirutal (Arene et al. 1993). The wide range of explants and experimental approaches that have been employed with different rose species and cultivars strongly suggest that it is difficult to develop a universal genotype-independent method for the production of embryogenic callus in rose (Marchant et al. 1996). Recent progress on rose regeneration has been reviewed by Rout et al. (1999). However, in this chapter we will provide detailed protocols for initiation of embryogenic cultures of rose as well as review some of the applications for these embryogenic cultures for genetic improvement and/or manipulation of roses.

2
Embryogenic Culture Initiation

2.1
Explant Preparation

Among all different tissues used for induction of somatic embryogenesis, it is apparent that in vitro-grown leaves provide the most reliable source of explants for induction of somatic embryogenic cultures from various genotypes of rose.

2.2
Establishing Proliferating Shoot Cultures

To begin with, proliferating shoot cultures of rose must be first established. On the basis of our own experience with various genotypes of *R. hybrida* and *R. chinenesis* minima, nodal stem segments (2 cm in length) that are closest to the apical meristem must be collected from healthy and vigorously growing greenhouse-grown plants. Once cut from actively growing donor plants, all leaves must be removed from stem segments, but retaining the apical meristem intact.

Stem segments (1.5 cm in length) are surface-sterilized with 0.525% sodium hypochlorite solution (10% Clorox commercial bleach) for 10 min, and rinsed three times with sterilized-distilled water (5 min per rinse). Nodal stem sections are then given a fresh cut (along the basal end), and placed in 25 × 150 mm culture tubes containing the medium listed in Table 1. It is important to point out that stem segments with relatively large diameter (0.6–0.8 mm) and long internodes (> 2 cm) are preferred. Cultures should be incubated under a 16 h photoperiod provided by cool-white fluorescent light (60 mmol m^{-2} s^{-1}).

Table 1 Composition of media for establishment and proliferation of shoot cultures of rose using nodal stem segments

Medium component	Culture establishment (per liter)	Shoot proliferation (per liter)
MS salts	4.30 g	4.43 g (salts + MS vitamins)
BA	4.44 mM	2.22 mM
NAA	0.54 mM	0.27 mM
Sucrose	30.00 g/l	30.00 g/l
Agar	7.00 g (Difco-bacto)	2.5 g (gelrite)
pH	5.7	5.7

Table 2 Composition of media for induction of callogenesis followed by embryogenesis from rose leaf explants

Medium component	Callus induction (per liter)	Induction of embryogenesis (per liter)
MS salts + MS vitamins	4.43 g	2.25 g (1/2 MS salts + full MS vitamins)
2,4-D	11.3 mM	0.00 mM
TDZ	0.00 mM	2.30 mM
GA3	0.00 mM	2.90 mM
Sucrose	30.00 g/l	30.00 g/l
Agar	2.50 g (gelrite)	2.5 g (gelrite)
pH	5.7	5.7

Within two weeks following culture establishment, shoots developing from buds should be excised and transferred to a fresh medium to promote shoot growth and proliferation. Proliferating shoot cultures should be periodically subcultured to fresh medium once every 4–5 weeks to maintain growth and proliferation of healthy and vigorous shoots.

2.3 Callus Induction

The top four vigorously growing leaves are excised from in vitro-grown proliferating shoots. Either whole leaves or leaflets should be used as explants for callus induction. All leaf explants should be placed with the abaxial surface in contact with the medium. The basal medium containing full-MS salts, MS vitamins, 30 g sucrose, is supplemented with 2,4-D, and solidified with 2.5 g gelrite. Concentrations of 2,4-D of either 11.3 or 45.2 mM are recommended. pH of the medium is adjusted to 5.7. However, the concentrations of 2,4-D may have to be amended depending on the rose genotype used. Cultures are then incubated in the dark for 4 weeks at a temperature of 23 ± 1 °C.

2.4 Induction of Somatic Embryogenesis

Explants with callus, previously incubated on medium containing 2,4-D, are transferred to a 1/2 MS basal medium, full-strength MS vitamins, 30 g sucrose, and containing either no PGRs, 2.9 mM gibberellic acid (GA_3) alone, or 2.9 mM GA_3 with either 2.2 mM BA or 2.3 mM thidiazuron (TDZ). The medium is solidified with 2.5 g gelrite gellan gum (PhytoTechnology), and pH is adjusted to 5.7. Cultures are grown under light conditions as described above.

Fig. 1 Somatic embryogenesis and plantlet regeneration in rose. **a** Primary somatic embryos induced from embryogenic callus; **b** whole plantlets regenerated from somatic embryos

Callus explants are incubated for a period of two months on the above media and then subcultured to a PGR-free medium for an additional two months. The development of embryogenic callus should be observed throughout this incubation regime. An asynchronous development of embryogenic callus with globular, heart-shaped, and cotyledonary stages are observed throughout this period (Fig. 1a). Embryogenic callus is soft, friable, and opaque-white in color. At times, explants might turn brownish in color (especially those continuously incubated on PGR-free medium), but this callus can still produce somatic embryos. However, if hard, compact, and green-colored callus is observed, it is most likely to be either a nondifferentiating callus or organogenic callus.

2.5
Induction and Proliferation of Secondary Somatic Embryogenesis

Induction of secondary embryogenesis is highly desirable for both micropropagation and genetic improvement (e.g., via transformation) efforts. Inducing secondary embryogenesis from primary somatic embryos can be accomplished by transferring primary embryogenic callus onto petri plates containing 1/2 MS basal salts, full-strength MS vitamins, and solidified with 2.5 g gelrite gellan gum for a period of one month. These are then transferred onto a PGR-free medium with monthly subcultures. All cultures are maintained under light conditions as described above. Proliferation of somatic embryos can be maintained for at least 1 year.

2.6
Maturation and Germination of Somatic Embryos

Maturation and germination of somatic embryos is achieved by transferring individual clumps of somatic embryos onto a similar medium as described above, but with a slight modification. Essentially, the medium consists of 1/2 MS basal salts, full-strength MS vitamins, 30 g sucrose, 3.8 mM abscisic acid (ABA), and solidified with 2.5 g gelrite gellan gum. Bipolar plantlets are then excised, and individually transferred to a PGR-free shoot elongation medium consisting of 1/2 MS medium, full-strength vitamins, and 30 g sucrose for a period of one month. This medium also promotes shoot elongation and root development.

2.7 Plantlet Development, Acclimatization, and Transfer to the Greenhouse or Field

Rooted plantlets are transferred to a soil mix (1 : 1 : 1 of soil, peat, and perlite) in 4 cm plastic pots for a period of two weeks, and covered with a clear plastic bag. If the plantlets are in flats, then a clear plastic cover can be used instead. The top of the plastic bag/cover is gradually removed/opened to allow for plantlets to be acclimatized. This process can take anywhere from two to three weeks.

Acclimatized plants are then transferred to the greenhouse and grown at 23 °C. Plants are watered daily using a drip-irrigation system, and fertilized once every 2 weeks with 250 ppm of a 20-20-20 NPK fertilizer solution. Once the plants are well established in the greenhouse (Fig. 1b), then these can be transferred to the field.

3 Regulation of Gene Expression During Somatic Embryogenesis

As plant cells grown in vitro undergo the process of somatic embryogenesis, these are accompanied by changes in DNA methylation that are associated with regulation of gene expression (Finnegan 2001). In higher plants, the 5-methylcytosine (5 mC) is predominantly modified, and among all CpG sequences in a plant genome, 60–90% of those are methylated, while unmethylated CpG sequences are clustered as CpG islands (Ng and Bird 1999). DNA methylation can inhibit transcription by modifying target sites of transcriptional factors thus blocking their binding to these sites, but also changes occurring in the chromatin of a methylated template also contribute to the observed inhibition of transcription (Finnegan 2001). In plant genomes, methylation is not only restricted to CpG sequences as significant levels of cytosine methylation are also observed in nonCG sequences, which include symmetrical CNG and asymmetrical CNN sequences (Tariq and Paszkowski 2004).

The presence of 5 mC is a feature of transcriptionally silenced chromatin, and provides a plant genome with a mechanism to defend itself against transposable elements and retroviruses (Martinesen and Colot 2001; Bird 2002). Genetic alterations that reduce methylation levels result in various pleiotropic phenotypes in plants (Bird 2002). The *Arabidopsis thaliana* genome contains at least 10 genes encoding DNA methyltransferases (Finnegan and Kovac 2000; Kankel et al. 2003). Among those, the *Arabidopsis* MET1 has been extensively investigated, and found to have a complex role in various developmental processes (Finnegan and Kovac 2000). Screening plants with reduced methylation of repetitive sequences, *MET1* missense mutations (*met1-1* and *met1-2*) have been isolated exhibiting delayed flowering and loss of gene silencing (Kankel et al. 2003). Methylation in nonCG sequences, which is a common modification in plant DNA, is also catalyzed by a domain containing plant-specific methyltransferase *CHROMOMETHYLASE3* (*CMT3*) (Bartee et al. 2001). Moreover, *CMT3*, is a key determinant in CpXpG methylation (Bartee et al. 2001).

Recently, Xu et al. (2004) have conducted a detailed investigation of DNA methylation alterations during reprogramming events in somatic tissues of *R. hybrida* using the amplified fragment-length polymophism (AFLP) technique. On the basis of banding patterns, it has been observed that the highest numbers of AFLP bands are observed in embryogenic callus and in regenerants from embryogenic callus. This indicates that a number of internal cytosines are methylated during the processes of somatic embryogenesis and subsequent regeneration of somatic embryos into whole plantlets. Moreover, methylation alterations during somatic embryogenesis have been found to be characterized by extensive demethylation of outer cytosines in $5'$-$_{m}$CCGG-$3'$ sequences, and these are passed along to their regenerants. These findings provide support to the hypothesis that modified cytosines are likely essential for the acquisition of embryogenic potential in somatic cells of rose, and that these are then passed on to subsequent regenerants from somatic embryos (Xu et al. 2004).

Among methylation-related bands that have been sequenced, some have been found to be tissue-specific, and more specifically these are associated with embryogenic callus and regenerants of somatic embryos (Xu et al. 2004). The amino acid sequence of one such embryogenesis-specific band appears to be derived from the Deetiolated 1 (DET1) protein in rose. Although the function of this protein is not clearly identified in rose, it has been reported to be a regulatory gene that represses several signaling pathways controlled by light (Schafer and Bowler 2002). Moreover, some clues as to the function of this gene can be discerned from extensive studies in tomato. It has been reported that mutations in this gene are responsible for *high pigment-2* (*hp-2*) phenotypes in tomato that are characterized by exaggerated photo-responsiveness (Mustilli et al. 1999). Light-grown *hp-2* mutants display high levels of anthocyanins, are short, and more deeply-pigmented than wild-type plants. The higher pigmentation of mature fruits from these mutants is due to elevated levels of both flavonoids and carotenoids (Mustilli et al. 1999; Levin et al. 2003). Therefore, it is likely to expect that the DET1 in rose is also associated with anthocyanin content as well.

4 Genetic Transformation of Somatic Embryos

One of the most successful applications of somatic embryogenesis in rose has been the use of this cellular differentiation pathway for developing a genetic transformation system for roses. The ability to introduce and express diverse foreign genes into plants has long been employed for genetic improvement of various plant species, and it has become an important strategy for genetic improvement of roses as well. The promise of genetic transformation of roses is slowly being realized with opportunities for developing genotypes with enhanced and desirable traits coming along as recent advances are made in both somatic embryogenesis and genetic transformation protocols of rose.

Generally, plant transformation is achieved either via *Agrobacterium*-mediated transformation or via microprojectile bombardment. However, a small number of target cells typically receive the foreign DNA during these transformation events, and even a smaller number of these cells survive selection and subsequent regeneration of stable transformants. Therefore, efforts have been made to develop transformation protocols for rose using *Agrobacterium*-mediated transformation, and to a lesser extent via microprojectile bombardment.

Over a decade ago, Firoozabady et al. (1991) published the first report on successful *Agrobacterium*-mediated transformation of *R. hybrida* cv. Royalty. Later, transgenic rose plants were obtained by transforming friable embryogenic tissues of rose, recovered from filament cultures, with either *Agrobacterium tumefaciens* or *A. rhizogenes* (Firoozabady et al. 1994). Mathews et al. (1994) regenerated transgenic rose from protoplasts of embryogenic cell lines.

Van der Salm et al. (1997) obtained transgenic plants from roots derived from stem slices of the rootstock *R. hybrida* cv. Moneyway following co-cultivation with A. tumefaciens strain GV3101 containing an *nptII* gene and individual *rol* genes from *A. rhizogenes*. Grafting the transformed rootstock resulted in stimulation of both root development of the rootstock and axillary-bud break of the untransformed scion (Van der Salm et al. 1998). Marchant et al. (1998a) regenerated transgenic plants from embryogenic callus of *R. hybrida* following microprojectile bombardment with the biolistic gene gun. Subsequently, Marchant et al. (1998b) successfully introduced a chitinase gene into *R. hybrida* cv. Glad Tiding, and found that expression of the chitinase transgene reduced the severity of black spot (*Diplocarpon rosae* Wolf.) development by 13–43%.

Recently, Li et al. (2002b) have reported on an enhanced efficiency of *Agrobacterium*-mediated transformation of embryogenic cultures of *R. hybrida* cv. Carefree Beauty by taking advantage of induced secondary somatic embryogenesis (Li et al. 2002a). As transformed embryogenic cells act independently from neighboring cells, these develop into somatic embryos that further undergo secondary embryogenesis. It is observed that transgenic lines with similar Southern hybridization profiles exhibit the same level of transcription as demonstrated by similar band intensities in Northern blots. Therefore, the transformation efficiency is estimated to be at least 9%. As the number of transgenic plants developing from the same transformation event is high (having undergone secondary somatic embryogenesis), this approach avoids the recovery of chimeric transgenic plants. This finding is especially important for plant species that rely on vegetative propagation.

In a later study (Li et al. 2003), this transformation protocol was used to introduce an antimicrobial protein encoding gene, *Ace-AMP1*, into *R. hybrida* cv. Carefree Beauty. Some of the recovered transgenic plants exhibited enhanced resistance to the fungal pathogen powdery mildew [*Sphaerotheca pannosa* (Wallr.: Fr.) Lev. var. *rosae*]. This was demonstrated in both a detached leaf assay and an in vivo greenhouse assay of whole plants. These promising findings offer new opportunities for developing roses with resistance to various economic diseases, among other useful and desirable traits such as flowering habit, growth habit, and flower quality and longevity.

5 Conclusions

Somatic embryogenesis has been successfully achieved in a number of rose genotypes. Various efforts have been made to induce somatic embryos from different tissues of rose plants as well. Recent efforts to induce secondary somatic embryogenesis have been quite promising and encouraging. However, it is important to note that plant cells may undergo some genetic changes while they undergo cellular differentiation, such as somatic embryogenesis, in vitro. As a result, it is important to monitor those changes in gene regulation that are often attributed to changes in DNA methylation. These changes in DNA methylation may contribute to tissue culture-induced mutagenesis, and can also lead to chromatin structure alternations, and changes in gene expression.

However, it is important to point out that the success in inducing somatic embryogenesis in roses has been critical for the successful development of transformation systems for roses. So far, these transformation protocols have resulted in the recovery of transgenic rose lines either with enhanced rooting, bud break, or disease resistance. Further opportunities for developing transgenic roses with other desirable horticultural traits will certainly arise in the near future.

References

Arene L, Pellegrino C, Gudin S (1993) A comparison of the somaclonal variation level of *Rosa hybrida* L. cv. Meirutral plants regenerated from callus of direct induction from different vegetative and embryonic tissues. Euphytica 71:83–90

Bird A (2002) DNA methylation patterns and epigenetic memory. Genes Dev 16:6–21
Davies PJ (1995) The plant hormones: their nature, occurrence, and functions. In: Davies PJ (ed) Plant Hormones: Physiology, Biochemistry, and Molecular Biology. Kluwer, Dordrecht, pp 1–12
De Wit JC, Esendam HF, Honkanen JJ, Tuominen U (1990) Somatic embryogenesis and regeneration of flowering plants in rose. Plant Cell Rep 9:456–458
Finnegan EJ (2001) Is plant gene expression regulated globally? Trends Genet 17:361–365
Finnegan EJ, Kovac KA (2000) Plant DNA methyltransferases. Plant Mol Biol 43:189–201
Firoozabady E, Lemieux CS, Moy YS, Moll B, Nicholas JA, Robinson KEP (1991) Genetic engineering of ornamental crops. In vitro 27:96A
Firoozabady E, Moy Y, Courtney-Gutterson N, Robinson K (1994) Regeneration of transgenic rose (*Rosa hybrida*) plants from embryogenic tissue. Bio/Tech 12:609–613
Hill GP (1967) Morphogenesis of shoot primordia in cultured stem tissue of a garden rose. Nature 216:596–597
Hsia C, Korban SS (1996) Organogenesis and somatic embryogenesis in callus cultures of *Rosa hybrida* and *Rosa chinensis minima*. Plant Cell Tiss Org Cult 44:1–6
Kankel MW, Ramsey DE, Stokes TL, Flowers SK, Haag JR, Jeddeloh JA, Riddle NC, Verbsky ML, Richards EJ (2003) *Arabidopsis* MET1 cytosine methyltransferase mutants. Genetics 163:1109–1122
Kintzois S, Manos C, Makri O (1999) Somatic embrogenesis from mature leaves of rose (*Rosa* sp.). Plant Cell Rep 18:467–472
Kunitake H, Imamizo H, Mii M (1993) Somatic embryogenesis and plant regeneration from immature seed-derived calli of rugosa rose (*Rosa rugosa* Thumb). Plant Sci 90:187–194
Levin I, Frankel P, Gilboa N, Tanny S, Lalazar A (2003) The tomato *dark green* mutation is a novel allele of the tomato homolog of the *DEETIOLATED1* gene. Theor Appl Genet 106:454–460
Li, X, Krasnyanski S, Korban SS (2002a) Somatic embryogenesis, secondary somatic embryogenesis, and shoot organogenesis in Rosa. J Plant Physiol 159:313–319
Li X, Krasnyanski S, Korban SS (2002b) Optimization of the *uidA* gene transfer into somatic embryos of rose via *Agrobacterium tumefaciens*. Plant Physiol Biochem 40:453–459
Li, X, Gasic K, Cammue B, Broekaert W, Korban SS (2003) Transgenic rose lines harboring an antimicrobial protein gene, Ace-AMP1, demonstrate enhanced resistance to powdery mildew (*Sphaerotheca pannosa*). Planta 218:226–232
Litz RE, Gray DJ (1995) Somatic embryogenesis for agricultural improvement. World J Microbiol Biotech 11:416–425
Marchant R, Davey MR, Lucas JA, Power JB (1996) Somatic embryogenesis and plant regeneration in floribunda rose (*Rosa hybrida* L. cvs. Trumpeter and Glad Tidings). Plant Sci 120:95–105
Marchant R, Power JB, Lucas JA, Davey MR (1998a) Biolistic transformation of Rose (*Rosa hybrida* L.). Ann Bot 81:109–114
Marchant R, Davey MR, Lucas JA, Lamb CJ, Dixon RA, Power JB (1998b) Expression of a chitinase in rose (*Rosa hybrida* L) reduces development of black spot disease (*Diplocarpon rosae* Wolf). Mol Breed 4:187–194
Martienssen RA, Colot V (2001) DNA methylation and epigenetic inheritance in plants and filamentous fungi. Science 293:1070–1074
Murli S, Sreedhar D, Lokeswari TS (1996) Regeneration through somatic embryogenesis from petal-derived calli of *Rosa hybrida* L. Arizona (hybrid tea). Euphytica 91:271–275
Mustilli AC, Fenzi F, Ciliento R, Alfano F, Bowler C (1999) Phenotype of the tomato *high pigment-2* mutant is caused by a mutation in the tomato homolog of *DEETIOLATED1*. Plant Cell 11:145–157

Ng HH, Bird AP (1999) DNA methylation and chromatin modification. Curr Opin Genet Dev 9:158–163

Noriega C, Sondahl MR (1991) Somatic embryogenesis in hybrid tea roses. Bio/Tech 9:991–993

Raemakers, CJJM, Jacobsen E, Visser RGF (1995) Secondary somatic embryogenesis and applications in plant breeding. Euphytica 81:93–107

Rout GR, Debata BK, Das P (1991) Somatic embryogenesis in callus culture of *Rosa hybrida* L. cv. Landora. Plant Cell Tiss Org Cult 27:65–69

Rout GR, Samantaray S, Mottey J, Das P (1999) Biotechnology of the rose: a review of recent progress. Scient Hort 81:201–228

Sarasan V, Roberts AV, Rout GR (2001) Methyl laurate and 6-benzyladenine promote the germination of somatic embryos of a hybrid rose. Plant Cell Rep 20:183–186

Schafer E, Bowler C (2002) Phytochrome-mediated photoperception and signal transduction in higher plants. EMBO Rep 3:1042–1048

Tariq M, Paszkowski J (2004) DNA and histone methylation in plants. Trends Genet 20:244–251

Van der Salm TPM, Bouwer R, van Dijk AJ, Keizer LCP, Hanish Ten Cate CH, Van Der Plas LHW, Dons JJM (1998) Stimulation of scion bud release by *rol* gene transformed rootstocks of *Rosa hybrida* L. J Exp Bot 49:847–852

Van der Salm TPM, van der Toorn CJG, Hanischten cate CH, Dons HJM (1996) Somatic embryogenesis and shoot regeneration from excised adventitious roots of the rootstock *Rosa hybrida* cv. Money Way. Plant Cell Rep 15:522–526

Van der Salm TPM, van der Toorn CJG, Bouwer R, Don HJM (1997) Production of *rol* gene transformed plants *Rosa hybrida* L. and characterisation of their rooting ability. Mol Breed 3:39–47

Visessuwan R, Kawai T, Mii M (1997) Plant regeneration systems from leaf segment culture through embryogenic callus formation of *Rosa hybrida* and *R. canina*. Breed Sci 47:217–222

Xu ML, Li X, Korban SS (2004) DNA-methylation alterations and exchanges during in vitro cellular differentiation in rose (Rosa hybrida L.). Theor Appl Genet 109:899–910

Plant Cell Monogr (2)
A. Mujib · J. Šamaj: Somatic Embryogenesis
DOI 10.1007/7089_030/Published online: 20 December 2005

Embryogenesis in *Catharanthus roseus*: Roles of Some External Factors in Proliferation, Maturation and Germination of Embryos

A. Junaid (✉) · A. Mujib · M. A. Bhat · A. Ilah · M. P. Sharma

Cellular Differentiation and Molecular Genetics Section, Department of Botany, Hamdard University, 110062 New Delhi, India

Abstract *Catharanthus roseus* is an important medicinal plant that contains two well-known anticancerous alkaloids, vincristine and vinblastine. Cell culture technology has been employed for a long time to improve the alkaloid yield. In this chapter, various processes of somatic embryogenesis such as embryo induction, proliferation, maturation and germination are described. In this embryogenic system, embryos showed irregularities in structure and registered poor conversion frequency. Several carbon sources were added in order to improve the embryo quality before germination: 3% fructose or 3–6% maltose were found to be effective during maturation. Plantlet conversion was high on 3–6% maltose and 3% fructose. In addition, suspension culture, indirect embryogenesis and loss of embryogenic potential with time are discussed in brief. The authors felt that the low yields of vincristine and vinblastine may be improved if the single cell embryo origin concept is utilized in a genetic modification program.

1 Introduction

Catharanthus roseus is a fleshy perennial, growing up to 32-in. (80-cm) high. It has glossy, dark green, oval leaves and flowers all summer long. *C. roseus* is native to the Indian Ocean island of Madagascar. This herb is commoner in many tropical and subtropical regions worldwide, including the southern USA.

Extracts of entire dried plant contain many alkaloids of medicinal use. The principal alkaloid is vinblastine or vincaleukoblastine (vinblastine sulfate), sold as Velban. The alkaloid has a growth inhibition effect in certain human tumors. Vinblastine is used experimentally for treatment of neoplasms and is recommended for generalized Hodgkin's disease and resistant choricarcinoma. Another pharmacologically important alkaloid is vincristine sulfate or vincristine. Vincristine is used against leukemia in children. Vinblastine and vincristine in combination has resulted in 80% remission in Hodgkin's disease, 99% in acute lymphocitic leukemia, 80% in Wilm's tumor, 70% in gestational choricarcinoma and 50% in Burkett's lymphoma. There are over 100 other alkaloids in addition to vinblastine and vincristine. Synthetic vincristine also used to treat leukemia is, however, only 20% effective as compared with the natural product derived from *C. roseus*.

Since 1950, cell culture techniques have been used to improve alkaloid content in *Catharanthus*. The process has been divided into two phases:

1. Establishment of culture
2. Extraction of alkaloids

For establishing culture, various plant parts, i.e., explants (shoot, root, callus, organs, suspension, etc.), have been used. Several key factors that have major control over the biosynthesis of alkaloids have been optimized and were reviewed (Moreno et al. 1995; Mujib et al. 2003). However, the study of somatic embryogenesis has not yet been reported and its importance to enhance yield not assessed. It is a remarkable process by which plant cells are transformed into embryos in culture. Although, the process has been reported in a wide range of plants, plantlet recovery is not always satisfactory. This is partly due to the absence of an optimized system which induces rapid embryo formation and proliferation. The induced somatic embryos also show a range of abnormalities in structure, secondary/adventive embryo formation on primary structures and a higher degree of heterogeneity (Akula et al. 2000; Cho et al. 1998; Ilah et al. 2002). The quality of somatic embryo, in turn, determines the success of maturation and in vitro germination. The low rate of embryo germination and subsequent poor conversion is one of the major challenges in embryogenic research. Somatic embryos with normal morphology also behave differently in different cultural conditions (Soh et al. 2000). A variety of studies have recently been conducted to enhance proliferation rate and plant recovery (Saito et al. 1991; Etlenne et al. 1997; Afreen et al. 2002; Lee et al. 2001). The present chapter describes the role of plant growth regulators in *Catharanthus* and the involvement of external factors like carbohydrate and pH is assessed at different stages of development.

2 Establishment of Somatic Embryo in *Catharanthus*

The process of somatic embryogenesis is a complex multistep process which is divided into the following stages: (1) establishment and maintenance of embryogenic tissues from explant, (2) proliferation of embryos and (3) embryo maturation and germination.

2.1 Induction and Maintenance of Embryogenic Tissue

Initiation of callus tissues in *Catharanthus* was induced from various tissues like stem, leaf and root; however, induction of embryogenic callus was only achieved from hypocotyl tissue derived from in vitro germinated seeds. Two media, namely, Murashige and Skoog (MS) and White, were used, and both proved to be effective in establishing culture. The production of somatic embryos is controlled by various external factors such as carbon sources, nitrogen source, dissolved oxygen and pH. Since Skoog and Miller (1957) the role of auxins in tissue culture especially in somatic embryogenesis has been well established (Dudits et al. 1991; Davletova et al. 2001). A number of auxins, both natural (indole-3-acetic acid, indole-3-butyric acid) and synthetic (such as naphthalene acetic acid, NAA, 2,4-dichlorophenoxyacetic acid, 2,4-D, chlorophenoxyacetic acid, CPA, and 2,4,5-trichlorophenoxyacetic acid) have been regularly added to the culture media for somatic embryogenesis. However, auxin involvement in triggering embryogenesis has been only noted at the early stage of embryogenesis; later on auxins inhibit embryo growth. The removal or addition of lower concentrations of auxins was thus necessary. Auxins like 2,4-D and CPA are also required for the formation of the callus on which the embryo originates from "induced embryogenic cells" (Sharp et al. 1980). The rapid uptake of auxins results in depletion of the medium and in liquid medium they disappear early and eventually increase the plant growth regulator (PGR) level within the tissues. In *Catharanthus*, all the auxins (2,4-D, CPA and NAA) had a profound influence on callusing: the effective concentration only varies and generally lies within 0.5–2.0 mg/L. The hypocotyl callus was friable, light yellow, fast growing and the callus mass transformed into embryogenic tissue. The other explant sources (stem, leaf, etc.) induced calli which are non-embryogenic in nature, being characterized by their compact and nodular appearance and relatively slow growth. The embryogenic calli of hypocotyl origin were routinely maintained on medium supplemented with the same or a lower concentration of auxin alone or in combination with cytokinin (6-benzylaminopurine, BAP). Periodic transfer of tissues (3-week intervals) onto fresh nutrient media kept the callus mass growing and prevented necrosis. Subculture at extended intervals, however, reduced embryogenic ability; this temporary regenerative loss was resumed on restoration of normal cultural conditions.

2.1.1 Indirect Embryogenesis

Somatic embryogenesis has been reported in numerous plant genera where two distinct modes have been recognized. In some cultures, embryogenesis occurs directly without any callus phase, whereas in indirect embryogenesis the embryo develops from already induced meristematic callus clusters. In *Catharanthus* vigorous embryogenesis was established following a callus phase from hypocotyls. Many of the cultures also developed adventitive/secondary embryos. In such cases the growth of primary structures was significantly checked. Several embryos on solid medium were aggregated, laterally coalesced and showed ill-developed roots and other abnormalities.

2.1.2 Suspension Culture

Suspension culture was established by transferring 2–3-week-old embryogenic callus on MS liquid medium containing auxin alone or with cytokinin. Continuous agitation on a gyratory shaker at 120 rpm yielded rapid proliferation of embryogenic callus and released free cells with cell aggregates. The embryogenic cells were small and round, contained abundant starch and divided rapidly to form cell aggregates (two to five cells) of quadrilateral to hexagonal appearance. Some of the single cells were elongated and vacuolated; these cells showed limited cell divisions with transverse end-to-end attachment. After 1 week, a proembryo-like structure developed from cell aggregates and later transformed into globular and heart-shaped embryos. However, elongated and vacuolated cells did not participate in embryogenic processes.

The globular or heart-shaped embryos did not progress to maturity in liquid medium. Use of solid medium at this stage and onwards is important for embryo maturity. This strongly suggests a need of stability (which the semisolid agar provides) to establish a shoot–root axis/polarity at advanced stages of embryo development. However, a second round of callusing and embryogenesis was also simultaneously noted in solid media. A similar arrest of growth of somatic embryos in liquid medium was earlier noted in other plant systems (Soh et al. 2000).

2.2 Proliferation of Embryos

Four to five week old embryogenic calli differentiated into embryos in NAA (1.0 mg/L) added medium; other auxin sources were less effective for production of embryos. A heterogeneous mixture of somatic embryos (globular, heart and cotyledonary) was visible under a simple microscope. Embryos were induced generally in masses along with proliferating clumps of embryogenic callus. Addition of BAP in NAA-supplemented (1.0 mg/L) medium improved the embryo proliferation process (Fig. 1a, Table 1).

The pH of the medium, a key cultural condition, influences in vitro responses. Thus, a range of pH values (4.0–7.0) were tested to see their effect on embryogenesis. Table 4 shows the influence of the initial pH on the production of somatic embryos. The maximum embryo productivity was recorded in media with pH 5.5–6.0, adjusted before autoclaving. Wetherell and Dougall (1976) earlier observed the same pH range for somatic embryo production in carrot. However, the set pH generally changes in all the media after auto-

Table 1 Somatic embryogenesis in proliferation media. Murashige and Skoog (*MS*) medium contained naphthalene acetic acid (1.0 mg/L) with various 6-benzylaminopurine (*BAP*) concentrations (sources: 40–50 mg embryogenic callus, data scored after the seventh week of culture)

BAP concentration (mg/l)	Embryogenesis (%)	No. of somatic embryos/culture	Different stages of somatic embryos: Globular	Heart	Torpedo	Cotyledonary
0.5	43.75 ± 4.20^{d}	38.75 ± 2.27^{d}	18.25 ± 1.7^{c}	12.0 ± 1.6^{c}	6.25 ± 1.70^{b}	2.25 ± 1.2^{b}
1.0	61.75 ± 3.03^{ab}	82.5 ± 3.69^{b}	54.0 ± 1.87^{a}	18.5 ± 2.6^{b}	7.00 ± 1.75^{b}	3.00 ± 0.9^{b}
1.5	73.00 ± 3.67^{a}	99.25 ± 2.27^{a}	61.5 ± 1.18^{a}	22.5 ± 1.2^{a}	9.00 ± 0.80^{a}	6.25 ± 1.7^{a}
1.75	49.00 ± 5.52^{c}	46.25 ± 2.58^{d}	21.5 ± 1.29^{c}	18.5 ± 1.3^{b}	4.00 ± 0.81^{c}	2.25 ± 1.2^{a}
2.0	41.25 ± 4.60^{d}	64.75 ± 3.69^{c}	39.25 ± 1.7^{b}	12.2 ± 1.7^{c}	10.2 ± 1.70^{a}	3.00 ± 0.8^{a}
ANOVA						
F	9.031	11.310	2.346	0.759	1.102	2.011
P^{+}	0.000***	0.000***	0.88 ns	0.44*	0.550 ns	0.492*
LSD 5%	5.102	3.909	2.536	2.176	1.182	1.075

Values are means ± standard errors of at least three replicates. Within each column, values followed by the *same superscript letter* are not significantly different at the $P = 0.05$ level according to the least significant difference (*LSD*) test.
F test significant at $^{***}P < 0.001$, $^{*}P < 0.05$

Fig. 1 Somatic embryogenesis and plant regeneration in *Catharanthus roseus* **a** A heterogeneous mixture of somatic embryos in proliferation medium. **b** Large green somatic embryo in maturation medium. **c** Somatic embryo with black necrotic zone at shoot-root axis. **d** Somatic embryo regenerated plantlet

claving and during the culture period (Smith and Krikorian 1990; Owen et al. 1991; Huang et al. 1993; Sakano 1997) and may alter embryo production ability. The mature embryo productivity was also similarly high (Table 5) when the initial pH was adjusted to 5.5–6.0.

2.3 Embryo Maturation, Germination and Role of Carbon Sources

In in vitro culture plant cells or tissues show little autotrophic property, even the apparently green tissues are not fully autotropic and need external carbon for energy. The addition of various carbon sources in the media enhances cell growth, regeneration and also influences somatic embryogenesis (Verma and Dougall 1977). However, poor embryo quality limits plantlet conversion frequency. Recently, a number of treatments have been adapted to embryos involving the use of abscisic acid, sugar, sugar–alcohol, poly(ethylene glycol), etc. during maturation and germination (Xing et al. 1999; Lipavska and Konradova 2004; Robichaud et al. 2004). Sucrose is generally the carbon source of choice; however, other sugars are used frequently in tissue culture. In this chapter the roles of various carbon sources are evaluated at different stages of embryogenesis.

Individual white-opaque cotyledonary somatic embryos were cultured on MS medium fortified with gibberellic acid (1.0 mg/L) for maturation. Somatic embryos turned green (Fig. 1b), increased in length and occasionally became coiled but did not germinate into plantlets. However, the green embryos germinated well (Fig. 1d) in media supplemented with BAP (0.5 mg/L). The maturation and germination were influenced by carbohydrate sources (Tables 2, 3) as the somatic embryos increased in size in all the sugar sources tested and maintained steady growth up to the seventh week of culture. The 3% level of carbohydrate is more active than the 6% level in which embryo growth was slow and this tendency was noted for all carbon sources, such as maltose, glucose, fructose and even sucrose. Germination, i.e., plantlet conversion, is high in 3–6% maltose and 3% fructose, whereas 3% glucose and 6% sucrose/fructose had little effect on germination. In some of the sugars

Table 2 Somatic embryo in maturation media (MS + 1.0 mg/L gibberellic acid), added with different carbohydrates

Treatment	Initial length of embryos (mm)	Length after 5 weeks (mm)	Length after 7 weeks (mm)
Sucrose 3%	5.70 ± 0.5^{bc}	8.50 ± 0.3^{b}	10.05 ± 0.2^{c}
Sucrose 6%	5.52 ± 0.2^{bc}	7.275 ± 0.2^{a}	9.05 ± 0.3^{dc}
Maltose 3%	6.57 ± 0.3^{a}	9.675 ± 0.2^{a}	11.47 ± 0.3^{a}
Maltose 6%	5.80 ± 0.6^{b}	8.300 ± 0.2^{bc}	9.54 ± 0.3^{cd}
Glucose 3%	6.37 ± 0.2^{a}	8.750 ± 0.2^{b}	10.725 ± 0.5^{b}
Glucose 6%	6.00 ± 0.4^{ab}	7.533 ± 0.2^{d}	8.625 ± 0.4^{c}
Fructose 3%	5.725 ± 0.5^{b}	8.675 ± 0.3^{b}	10.625 ± 0.4^{b}
Fructose 6%	5.175 ± 0.3^{c}	7.42 ± 0.8^{cd}	9.70 ± 0.2^{c}
ANOVA			
F	8.416	12.262	13.12
P^{+}	0.000***	0.000***	0.002 ns
LSD 5%	0.542	0.657	0.559

Values are means ± standard errors of five replicates with six embryos in each replicate. Within each column, values followed by the *same superscript letter* are not significantly different at the $P = 0.05$ level according to the LSD test.
F test significant at $^{***}P < 0.001$

Table 3 Somatic embryo germination (plantlet conversion) in BAP (0.5 mg/L). MS medium also contained different sugars and concentrations

Treatments	Plant conversion (RL + SL) RL (mm)	SL (mm)	Only root RL (mm)	Only shoot SL (mm)	Leaf number LN
Sucrose 3%	5.80 ± 0.7^{d}	8.57 ± 0.4^{d}	–	11.80 ± 0.4^{b}	3.5 ± 1.2^{a}
Sucrose 6%	–	–	11.92 ± 0.5^{a}	9.275 ± 0.4^{d}	2.0 ± 1.8^{a}
Maltose 3%	8.17 ± 0.2^{b}	11.6 ± 0.4^{b}	–	11.00 ± 0.1^{bc}	2.5 ± 1.2^{a}
Maltose 6%	9.36 ± 0.4^{a}	12.80 ± 0.1^{a}	9.35 ± 0.5^{b}	12.77 ± 0.7^{a}	3.0 ± 2.1^{a}
Glucose 3%	–	–	–	9.625 ± 0.6^{d}	1.5 ± 1.2^{a}
Glucose 6%	5.30 ± 0.4^{d}	8.55 ± 0.5^{d}	–	11.00 ± 0.4^{c}	2.5 ± 2.3^{a}
Fructose 3%	7.26 ± 0.3^{c}	9.55 ± 0.5^{c}	–	11.75 ± 0.1^{bc}	2.5 ± 1.2^{a}
Fructose 6%	–	–	7.30 ± 2.6^{b}	7.925 ± 0.3^{c}	2.2 ± 2.6^{a}
ANOVA	0.752	0.267	2.646	0.379	0.567
F	0.043*	0.0674 ns	0.048*	0.820 ns	0.345*
P^{+}	0.536	0.554	0.436	0.700	2.455
LSD 5%					

Data were recorded after 10 weeks on germinating medium following 7 weeks in maturation media. Values are means ± standard errors of five replicates with six embryos per replicate per treatment. Values followed by the *same superscript letter* are not significantly different at the $P = 0.05$ level according to the LSD test.

Dashes indicate that there were no converted plantlets and and there was no root development.

F test significant at $^{*}P < 0.05$

Table 4 Effect of pH on somatic embryo proliferation in *Catharanthus roseus*

pH values	No. of somatic embryo/culture	Different stages of somatic embryos Globular	Heart	Torpedo	Cotyledonary
4.0	53.66 ± 2.4	31.34 ± 2.6	16.00 ± 1.7	4.66 ± 0.5	2.33 ± 0.5
4.5	61.00 ± 2.4	32.00 ± 1.6	21.00 ± 1.0	5.00 ± 2.0	3.00 ± 2.6
5.0	61.33 ± 2.6	34.67 ± 2.0	16.00 ± 3.6	6.33 ± 2.0	4.33 ± 2.0
5.5	69.66 ± 3.2	40.33 ± 1.2	16.67 ± 3.7	7.00 ± 2.0	5.66 ± 1.5
5.8	80.33 ± 2.9	49.34 ± 2.3	21.67 ± 4.0	9.33 ± 2.1	7.00 ± 2.4
6.0	99.25 ± 2.2	61.50 ± 1.1	22.50 ± 1.2	9.00 ± 0.8	6.25 ± 1.7
6.5	59.00 ± 7.4	30.66 ± 1.6	20.33 ± 2.0	4.00 ± 2.6	4.00 ± 2.0
7.0	24.00 ± 3.5	15.33 ± 1.2	8.66 ± 3.7	-	

- Values are means ± standard errors of at least 3 replicates.
- Hormones for proliferation (MS+NAA 1.0 mg/L)+BAP (1.5 mg/L)
- Incubation period: 6th weeks of culture.
- Sugar: Maltose 6%
- Inoculam: Embryogenic callus

Table 5 Effect of pH on somatic embryo maturation

pH values	Matured embryo/culture	Forms of embryos Normal embryo (%)	Abnormality (%)
4.0	18.00 ± 3.00	10.00	50.00
4.5	18.67 ± 3.51	33.99	21.94
5.0	19.66 ± 3.05	51.00	7.98
5.5	21.65 ± 0.57	60.84	3.95
5.8	22.48 ± 0.34	73.42	3.08
6.0	24.00 ± 2.0	78.34	2.20
6.5	29.00 ± 0.81	84.00	2.34
7.0	6.33 ± 1.53	7.98	10.32

- Values are means ± standard errors of at least 3 replicates.
- Hormones for plant maturation: MS + GA_3 (1.0 mg/L)
- Incubation period: 6 weeks
- Sugar: Maltose 6%
- Inoculum: 30 embryo/culture

tested, a black necrotic zone developed at the shoot-root junction as a mark of an adverse effect (Fig. 1c).

Except for glucose, the sugar level only induced primary roots without any visible shoot and has little importance in a plantlet multiplication program. The involvement of carbohydrate sources on embryo maturation and germination was observed earlier in some systems (Alemanno et al. 1997; Li et al.

1998; Corredoira et al. 2003). But the entire physiology is still very complex to understand fully.

3 Loss of Embryogenic Potential

In *Catharanthus*, embryogenesis is very fast and readily induced from hypocotyl. The potentiality decreases with the age of the culture. The plant growth regulator that was active previously is less effective with the age of the cultures. The lost potentiality was recovered at least partially, where the combination and level of PGR was replaced with a new set of combinations. Early subculturing (2-week interval) has proved to be effective also to some degree. This incidence, however, is common in tissue culture; changes in ploidy of the culture cells and inhibitors released by the aging tissues were previously described as some of the reasons responsible for this embryogenic loss.

4 Conclusion and Some Areas of Interest

C. roseus is a medicinal plant well known for its anticancerous properties. In cell culture techniques several tissues/explants have been used to establish culture; however, the importance of somatic embryogenesis has not been realized fully in an alkaloid improvement program. The present study indicates that embryos were produced in large numbers in solid media; however, in some cases embryogenesis is associated with embryo abnormalities like aggregation of proembryos/embryos, ill-developed roots, secondary callusing and embryogenesis, and root degeneration. Use of bioreactors may minimize such irregularities (Denchev et at. 1992; Hvoslef-Eide et al. 2002) and it also has the ability to improve biomass growth and to increase differentiation and plantlet production. Despite its many promises, the use of a vessel or bioreactor is still not integrated in alkaloid research.

Two different pathways of somatic embryogenesis have been discussed in plant systems, i.e., direct embryogenesis on explant and indirect embryogenesis via a callus phase. In both cases, the origin of the embryo is said to be from a single cell, which is easily amenable to genetic modification. The approaches like *Agrobacterium tumefaciens* mediated genetic alteration, T-DNA insertional mutagenesis, in vitro mutagenesis and selection of induced mutants, and protoplast fusion may generate new cell lines/plants with improved yield.

The process of embryogeny, particularly the aspect of maturation, germination or plantlet conversion, is a complex mechanism of interdisciplinary nature involving embryology, physiology, biochemistry and other subjects. Although many of the facts have been addressed quite successfully in recent times, there are still questions that remain unanswered. Reduction in structural abnormalities will definitely increase the regenerability of somatic embryos. Besides, proper embryo selection and their transfer to optimized germination medium, selection of germinated rooted plantlets and their transfer to soil for acclimatization are some of the important stages and/or cultural practices that need more attention for success and reproducibility of plantlet production.

Embryonal masses have been preserved for many purposes. In *Catharanthus* the cryopreservation method has recently been established where the pretreatment, cryoprotectants, cooling and thawing processes have been optimized (Mannonen et al. 1990). Storage in liquid nitrogen and mineral oil is also used for the preservation of genetically engineered cells. On receiving appropriate cultural conditions, superior cell lines with high alkaloid producing ability will resume normal growth (Bacchiri 1995), but the information is still not enough in *Catharanthus*.

References

Afreen F, Zobayed SMA, Kozai T (2002) Photoautotropic culture of arabusta somatic embryos: Development of a bioreactor for large scale plantlet conversion from cotyledonary embryos. Ann Bot 90:21–29

Akula A, Becker D, Bateson M (2000) High yielding repetitive somatic embryogenesis and plant recovery in a selected tea clone, "TRI-2025" by temporary immersion. Plant Cell Rep 19:1140–1145

Alemanno L, Berthouly M, Michaux-Ferriere N (1997) A comparison between *Theobroma cacao* L. zygotic embryogenesis and somatic embryogenesis from floral explants. In Vitro Cell Dev Biol Plant 33:163–172

Bachiri Y, Gazeau C, Hansz J, Morisset C, Dereuddre J (1995) Successful cryopreservation of suspension cells by encapsulation dehydration. Plant Cell Tissue Org Cult 43:241–248

Cho DY, Lee EK, Soh WY (1998) Anomalous structure of somatic embryos developed from leaf explant cultures of *Angelica gigas* Nakai. Korean J Plant Tissue Cult 25:1–5

Corredoira E, Ballester A, Vieitez AM (2003) Proliferation, maturation and germination of *Castanea sativa* Mill. somatic embryos originated from leaf explants. Ann Bot 92:129–136

Davletova S, Meszaros T, Miskolezi P, Oberschall A, Torok K, Magyar Z, Dudits D, Deak M (2001) Auxins and heat shock activation of a noval member of the calmodulin like domain protein kinase gene family in cultured alfalfa cell. J Exp Bot 52:215–221

Denchev PD, Kullin AI, Scragg AH (1992) Somatic embryo production in bioreactors. J Biotechnol 26:99–109

Dudits D, Bogre L, Gyorgyey J (1991) Molecular and cellular approaches to the analysis of plant embryo development from somatic cells in vitro. J Cell Sci 99:475–484

Etlenne H, Lartaud M, Michaux-Ferriere N, Carron MP, Berthouly M, Teisson C (1997) Improvement of somatic embryogenesis in *Hevea brasiliensis* (Mull. ARG.) using the temporary immersion technique. In Vitro Cell Dev Biol 33:81–87

Huang L, Chi C, Vits H, Sataba EJ, Cooke TJ, Hu W (1993) Population and biomass kinetics in fed-batch cultures of *Daucus carota* L. somatic embryos. Biotechnol Bioeng 41:811–818

Hvoslef-Eide AK, Olsen ORS, Lyngved R, Heyerdahl PH (2002) Bioreactor design for clonal propagation of somatic embryo. In: Abstracts of the 1st international symposium on liquid system for in vitro mass propagation of plants, Oslo, Norway, 30 May–1 June

Ilah A, Abdin MZ, Mujib A (2002) Somatic embryo irregularities in vitro cloning of Sandal (*Santalum album*). Sandalwood Res Newslett 15:2–3

Lee EK, Cho DY, Soh WY (2001) Enhanced production and germination of somatic embryos by temporary starvation in tissue culture of *Daucus carota*. Plant Cell Rep 20:408–415

Li XY, Feng H, Huang H, Murphy JB, Gbur EE Jr (1998) Polyethylene glycol and maltose enhance somatic embryo maturation in loblolly pine (*Pinus teada* L.). In vitro Cell Dev Biol Plant 34:22–26

Lipavska H, Konradova H (2004) Somatic embryogenesis in conifers: the role of carbohydrate metabolism. In Vitro Cell Dev Biol Plant 40:23–30

Mannonen L, Toivonen L, Kauppinen VC (1990) Effect of long term preservation on growth and productivity of *Panax ginseng* and *Catharanthus roseus* cell culture. Plant Cell Rep 9:173–177

Moreno PRH, Van der Heijden R, Verpoorte R, Vander-Heijden R (1995) Cell and tissue culture of *Catharanthus roseus*, a literature survey II. Updating from 1988–1993. Plant Cell Tissue Org Cult 42:1–25

Mujib A, Ilah A, Gandotra N, Abdin MZ (2003) In vitro application to improve alkaloid yield in *Catharanthus roseus*. In: Govil JN, Kumar PA, Singh VK (eds) Biotechnology and genetic engineering. Recent progress in medicinal plants, vol. IV. Sci Tech, Houston, USA, pp 415–440

Owen HR, Wengerd D, Miller AR (1991) Culture medium pH is influenced by basal medium, carbohydrate source, gelling agent, active charcoal and medium storage method. Plant Cell Rep 10:583–586

Robichaud RL, Lesser VC, Merkle SA (2004) Treatments affecting maturation and germination of American chestnut somatic embryos. J Plant Physiol 161:957–969

Saito T, Nishizawa S, Nishimura S (1991) Improved culture conditions for somatic embryogenesis from *Asparagus officinalis* L. using an aseptic ventilative filter. Plant Cell Rep 10:85–89

Sakano K, Kiyota S, Yazaki Y (1997) Acidification and alkalinization of culture medium by *Catharanthus roseus* cells is anoxic production of lactate a cause of cytoplasmic acidification? Plant Cell Physiol 38:1053–1059

Sharp WR, Sondahl MR, Caldas LS, Maraffa SB (1980) The physiology of in vitro asexual embryogenesis. Hortic Rev 2:268–310

Skoog F, Miller CO (1957) Chemical regulation of growth and organ formation in plant tissues culture in vitro. Symp Soc Exp Biol 11:118–130

Smith DL, Krikorian AD (1990) Somatic embryogenesis of carrot in hormone-free medium: external pH control over morphogenesis. Am J Bot 77:1634–1647

Soh WY, Lee EK, Cho DY (2000) Germination arrest of carrot somatic embryos cultured in MS basal liquid medium. Korean J Plant Tissue Cult 27:175–180

Verma DC, Dougall DK (1977) Influence of carbohydrates on quantitative aspects of growth and embryo formation in wild carrot suspension cultures. Plant Physiol 59:81–85

Wetherell DF, Dougall DK (1976) Sources of nitrogen supporting growth and embryogenesis in cultured wild carrot tissue. Physiol Plant 37:97–103

Xing Z, Powell WA, Maynard CA (1999) Development and germination of American Chesnut somatic embryos. Plant Cell Tissue Org Cult 57:47–55

Plant Cell Monogr (2)
A. Mujib · J. Šamaj: Somatic Embryogenesis
DOI 10.1007/7089_026/Published online: 22 December 2005

Somatic and Zygotic Embryogenesis in Avocado

C. Sánchez-Romero[1] · B. Márquez-Martín[1] · F. Pliego-Alfaro[2] (✉)

[1]IFAPA, CIFA Málaga, Cortijo de la Cruz s/n, 29140 Churriana-Málaga, Spain
csanrome@olinet.es

[2]Dpto. Biología Vegetal, Facultad de Ciencias, Campus de Teatinos s/n, 29071 Málaga, Spain
ferpliego@uma.es

Abstract Avocado is a species widely cultivated for its highly nutritious fruit. Currently, soil-borne diseases such as *Phytophthora* root rot are severe threats for commercial plantings and breeding programmes by conventional means to select genotypes tolerant to this disease are under way in different countries. Use of biotechnological tools would be very useful for improvement of this crop. Somatic embryogenesis could be used to generate variability in vitro as well as for genetic transformation. At present, somatic embryos can be obtained by culturing immature zygotic embryos in Murashige and Skoog's (MS) medium supplemented with 0.1 $mg\,l^{-1}$ picloram. Proliferation of embryogenic cultures can take place under the same conditions used for culture initiation; however, development of white-opaque somatic embryos requires cultivation of previously synchronized globular-stage embryos in a B5 formulation based medium solidified with agar at 10 $g\,l^{-1}$. Conversion of these embryos takes place at a 10–20% rate in MS medium at half strength with a 0.5 $mg\,l^{-1}$ benzylaminopurine supplement. Somatic embryos from adult explants have occasionally been obtained; however, methods are needed to improve the induction process as well as to properly mature the resulting embryos, before somatic embryogenesis can be widely used in avocado breeding programmes.

1 Introduction

Avocado (*Persea americana* Mill.) is the most important species within the Lauraceae family, which comprises 50 genera and 2500 or more species, distributed mainly in tropical and subtropical areas (Rohwer 1993). The genus *Persea* includes about 50 species, predominantly of American origin (Bergh 1969).

Avocado is a species widely cultivated for its fruits, which present a high oil content and constitute a well-balanced supply of proteins, carbohydrates, minerals and vitamins (Ray 2002). It is a vigorous and evergreen tree that may reach, at adult age, 20 m in height. Flowering occurs in winter and spring and avocado flowers are pubescent, regular, complete and trimerous. Hundreds of flowers appear grouped in a compound panicle of racemes. The fruit is a single-seeded berry whose size, shape, colour and oil content can vary widely among cultivars. The mature seed is large, fleshy and exalbuminous and is surrounded by a thick, fleshy and buttery pulp. Because of seed recalcitrance, they are cold- and desiccation-sensitive and lose viability shortly after harvest. The embryo has two massive, straight cotyledons (Bergh 1969, 2000.)

The species *Persea americana* has traditionally been subdivided into three horticultural races (Popenoe 1934) which have recently been recognized as botanical varieties: *P. americana* var. *americana* (West Indian race), var. *guatemalensis* (Guatemalan race) and var. *drymifolia* (Mexican race) (Bergh 2000). Botanical varieties appear to have evolved in different geographical locations from the centre of origin of the species, the Chiapas (Mexico)–Guatemala–Honduras area (Kopp 1966). The three varieties have distinctive adaptations and evident distinguishing horticultural and botanical features (Bergh 1975).

2 Crop Improvement Strategies and Applications of Biotechnology

Avocado is a commercially important tropical crop, which is extending its geographical distribution from its origin to other tropical and subtropical areas. At present, major centres of avocado production are Mexico, Brazil, the Dominican Republic, Colombia, the USA, Chile and Indonesia (Ray 2002). Productivity of avocado orchards is limited by a series of problems that constitute the main objectives of avocado breeding programmes currently carried out in different countries, e.g. resistance to *Phytophthora* root rot, caused by the fungus *Phytophthora cinnamomi* Rands, the most important disease affecting avocado orchards throughout the world (Zilberstaine and Ben-Ya'ácov 1999), as well as resistance to other diseases such as *Rosellinia* root rot, *Verticillium* wilt, anthracnose, *Cercospora* spot, and sunblotch viroid. Additional breeding objectives include regular bearing and good fruit quality, dwarf plant stature, soil stress resistance and fruit ripening (Ray 2002; Litz et al. 2005).

Conventional breeding shows a series of inconvenience, when applied to tree species, that largely limits the transference and expression of genes of interest. Thus, biotechnological approaches could be useful for improvement of this crop. Somatic embryogenesis offers a high potential for use as a complementary technology to traditional breeding programmes. It can allow widening of the genetic basis by generating variability in vitro (somaclonal variation), through somatic hybridization or by direct introduction of genes (genetic transformation). Moreover, embryogenic systems are suitable for long-term cryopreservation of germplasm and can also become a means for mass clonal propagation of avocado selections.

Some of these applications are already being used in avocado, e.g. genetic transformation of embryogenic cultures with genes related to disease tolerance (antifungal protein, glucanase and chitinase) and fruit ripening (*S*-adenosylmethionine hydrolase) has recently been reported (Gómez-Lim and Litz 2004). Preliminary studies have also been carried out on the effect of γ-radiation on avocado embryogenic cultures to obtain somaclonal variants (Witjaksono and Litz 2004). Somatic hybridization has also been attempted by Witjaksono and Litz, who reported the fusion of protoplasts from avocado embryogenic cultures with non-embryogenic protoplasts of *Phytophthora* root rot resistant *Persea* species (Litz et al. 2005).

3 Zygotic Embryogenesis: Histological and Biochemical Aspects

The establishment of a pattern for zygotic embryogenesis is a very useful tool for evaluating somatic embryogenesis protocols and determining the way in which nutrients, hormones and other culture factors affect this developmental process under in vitro conditions (Wetzstein et al. 2000). In avocado, zygotic embryogenesis has been exhaustively studied considering physiological, histological and biochemical aspects (Perán-Quesada 2001; Perán-Quesada et al. 2005; Sánchez-Romero et al. 2002).

The avocado zygotic embryo, as others of the recalcitrant type, has a prolonged growth period, which ranges from 6 to 12 months depending on the cultivar (Whiley 1992). In cultivar Hass, histodifferentiation has been shown to occur until about 100 days after pollination (DAP) (16–18 mm long embryos), while the beginning of the maturation phase was evident at 125 DAP, in embryos 24–26 mm in length; this event was revealed by a massive starch granule accumulation and the visualization of protein bodies for the first time (Perán-Quesada et al. 2005). The initiation of this stage has also been associated with a significant increase in fresh and dry weights as well as a decrease in water content and hexoses to sucrose ratio; afterwards, the synthesis of specific storage proteins takes place (Sánchez-Romero et al. 2002). Avocado zygotic embryo development continues until approximately 305 DAP, when embryos reach 38–40 mm in length; at physiological maturity, embryos present maximum fresh and dry weights, minimum water content and an albumin of 49 kDa whose expression has been associated with the latest maturation stage (Sánchez-Romero et al. 2002).

The developmental pattern established for the avocado zygotic embryo reveals an important role of the maturation phase. During this period, a sequential and ordered accumulation of specific storage products as well as a series of physiological changes occur, which appear to be required for achieving physiological maturity, at which virtually 100% of the embryos can be converted into vigorous and healthy plants (Perán-Quesada et al. 2005).

4 Somatic Embryogenesis

Somatic embryogenesis in avocado was first reported by Pliego-Alfaro (1981) using immature zygotic embryos, cultivar Hass. Since then, it has been accomplished from zygotic embryos of different cultivars (Mooney and Van Staden 1987; Raviv et al. 1998; Witjaksono and Litz 1999a; Perán-Quesada et al. 2000). More recently, induction of embryogenic cultures from the nucellus of immature fruits (Witjaksono 1997; Vidales et al. 2003; Márquez-Martín, unpublished results) and different floral structures (Chaparro and Sánchez-Romero, unpublished results) has also been reported. Nevertheless, somatic embryogenesis from adult explants is genotype-dependent and the majority of studies are currently carried out using immature zygotic embryos as the initial explant.

4.1 Preparation of Explants

Immature avocado fruits are surface-sterilized according to Pliego-Alfaro (1981) and Pliego-Alfaro and Murashige (1988) by immersion in a 0.5% (v/v) sodium hypochlorite solution supplemented with 0.1% (v/v) Tween 20 for 10 min, and rinsed three times with sterilized distilled water. Afterwards, fruits are carefully cut lengthways under sterile conditions and the embryos are found embedded in a gelatinous endosperm in the ovule's micropylar end. The embryos are isolated with care and placed individually on the surface of the nutrient medium.

4.2 Initiation of Embryogenic Cultures

Pliego-Alfaro (1981) and Pliego-Alfaro and Murashige (1988) established the culture medium components required for inducing embryogenic cultures from immature zygotic embryos. Induction medium (MS+0.1P) consisted of MS salts and vitamins (Murashige and Skoog 1962), 30 $g\,l^{-1}$ sucrose, 0.1 $mg\,l^{-1}$ picloram and 8 $g\,l^{-1}$ agar. The 0.1 $mg\,l^{-1}$ picloram supplement appeared to be critical since lower (0.001–0.01 $mg\,l^{-1}$) or higher (1 $mg\,l^{-1}$) concentrations failed to induce embryogenic callus. These results have also been confirmed by Mooney and Van Staden (1987).

In the first reports, explants for somatic embryogenesis induction were incubated either in darkness or under light conditions (32–35 $\mu mol\,m^{-2}\,s^{-1}$) (Pliego-Alfaro 1981; Mooney and Van Staden 1987; Pliego-Alfaro and Murashige 1988). Nowadays, however, the initiation of this type of cultures is usually carried out under constant darkness (Witjaksono and Litz 1999a; Perán-Quesada et al. 2004).

An important factor affecting the induction of embryogenic cultures in avocado is the developmental stage of the zygotic embryo used as explant. The optimum size for embryogenic response is 0.6–0.8 mm, approximately 21–28 DAP (Pliego-Alfaro and Murashige 1988).

The embryogenic response occurs 18–40 days after explanting (Witjaksono and Litz 1999a). Cultures initiate mainly from the embryo's hipocotyl region (Fig. 1) (Perán-Quesada 2001), although Witjaksono and Litz (1999a) state that very immature zygotic embryos are totally responsive. The initiation frequency ranges between 0 and 25%, varying clearly among cultivars (Witjaksono and Litz 1999a; Perán-Quesada et al. 2000).

Established avocado embryogenic cultures are quite heterogeneous and are composed by nodular structures, proembryogenic masses (PEMs) and somatic embryos at different developmental stages. Since the initiation, the morphology of avocado embryogenic cultures varies greatly depending upon the genotype. Non-embryogenic white to grey and amorphous callus is also observed on induction medium although it can be easily distinguished from the light creamy-pale yellow and friable embryogenic callus (Pliego-Alfaro 1981; Pliego-Alfaro and Murashige 1988).

Witjaksono and Litz (1999a) studied the effect of different mineral formulations on the induction of avocado embryogenic cultures. Although B5 (Gamborg et al. 1968) major salts induced somatic embryogenesis in more

Fig. 1 Initiation of avocado embryogenic cultures from immature zygotic embryos on MS+0.1P medium

avocado genotypes than standard or modified MS major salts, no significant differences among treatments could be inferred.

4.3 Culture Maintenance and Proliferation

Maintenance and proliferation of avocado embryogenic cultures are carried out in the same culture conditions indicated for the induction phase: MS+0.1P medium solidified with 6–8 $g\,l^{-1}$ agar and incubation in darkness at $25 \pm 1\,°C$ (Pliego-Alfaro and Murashige 1988; Witjaksono et al. 1999; Perán-Quesada et al. 2004). Under these conditions, embryogenic cultures continually sector into embryogenic and non-embryogenic portions with a frequency clearly dependent upon the genotype. Therefore, subculturing to fresh medium implies careful discrimination of non-embryogenic callus.

According to Witjaksono and Litz (1999a), two types of embryogenic cultures can morphologically be distinguished in avocado (Fig. 2):

- PEM-type cultures, consisting of PEMs with occasional development of proembryos and somatic embryos at heart and later stages
- SE-type cultures, consisting of somatic embryos at different developmental stages, from the globular to the cotyledonary stage, and low frequency of PEMs and proembyos

Cultures appearance is genotype-dependent with most genotypes showing a SE-type response (Witjaksono and Litz 1999a).

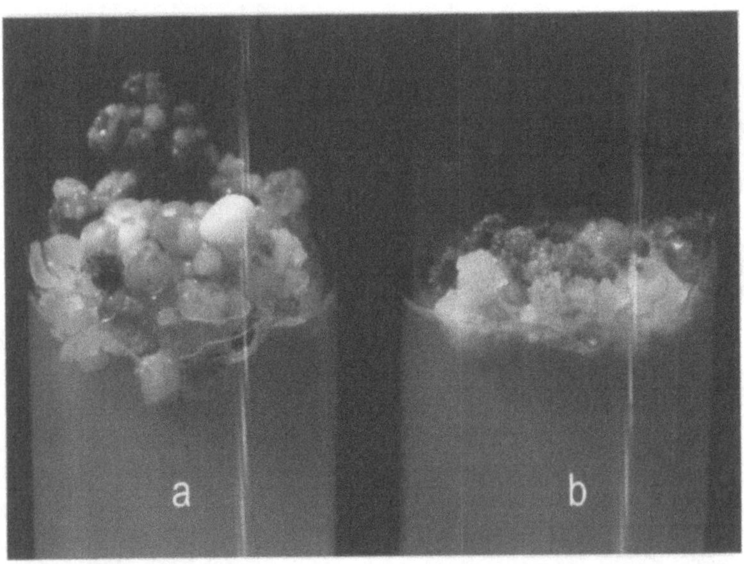

Fig. 2 Avocado embryogenic cultures on solid maintenance medium: SE-type culture (*a*) and PEM-type culture (*b*)

Different factors have been studied in relation to maintenance of avocado embryogenic cultures with special attention to the basal medium formulation and the gelling agent (Witjaksono and Litz 1999a; Perán-Quesada 2001). When testing the effect of different formulations [MS major salts, B5 major salts without $(NH_4)_2SO_4$ ($B5^-$) and $B5^-$ supplemented with 400 mg l^{-1} glutamine (B5-G)] and gelling agents (8 g l^{-1} agar or 2 g l^{-1} gellan gum), Witjaksono and Litz (1999a) found that medium formulation, gelling agent and their interaction showed a significant effect on the morphology of avocado embryogenic cultures. MS medium (Murashige and Skoog 1962) gelled with agar resulted in maximum PEM proliferation, while modifications of B5 major salts solidified with gellan gum produced a larger number of globular and cotyledonary (smaller than 5 mm) embryos. Similar results were found by Perán-Quesada (2001), who also indicated that depending upon the genotype, prolonged maintenance in gellan gum solidified media causes an increase of the disorganized growth and a loss of embryogenic potential. Consequently, as stated before, the optimum conditions for proliferation and maintenance of embryogenic traits are the same as those indicated for the induction phase.

Avocado embryogenic cultures maintained in agar-gelled MS+0.1P medium continue to proliferate over several years; however, the appearance of the cultures changes with time and a trend to form smaller and more disorganized structures is observed (Fig. 3), e.g. SE-type cultures lose their ability to develop somatic embryos at advanced developmental stages and acquire, at the end, the PEM-type morphology. In PEM-type cultures, a disorganization of the embryogenic structures is also observed (Witjaksono and Litz 1999a).

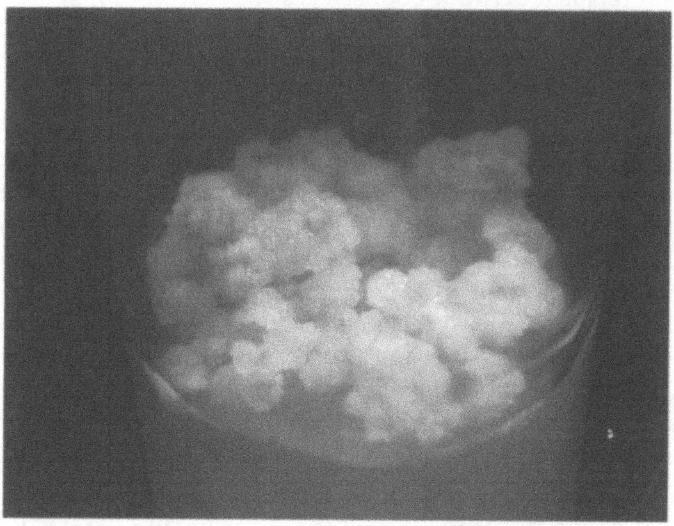

Fig. 3 Disorganization of avocado embryogenic cultures after a prolonged maintenance period

Maintenance of embryogenic cultures in liquid medium is also feasible. Suspension cultures can be established following the protocol of Witjaksono and Litz (1999a), e.g. inoculation of 0.4 g embryogenic cultures in 40 ml liquid MS+0.1P medium in 100 ml Erlenmeyer flaks. Cultures are maintained on a rotary shaker at 120 rpm and 25 ± 1 °C in semidarkness conditions with subculturing onto fresh medium every 2 weeks. However, liquid conditions are not adequate for maintaining all types of avocado embryogenic cultures. Disorganization is generally quicker in SE-type cultures and, in fact, some SE-type genotypes cannot be maintained in liquid medium for prolonged periods (Witjaksono and Litz 1999a). Consequently, avocado embryogenic cultures are usually maintained in solid medium and embryogenic suspensions are only occasionally established.

4.4 Somatic Embryo Development and Maturation

Maturation involves accumulation of storage products and, as a consequence, translucent somatic embryos turn white-opaque (w-o) (Cailloux et al. 1996). This morphological change has been used as an indicator of the efficiency of treatments employed for inducing development and maturation of avocado somatic embryos (Fig. 4) (Witjaksono and Litz 1999b; Perán-Quesada et al. 2004).

Witjaksono and Litz (1999b) used embryogenic suspensions as source material for inducing avocado somatic embryo development. Embryogenic suspensions were successively filtrated through 1.8 and 0.8 mm screens and the fraction retained between both screens was selected to induce development of w-o embryos. Culturing in liquid medium and subsequent selective sieving allowed synchronization of cultures, a very important aspect since it allows the application of maturation treatments to developmentally uniform embryogenic material.

Investigations carried out by Márquez-Martín et al. (2003) have shown that factors related to the embryogenic suspension, such as inoculum size and time in culture, have an important influence on the subsequent capacity for developing w-o somatic embryos. Nine days in liquid medium was optimum for SE-type cultures, while for PEM-type cultures, growth for 14 days gave better results. In both cases, maximum w-o somatic embryo development occurred when the inocula derived from suspensions in the linear growth phase. In relation to culture initial density, standard values (0.4 g per 40 ml) gave good results in SE-type cultures, while for PEM-type cultures, better results were obtained at much higher initial densities (4.0 g per 40 ml).

The ability of avocado embryogenic cultures to produce w-o somatic embryos varied significantly with the type of culture (SE or PEM type). A higher recovery of w-o somatic embryos as well as a higher frequency of w-o embryos at advanced developmental stages (5 mm or larger) could only be ob-

Fig. 4 White-opaque somatic embryos obtained after 4–5 weeks on B5m medium solidified with 10 g l^{-1} agar

tained from SE-type cultures (Márquez-Martín et al. 2003). This result is in accordance with previous observations of Witjaksono and Litz (1999a), who reported that a low-frequency somatic embryo production was associated with PEM-type cultures.

Generally, auxin removal is a critical step to induce somatic embryo development (von Arnold et al. 2002). In avocado, embryos at advanced developmental stages can be observed in the presence of auxin, although its removal enhances the process. Besides auxin, the influence of other factors affecting somatic embryo development has also been studied; Witjaksono and Litz (1999b) recommended the MS formulation; however, significantly better results have been obtained with B5 major salts by Perán-Quesada et al. (2004). According to these authors, while MS formulation favoured the formation of PEMs and somatic embryos at early developmental stages, B5 macronutrients stimulated the development of w-o cotyledonary somatic embryos.

Gelling agent type and concentration also appeared to be critical factors for development of avocado somatic embryos. Witjaksono and Litz (1999b) recommended the use of gellan gum at 6 g l^{-1}; however, Márquez-Martín et al. (2001), obtained better results when using agar at 10–12 g l^{-1} in comparison with agargel (3.5–10 g l^{-1}) or gellan gum (1.7–6.8 g l^{-1}).

In relation to the sugar effects, although w-o somatic embryo production was significantly stimulated when using high sucrose concentrations, alone (Perán-Quesada et al. 2004) or in combination with different agar concentrations (Márquez-Martín, unpublished results), the quality of the resulting w-o embryos was worse than that obtained when only high agar concentrations were used, e.g. in the presence of sucrose, w-o embryos appeared partially beige to tan-coloured and their surface was quite irregular.

Abscisic acid (ABA) has been repetitively used on somatic embryo maturation; however, its role in avocado is not clear. On average, higher w-o somatic embryo production has been recorded in ABA-supplemented media; nevertheless, this effect appeared to be a consequence of a previous increase in the production of globular translucent somatic embryos caused by ABA (Perán-Quesada et al. 2004).

Other supplements tested, e.g. organic nitrogen sources, osmotic agents or active charcoal, have not improved the results obtained when using B5m basal medium (B5 major salts with MS minor salts and vitamins) gelled with $10\,g\,l^{-1}$ agar (Márquez-Martín, unpublished results). Nevertheless, only a low percentage of w-o somatic embryos developed under these conditions can be converted into plants (Sánchez-Romero, unpublished results).

Two causes have generally been indicated as limiting somatic embryo conversion: morphological abnormalities and deficient maturation (Ammirato 1987). Problems related to the correct development of shoot and root meristems have repeatedly been reported in avocado somatic embryogenesis (Mooney and Van Staden 1987; Pliego-Alfaro and Murashige 1988). However, considering that full maturity is needed in avocado zygotic embryos for achieving high germination percentages (Perán-Quesada et al. 2005), the low conversion rates observed in the somatic embryos could also be due to lack of maturation. Therefore, the introduction into the culture sequence of an additional maturation phase appears to be advisable.

4.5
Somatic Embryo Conversion

Different culture media and conditions have been tested to induce conversion of avocado somatic embryos, e.g. Witjaksono and Litz (1999b) used MS medium supplemented with $1\,mg\,l^{-1}$ benzylaminopurine and $1\,mg\,l^{-1}$ gibberellic acid and solidified with $8\,g\,l^{-1}$ agar, while Perán-Quesada et al. (2004) induced germination of embryos following partial cotyledon removal and culture on M1 medium (Skene and Barlass 1983) gelled with $1.7\,g\,l^{-1}$ gellan gum; however, the reported germination percentages were generally low (0–11.11%) and very dependent on the genotype. Treatments that have significantly improved zygotic embryo conversion, such as germination in liquid medium (Peran-Quesada 2001) and desiccation at high relative humidity (Sánchez-Romero et al. 2003), have also been applied to w-o somatic embryos; however, no positive results were obtained (Marquez-Martin, unpublished results).

Somatic seedlings obtained after germination (Fig. 5) are generally weaker and smaller than those derived from mature zygotic embryos, with shoots

Fig. 5 Avocado somatic embryo germinated on solid M1 medium

2–3 mm in length in some cases. Skene and Barlass (1983) used micrografting on seedlings to recover weak shoots obtained after germination of immature avocado zygotic embryos. This technique was also used successfully by Raharjo and Litz (2003) to recover weak shoots derived from somatic embryos.

4.6 Acclimatization

Somatic embryo derived plantlets have successfully been transferred to ex vitro conditions (Fig. 6) (Perán-Quesada et al. 2004). When somatic embryo derived shoots reached a minimum size (approximately 0.5 cm), they could be micropropagated following the procedure of Barceló-Muñoz et al. (1990) for juvenile avocado. Micropropagated shoots longer than 1.5 cm rooted at an 80% rate following a 3-day exposure to liquid MS medium with macroelements at 0.3X and supplemented with 1 mg l^{-1} indolebutyric acid. Somatic plantlets were transplanted to trays containing a mix of peat, coconut fibre and perlite and maintained under polyethylene tunnels with 100% relative humidity and light irradiance of 110–120 μmol m^{-2} s^{-1}. After 8 weeks under these conditions, the survival rate was 92%.

Fig. 6 Somatic embryo derived plantlet under ex vitro conditions

5
Conclusions

Since Pliego-Alfaro (1981) reported somatic embryogenesis in avocado, much progress has been made in this area. Induction of embryogenic cultures and plantlet regeneration has been reported for different avocado cultivars; although the most commonly used explant is still the immature embryo. Moreover, somatic embryo conversion occurs at low rates. For somatic embryogenesis to be applied in breeding programmes, further research needs to be carried out on the induction of embryogenic cultures from mature explants of selected trees as well as on the improvement of maturation conditions to promote the optimal accumulation of storage products and, ultimately, the maximal conversion into quality plants.

Acknowledgement The authors are grateful for the support provided by the Comisión Interministerial de Ciencia y Tecnología, Spain (grant no. AGL 2004-07028-C03-03/AGR).

References

Ammirato PV (1987) Organizational events during somatic embryogenesis. In: Green CD, Somer DA, Hackett WP, Biesboer DD (eds) Plant tissue and cell culture. Liss, New York, pp 57–81

Barceló-Muñoz A, Pliego-Alfaro F, Barea JM (1990) Micropropagación de aguacate (*Persea americana* Mill) en fase juvenil. Actas de Horticultura 1:503–506

Bergh BO (1969) Avocado (*Persea americana* Miller). In: Ferwerda FP, Witt F (eds) Outlines of perennial crop breeding in the tropics. Miscellaneous papers no 4. Landbouwhogeschool (Agricultural University), Wageningen, pp 23–51

Bergh BO (1975) Avocados. In: Janick J, Moore JN (eds) Advances in fruit breeding. Purdue University Press, West Lafayette, p 541–567

Bergh BO (2000) *Persea americana*. In: Halevy AH (ed) Handbook of flowering, vol 5. CRC, Boca Raton, pp 253–268

Cailloux F, Julien-Guerrier J, Linossier L, Coudret A (1996) Long-term somatic embryogenesis and maturation of somatic embryos in *Hevea brasiliensis*. Plant Sci 120:185–196

Gamborg OL, Muller RA, Ojima K (1968) Nutrient requirements of suspension cultures of soybean root cells. Exp Cell Res 50:151–158

Gómez-Lim MA, Litz RE (2004) Genetic transformation of perennial tropical fruits. In Vitro Cell Dev Biol Plant 40:442–449

Kopp LE (1966) A taxonomic revision of the genus *Persea* in the western hemisphere (*Persea*-Lauraceae). Mem NY Bot Gard 14:1–120

Litz RE, Witjaksono, Raharjo S, Efendi D, Pliego-Alfaro F, Barceló-Muñoz A (2005) *Persea americana* Avocado. In: Litz RE (ed) Biotechnology of fruit and nut crops. Biotechnology in agriculture series, no 29. CABI, Wallingford, pp 326–347

Márquez-Martín B, Sánchez-Romero C, Barceló-Muñoz A, Pliego-Alfaro F (2001) Efecto del agente gelificante sobre el desarrollo de embriones somáticos de aguacate. Abstracts book. IV Reunión de la Sociedad Española de Cultivo in vitro de Tejidos Vegetales, Santiago de Compostela, Spain, p 29

Márquez-Martín B, Sánchez-Romero C, Perán-Quesada R, Barceló-Muñoz A, Pliego-Alfaro F (2003) Efecto del tipo de callo, tiempo de precultivo en suspensión y densidad de inóculo en el desarrollo de embriones somáticos de aguacate (*Persea americana* Mill.). Abstracts book. V Reunión de la Sociedad Española de Cultivo in vitro de Tejidos Vegetales, Pamplona, Spain, p 13

Mooney PA, Van Staden J (1987) Induction of embryogenesis in callus from immature embryos of *Persea americana*. Can J Bot 65:622–626

Murashige T, Skoog F (1962) A revised medium for rapid growth and bioassays with tobacco tissue cultures. Physiol Plant 15:473–497

Perán-Quesada R (2001) Embriogénesis in vitro de aguacate (*Persea americana* Mill.). PhD thesis, University of Málaga, Spain

Perán-Quesada R, Sánchez-Romero C, Barceló-Muñoz A, Simón-Pérez E, Pliego-Alfaro F (2000) Somatic embryogenesis in different avocado (*Persea americana* Mill) cultivars. In: Ríordáin FO (ed) Development of integrated systems for large-scale propagation of elite plants using in vitro techniques (EUR 19237-COST action 822). Office for Official Publications of the European Communities, Luxembourg, p 125

Perán-Quesada R, Sánchez-Romero C, Barceló-Muñoz A, Pliego-Alfaro F (2004) Factors affecting maturation of avocado somatic embryos. Sci Hortic 102:61–73

Perán-Quesada R, Sánchez-Romero C, Pliego-Alfaro F, Barceló-Muñoz A (2005) Histological aspects of avocado embryo development and effect of developmental stages on germination. Seed Sci Res 15:125–132

Pliego-Alfaro F (1981) A morphogenetic study of the avocado (*Persea americana*, Mill.) in vitro. I. Development of a rooting bioassay and its application to studying restoration by grafting of rooting competence in adult shoots. II. Somatic embryogenesis in callus. PhD thesis, University of California

Pliego-Alfaro F, Murashige T (1988) Somatic embryogenesis in avocado (*Persea americana* Mill.) in vitro. Plant Cell Tissue Org Cult 12:61–66

Popenoe W (1934) Early history of the avocado. Calif Avocado Assoc Ybk 106–110

Raharjo S, Litz RE (2003) Rescue of genetically engineered avocado by micrografting. Proceedings V congreso mundial del aguacate, vol 2, Málaga, Spain, pp 119-122

Raviv A, Avenido RA, Tisalona LF, Damasco OP, Mendoza EMT, Pinkas Y, Zilkah S (1998) Callus and somatic embryogenesis of *Persea* species. Plant Tissue Cult Biotech 4:196–206

Ray PK (2002) Breeding tropical and subtropical fruits. Springer, Berlin

Rohwer JG (1993) Lauraceae. In: Kubitzi K, Rohwer JG, Bittrich V (eds) The families and genera of vascular plants. II. Flowering plants, dicotyledons. Springer, Berlin, pp 366–391

Sánchez-Romero C, Perán-Quesada R, Barceló-Muñoz A, Pliego-Alfaro F (2002) Variations in storage protein and carbohydrate levels during development of avocado zygotic embryos. Plant Physiol Biochem 40:1043–1049

Sánchez-Romero C, Perán-Quesada R, Márquez-Martín B, Barceló-Muñoz A, Pliego-Alfaro F (2003) Efecto de la desecación parcial sobre la germinación de embriones zigóticos inmaduros de aguacate. Proceedings V congreso mundial del aguacate, vol 1, Málaga, Spain, pp 83–87

Skene KGM, Barlass M (1983) In vitro culture of abscissed immature avocado embryos. Ann Bot 52:667–672

Vidales-Fernández I, Salgado-Garciglia R, Gómez-Lim MA, Angel-Palomares E, Guillén-Andrade H (2003) Embriogénesis somática de aguacate (*Persea americana* Mill. cv. Hass). Proceedings V Congreso Mundial del Aguacate, vol 1, Málaga, Spain, pp 89–95

Von Arnold S, Sabala I, Bozhkov P, Dyachok J, Filonova L (2002) Developmental pathways of somatic embryogenesis. Plant Cell Tissue Org Cult 69:233–249

Wetzstein HY, Jeyaretnam BS, Vendrame WA, Rodriguez APM (2000) Somatic embryogenesis in pecan (*Carya illinoinensis*). In: Jain SM, Gupta PK, Newton RJ (eds) Somatic embryogenesis in woody plants, vol 6. Kluwer, Dordrecht, pp 391–414

Whiley AW (1992) *Persea americana* Miller. In: Verheij EWM, Coronel RE (eds) Plant resources of South-East Asia. No 2. Edible fruits and nuts. Pudoc DLO, Wageningen, pp 249–254

Witjaksono (1997) Development of protocols for avocado tissue culture: somatic embryogenesis, protoplast culture, shoot proliferation and protoplast fusion. PhD thesis, University of Florida

Witjaksono, Litz RE (1999a) Induction and growth characteristics of embryogenic avocado cultures. Plant Cell Tissue Org Cult 58:19–29

Witjaksono, Litz RE (1999b) Maturation of avocado somatic embryos and plant recovery. Plant Cell Tissue Org Cult 58:141–148

Witjaksono, Litz RE (2004) Effect of gamma irradiation on embryogenic avocado cultures and somatic embryo development. Plant Cell Tissue Org Cult 77:139–147

Witjaksono, Litz RE, Pliego-Alfaro (1999) Somatic embryogenesis of avocado (*Persea americana* Mill.). In: Jain SM, Gupta PK, Newton RJ (eds) Somatic embryogenesis in woody plants, vol 5. Kluwer, Dordrecht, pp 197-214

Zilberstaine M, Ben-Ya'ácov A (1999) Integration of strategies for controlling root-rot in avocado in Israel. Rev Chapingo Ser Hortic 5:251–253

Plant Cell Monogr (2)
A. Mujib · J. Šamaj: Somatic Embryogenesis
DOI 10.1007/7089_025/Published online: 30 November 2005

Somatic Embryogenesis in Genera *Medicago*: an Overview

A. Iantcheva (✉) · M. Vlahova · A. Atanassov

AgroBoilnstitute, Bul. "Dragan Tzankov" 8, 1164 Sofia, Bulgaria
aneliaiancheva@abi.bg

Abstract This chapter outlines the details of somatic embryogenesis in genera *Medicago*. Various factors that influence the process of somatic embryo induction, development, maturation and conversion are discussed. The role of genotype, explant choice and preparation hormonal compositions and the origin of somatic embryos are also reviewed. Brief attention is paid to the regenerant's phenotype and fertility.

1 Introduction

The genus *Medicago* is composed of annual and perennial species. They are diploid, tetraploid and polyploid; wild and cultivated. The perennial species *M. sativa*, *M. falcata*, *M. varia*, *M. coerulea*, *M. arborea* and *M. glutinosa* are generally grouped as *M. sativa* complex. Alfalfa (*M. sativa*) is the most important forage crop cultivated on over 32 million hectares in the world (Michaud et al. 1988). For a long time it has been the object of genetic, cellular and molecular studies because of its good regeneration capacity in vitro. The first report of regeneration of *M. sativa* (Sanders and Bingham 1972) was via somatic embryogenesis. Since then, many reports on regeneration of this perennial species have been published, mostly by indirect somatic embryogenesis (Bingham et al. 1988; Arcioni et al. 1990; McKersier and Brown 1996; Barbulova et al. 2002). Regeneration via direct somatic embryogenesis was also reported in *M. sativa* (Maheswaran and Williams 1984) and *M. falcata* (Denchev et al. 1991). Annual *Medicagos* are closely related to alfalfa but they are diploid, self-pollinated and possess a short life cycle. The regeneration of annual *Medicagos* is more difficult than that of perennials. The first regeneration protocol of annual *M. truncatula* via indirect somatic embryogenesis was achieved by Nolan et al. (1989) and a few more have been reported since (Chabaud et al. 1996; Hoffmann et al. 1997; Trinh et al. 1998). Protocols for regeneration of other annuals have also been made in *M. polymorpha* (Scarpa et al. 1993), *M. littoralis* (Zafar et al. 1995), *M. suffruticosa* (Li and Demarly 1996) and *M. lupulina* (Li and Demarly 1995). Regeneration via direct somatic embryogenesis in liquid and solid media for *M. truncatula* (Iantcheva et al. 2001; Iantcheva et al. 2005) and for *M. littoralis*, *M. murex* and *M. polymorpha* also has been established (Iantcheva et al. 1999).

In this chapter, various factors that affect the process of somatic embryo induction, development, maturation and conversion are discussed. The genotype, explant choice and preparation, origin of somatic embryos and hormonal composition of culture media are also described. Brief attention is paid to the phenotype and fertility of the obtained regenerants.

2 Induction of Somatic Embryogenesis

2.1 Type of Somatic Embryogenesis

Somatic embryogenesis is a process whereby a cell or group of cells from somatic tissue forms an embryo. The development of somatic embryos nearly replicates the process of zygotic embryo formation. Somatic embryogenesis mostly occurs indirectly via an intervening callus phase or directly, i.e. embryos develop on the explant surface like epidermal or sub-epidermal layers, as in *M. falcata* (Denchev et al. 1991) and *M. truncatula* (Iantcheva et al. 2001). See Table 1 for a list of references on the induction of somatic embryogenesis.

2.2 Genotype, Choice of Explant and Type of Preparation

Genotype is the most important factor influencing embryogenic response. Variability in the induction and frequency of the obtained embryos is observed among different species of genera *Medicago* and within the cultivars (Brown and Atanassov 1985; Chen et al. 1987). Considerable variations in embryogenic capacity were also observed between individuals of one cultivar or species. Genotype-dependent embryogenic capability was widely reported especially in *M. sativa* (Seitz Kris and Bingham 1988; Mitten et al. 1984; Chen et al. 1987; Nagaradjan et al. 1986; Barbulova et al. 2002; Ivanova et al. 1994). The use of this species for in vitro experiments requires the isolation of a highly embryogenic genotype. In general, a regenerative genotype could be found in any alfalfa germplasm if enough genotypes are screened (Brown and Atanassov 1985; Mitten et al. 1984). Together with genotype there are other factors affecting embryogenic response: explant, culture condition and medium composition. Removal of the explant from the mother plant is a prerequisite for the acquisition of embryogenic competence (Finstad et al. 1993). Choice of explant is a factor that determines success in establishing embryogenic protocol. Besides, somatic embryo regeneration was achieved from different explant sources in perennial and annual *Medicagos*. Indirect somatic embryogenesis of *M. sativa* was induced from a broad range of explants such as leaf (Meijer and Brown 1987; Barbulova et al. 2002), petiole (Lai and McKersie 1994), internode (Parrott and Bailey 1993), immature embryo (Ninkovic et al. 1995), hypocotyl (Meijer and Brown 1987; Kim et al. 2004), suspension culture and mesophyll protoplast (Atanassov and Brown 1984). For this perennial species, direct embryo formation was achieved from immature embryos (Maheswaran and Williams 1984) and from leaves in *M. falcata* (Denchev et al. 1991). Immature inflorescence is a suitable explant source for embryo induction in *M. lupulina* (Li and Demarly 1995). Explants including meristematic active zones such as hypocotyl, cotyledon, petiole base and nodal stem segments are used for direct embryo formation in *M. truncatula*, *M. littoralis*, *M. murex* and *M. polymorpha* (Iantcheva et al. 1999). Recently, direct embryo formation from root explants was reported in *M. truncatula* (Iantcheva et al. 2005).

In order to select the appropriate explant as an initial material for induction of embryogenic potential, donor tissue has to be tested for ploidy level. Different tissues are mixtures of cells with different ploidy levels (polysomaty). Moreover, given that tissue polysomaty predisposes to ploidy variation in regenerants, possible sources of explant have to be checked for polysomaty. In the study of Iantcheva et al. (2001) the ploidy levels of leaf and petioles are examined to select more uniform monosomatic tissue, dominated by 2C nuclei as an initial explant for induction of embryogenic potential.

The age of explants, size, preparation and culture environment are important factors for the type of somatic embryogenesis—indirect or direct. The age of the in vitro plant and physiological stage are of great importance for induction of somatic embryogenesis. In a direct somatic embryogenesis system (in liquid medium) of *M. falcata* (Denchev et al. 1991) and *M. truncatula* (Iantcheva et al. 2001), the leaf explants were excised from 30-day-old in vitro plant material. The explants were chopped into small pieces by razor blade to a size of 2–4 mm. Such explant preparation with severe wounding and small size, together with liquid culture conditions and agitation on a rotary shaker (100 rpm), led to direct embryo formation on the surface of the explants, with the period of induction shortened to 15–20 days. Embryos emerged first on the cut edges of the explants. In perennial *M. falcata*, indirect embryo formation was noted when leaf squares were cultured on a solid medium with the same composition, which was reported earlier by Denchev et al. (1991). Obviously the age of the explant, preparation (severe wounding) and culture environment (liquid media, agitation) are of great importance for the type of somatic embryogenesis induced. Wounding of explants on a small scale probably triggers the expression of specific genes (wound-inducible genes), which were already identified and cloned (Dudits et al. 1995) for further acquisition of embryogenic competence towards cell division and differentiation. Wounding of the explants was found to be a key factor in *M. sativa* A70-34 embryogenesis (Piccioni and Valecchi 1996).

Application of an osmoticum as stress stimulus can also lead to acquisition of embryogenic competence. Osmotic pre-treatment with 1 M sucrose of the initial root explant of *M. truncatula* is important for the shortening of the regeneration period (induction, maturation and conversion) and a higher percentage conversion of somatic embryos to plants (Iantcheva et al. 2005). This information also confirmed that embryo induction and regeneration from root explant is also genotype specific, even after osmotic pre-treatment of the primary explant of *M. truncatula cv.* Jemalong and *cv.* R 108 1. Perhaps osmotic stress activates pre-determined embryogenic cells to switch them from the somatic to embryogenic type followed by cell division. Osmotic pre-treatment for only 1 h with 1 M sucrose activated cells for division in root tips of transgenic *M. falcata* plants expressing the *gus* gene under cell cycle promoters: *cyc* 3a (cyclin type A) and *cyc* 1a (cyclin type B) (Iantcheva et al. 2004). Short-term osmotic stress is found to be necessary for the accumulation of free proline (Gangopadhyay et al. 1997) and this could be connected with the improvement of somatic embryogenesis. The positive role of proline in the induction of somatic embryogenesis of alfalfa was similarly reported by Shetty and McKersie (1993).

The endogenous hormone level of the initial explant is essential for determining the ability of a particular genotype to induce somatic embryogenesis (Jiménez 2001). The performed comparative investigation (Ivanova et al. 1994) of two *M. falcata* lines (highly embryogenic 47/1/150 and non-embryogenic 47/1/165) confirms that the level of endogenous indole-3-acetic acid (IAA) in the initial explant was higher in the embryogenic line. In the same study, the negative correlation between endogenous ABA and acquisition of embryogenic potential was observed. The investigation of Pintos et al. (2002) on the endogenous cytokinin level of embryogenic and non-embryogenic calli of *M. arborea* established the higher endogenous cytokinin level in non-embryogenic than embryogenic callus. The above studies indicated that the processes of in vitro morphogenesis (organogenesis and somatic embryogenesis) are the results of a proper balance of plant growth regulators supplied to the culture medium and endogenous regulators in the tissue of the primary explant.

The acquisition of embryogenic competence and direct formation of somatic embryos are directly related to genome size. After examination of the genome size of several annual species of *Medicago*, it was found that the smallest genome size species formed somatic embryos for the shortest period of time, with a high number of embryos per explant compared to those with a larger genome size (Iantcheva et al. 2003).

2.3 Hormonal Composition of Culture Media

To date, in all reports of alfalfa and wild *Medicago*, the induction of somatic embryogenesis is accomplished on media supplemented with an auxin (2,4-D, dichlorophenoxyacetic acid, or NAA, α-naphthaleneacetic acid) alone or in combination with cytokinin (Sanders and Bingham 1975; McKersie and Brown 1996; Brown and Atanassov 1985; Nolan et al. 1989; Chabaud et al. 1996; Pintos et al. 2002). The embryogenic effect of 2,4-D is well known in legumes and in genera *Medicago* (Denchev et al. 1991; Trinh et al. 1998; Zafar et al. 1995). 2,4-D can reach the highest intracellular concentration and usually results in high-frequency embryo formation. The concentration of 2,4-D also plays an important role in the processes of de-differentiation and differentiation in vitro (Denchev and Atanassov 1988). In the study of Barbulova et al. (2002), a 2,4-D concentration of 5 or 2 mg/l produced a more dense, necrotic and less embryogenic callus compared to the white, soft and highly embryogenic callus obtained in a medium with 1 mg/l 2,4-D. For these cultivars, the lowest concentration of 2,4-D is the optimum. According to Vergana et al. (1990) the higher concentrations of 2,4-D, at some point, block the cell division and inactivate the cells that already possess embryogenic potential. A high frequency of direct somatic embryo formation was observed in liquid medium in perennial *M. falcata* (Denchev et al. 1991) and annual species of *M. truncatula* and *M. polymorpha* (Iantcheva et al. 2001) in the presence of 4 mg/l 2,4-D. The concentrations up to 11 mg/l 2,4-D are able to induce somatic embryogenesis, while higher levels prevent induction.

The addition of NAA is essential for somatic embryogenesis initiation for some annual species like *M. polymorpha* (Scarpa et al. 1993), *M. rigidula* and *M. orbicularis* (Ibragimova and Smolenskaya 1997), and *M. truncatula* (Nolan et al. 1989). However, the molecular mechanisms involved in the induction of this process are still not fully understood. Recently a somatic embryogenesis receptor kinase (SERK) gene from *M. truncatula* (*Mt*SERK 1) was cloned and its expression examined in culture (Nolan et al. 2003). An auxin stimulates *Mt*SERK 1 expression, but its expression is much higher when both auxin (NAA) and cytokinin (6-benzylaminopurine (BAP)) are present in the medium. The effect of cytokinin appears to be more promotive in indirect somatic embryogenesis systems. Enhancement in the production of callus tissue with following embryo formation is observed in *M. truncatula* and *M. sativa* when the induction medium is supplemented with BAP (Trinh et al. 1998). For induction of embryogenic potential and its expression among different species of genera *Medicago*, different cytokinins (kinetin, BAP, zeatin, thidiazuron (TDZ) are required (Denchev et al. 1991; Nolan et al. 1998; Ding et al. 2003; Chabaud et al. 2004; Kim et al. 2004).

Induction of somatic embryogenesis by cytokinin alone is relatively rare among legumes and especially in genera *Medicago*. In legumes, somatic embryogenesis induced by cytokinin is established in *Trifolium repence* (Maheswaran and Williams 1985) and *Phaseolus* (Malik and Saxena 1992). In the annual *Medicago* species *M. truncatula*, *M. littoralis*, *M. murex* and *M. polymorpha*, direct induction of somatic embryos was achieved on solid media in the presence of TDZ or BAP (Iantcheva et al. 1999). In this system the whole process of embryogenesis from induction to maturation was completed on a medium containing cytokinin as well, and this system was species independent. Actually, the embryogenic effect of TDZ was more pronounced than that of BAP in terms of embryo number. TDZ possesses cytokinin-like activity and induced high-frequency direct somatic embryogenesis in other legumes (Saxena et al. 1992; Murthy et al. 1995). This growth regulator is found to be more active than 2,4-D and BAP. After 1 h treatment with 1 mg/l TDZ, root-tip cells of transgenic *M. falcata* plants were activated for division and expressed the *gus* reporter gene (under promoters from cell cycle regulating genes—*cyc* A and *cyc* B) more strongly than 2,4-D and BAP (Iantcheva et al. 2004). The positive embryogenic response induced by TDZ suggests that it might influence the endogenous level of cytokinins, auxins and abscisic acid (ABA) (Murthy et al. 1995; Hutchinson et al. 1996). The above mentioned *Mt*SERK 1 gene (Nolan et al. 2003) was not expressed in the presence of cytokinin, or the cells that expressed *Mt*SERK 1 were few in number in the direct somatic embryogenesis system, as the level of *Mt*SERK 1 mRNA in the tissue was relatively low and was not detectable. In the direct somatic embryogenesis system of annual *Medicago* induced by TDZ (Iantcheva et al. 1999) the process started in a small number of meristematic cells.

2.4
Origin of Somatic Embryos in Direct Embryogenesis of Model M. falcata and M. truncatula Systems in Liquid and Solid Media

The indirect somatic embryogenesis systems in genera *Medicago* are characterized by a sequence of events that includes the stimulation of cell proliferation, dedifferentiation, acquisition of embryogenic competence and the induction of embryogenesis. Treatment with an auxin (usually 2,4-D) is a characteristic move for the early stages but subsequent embryo development requires removal of exogenous auxin. One feature of indirect systems is that the initial activation of cell proliferation is temporary and physically separated from the induction of embryo-specific cell division.

Direct somatic embryogenesis is characterized by the formation of embryos directly from differentiated tissue without the apparent requirement of the dedifferentiation stage involving disorganized cell proliferation. For example, the somatic embryogenesis system of *M. falcata* (Denchev et al. 1991) and *M. truncatula* (Iantcheva et al. 2001) involved direct formation of embryos from young alfalfa leaf explants and petioles in response to an induction treatment. There are two different models to explain this phenomenon. The first proposes that there are cells within the tissue that are already embryogenically competent and require the inductive signal to trigger direct embryo formation (Williams and Maheswaran 1986; Carman 1990). It has also been argued that in the direct system, embryogenesis does not differ significantly from the indirect procedure at the molecular level, and both proceed through similar stages of genetic re-programming at different rates

(De Jong et al. 1993). These models have different explanations for cell division activation in the process of direct somatic embryogenesis. In the first, the inductive signal acts as a mitotic trigger and re-activates cell division in cells that are already competent to switch from the somatic to embryogenic type and proceed into asymmetric cell division to form embryos. In the second model, the induction of cell proliferation is required for dedifferentiation which then permits the acquisition of embryogenic competence in certain cells, just as in the indirect system.

To distinguish these models, the investigation of induction of first cell division was studied in two single-cell suspension culture systems of *M. falcata* and *M. truncatula* for direct somatic embryogenesis. Initial embryogenic cell division and embryogenic competence might be linked to the expression of reporter *gus* gene under the control of promoters from cell cycle regulatory genes (*cyc* 3a, *cdc* 2a) and green fluorescent protein (*gfp*) reporter gene under 35 S promoters. The expression pattern of the studied reporter genes and the behaviour of single embryogenic cells in liquid culture confirm the asymmetry of first cell division, which starts the process of direct somatic embryogenesis.

Confocal microscopy observation of the 35 S *gfp M. truncatula* single-cell fraction confirmed that the fraction is composed of three types of cells: spheroid, ovoid (Fig. 1A) and elongated (Iantcheva et al. 2001). Transfer of these cells into a fresh induction medium supplemented with 2,4-D reactivates the cell for division. The *gfp* was detected strongly in the nucleus where it tends to accumulate slowly and the nucleus is situated at the cell periphery (Fig. 1B). The first asymmetric division is probably a consequence of nuclear migration from the central region to the periphery, which was also observed in *M. sativa* mesophyll protoplast (Dijak and Simmonds 1988). Further development of such asymmetrically divided cells (Fig. 1C) continued with the formation of a three-cell proembryo (Fig. 1E). Confocal software offers the possibility of depicting the *gfp* fluorescence profile in cells and structures. The peak indicated that the highest level of *gfp* expression is concentrated in the nucleus. Two peaks confirm the presence of two nuclei with separation of the cell into two unequal cell parts (Fig. 1D). Three peaks correspond to the three nuclei of a three-cell proembryo (Fig. 1F.)

In other systems of direct somatic embryogenesis (*M. falcata*), a single-cell fraction (expressing *gus* gene) is formed from the initial suspension culture after 10–15 days of induction (Iantcheva et al. 2004). These cells possess the potential to divide and form embryos and develop into a whole plant. In this fraction, three types of cells are also observed: spheroid, ovoid and elongated. Most of the spherical and ovoid cells are highly cytoplasmic, reduced in size and divided asymmetrically (Fig. 2A), and are capable of forming embryos and completing their development. In most cases, the smaller cell from this division tends to form a suspensor composed of two to five cells and the other cell continues to divide and form the embryo (Fig. 2B). In a later development, the suspensor aborts in well-shaped globular embryos (Fig. 2C). By

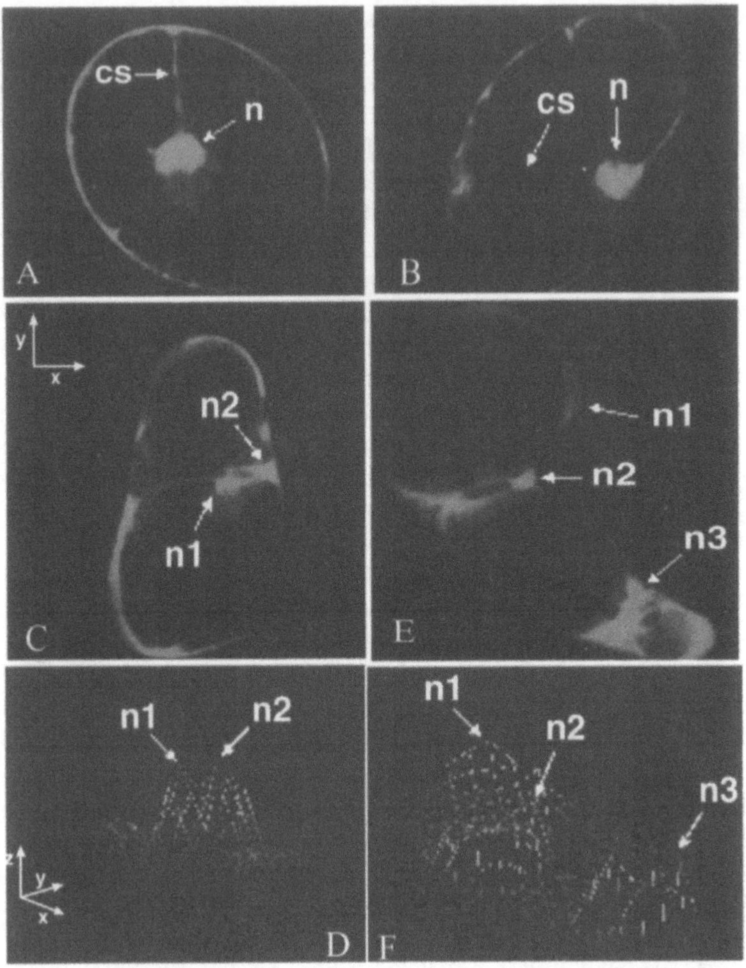

Fig. 1 Process of direct somatic embryo formation from a single cell in *Medicago truncatula*. **A** Spherical cell with a central nucleus (*n*) and cytoplasmic strands (*cs*) radiating to the cortical cytoplasm. **B** Ovoid cell with a nucleus in the cell periphery. **C** Asymmetric division with two nuclei *n1*, *n2*. **D** Level of fluorescence after first cell division; fishnet display of intensity (*z*) profiles for cell in (**C**). **E** Three-cell proembryo with nuclei *n1*, *n2*, *n3*. **F** Fishnet display of intensity profiles for proembryo in (**E**)

following the expression of *gus* gene under cyclin promoter it is possible to observe the aborted cells of suspensor which are not coloured blue, in contrast to cells of globular embryo that are still active for division (Fig. 2C). Further development of such a structure continued with the formation of torpedo and cotyledonary stage embryos, which eventually developed into plantlets and also formed secondary embryos on the surface of the primary structure (Fig. 2D,E).

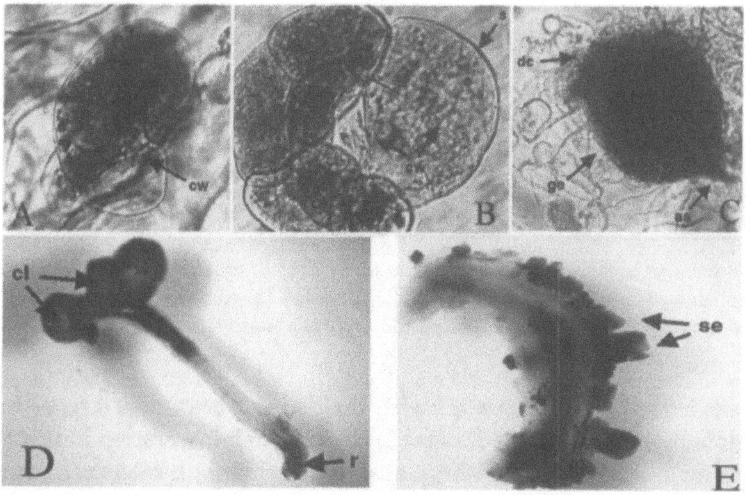

Fig. 2 Process of embryo formation from single cell to plantlet in *Medicago falcata*; *gus* activity is revealed by blue staining. **A** Asymmetric division in ovoid cell (*cw*=cell wall). **B** Proglobular embryo (*cw*=cell wall, *s*=suspensor). **C** Well-shaped globular embryo (*ge*=globular embryo, *dc*=divided cell peripheral cells, *as*=aborted suspensor). **D** Plantlet (*cl*=cotyledonary leaves, *r*=root). **E** Secondary embryo formation (*se*=secondary embryo)

2,4-D in induction medium acts as a mitotic trigger, which re-activates cell division as an inductive signal for cells in *M. falcata* and *M. truncatula* cell suspension culture. Asymmetry of the first cell division and establishment of cell polarity are the prerequisites for further embryo development. Similar results were observed in *M. varia* genotype A2 mesophyll protoplasts (Dudits et al. 1995). It is unclear what function is played by the suspensor, which develops on somatic embryos even in liquid media. It is perhaps essential for embryo polarity and serves as a channel for the nutrients and growth regulators to the developing embryo; however, it aborts later (Fig. 2C).

Single-cell suspension cultures of *M. falcata* and *M. truncatula* are particularly suitable for studying primary division and the induction of embryogenic potential of direct somatic embryos from single cells. They also confirm the asymmetry of the first cell division which starts the process of embryo formation.

Direct somatic embryogenesis of annual diploid *Medicago* on solid media supplemented with TDZ or BAP is characterized by the formation of embryos directly on the explants containing meristematic zones (Iantcheva et al. 1999). These somatic embryos develop without an intermediate callus phase. They are formed as independent units organized in clusters. The origin of somatic embryos is single cell or multicellular. The zones of embryo formation are characterized by groups of small meristematic cells with dense cytoplasm. In order to determine cell division and to reveal mitotic activity, $4',6-\text{diamidino}-2-\text{phenylindole}$ (DAPI) stained nuclei (unpublished

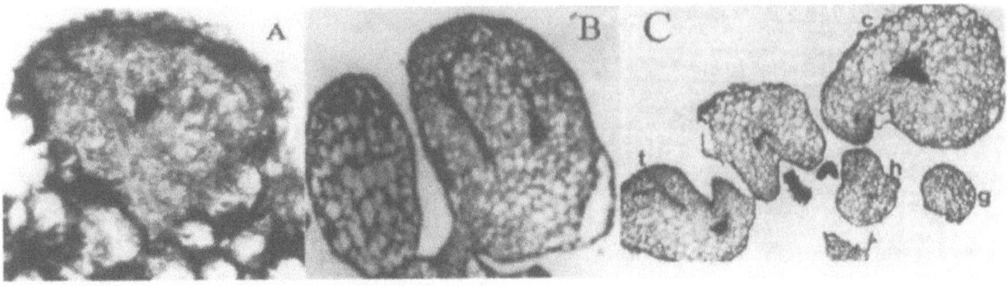

Fig. 3 Process of embryo formation promoted by TDZ in *Medicago truncatula*. **A** Primary late globular embryo. **B** Torpedo embryo connected by suspensor. **C** Stages in embryo formation (*g*=globular, *h*=heart, *t*=torpedo, *c*=cotyledonary)

data) confirmed that cells were highly divided and within 10 days of incubation globular embryos were observed. Histological observation indicated that somatic embryos develop without any connection with maternal tissue and in some cases they are connected with suspensor (Fig. 3A,B). Embryogenesis progresses through the stages typical of zygotic embryos: globular, heart, torpedo and cotyledonary (Fig. 3C). The appearance of an independent vascular system in the embryos indicated additionally that they develop as bipolar structures with apical and root parts, and possess the ability to convert to plantlets. The formation of secondary embryos on the surface of the primary structure is also detected. The origin of secondary embryos in most cases is a single cell that undergoes asymmetrical cell division. In this system and for direct induction of embryos, TDZ or BAP act as a mitotic trigger and start the process with activation of meristematic cells. Induction, development and maturation of somatic embryos proceed on the same medium in the presence of mitogene. The other advantage of the system is that it is species independent. The embryogenic capacities of the species used differ very slightly. The stable and positive embryogenic response could be due to the presence of meristematic cells in the explants, which are morphologically and physiologically more similar to each other than are differentiated somatic cells. The genotype-dependent embryogenic response which is typical for diploid *Medicago* is reduced. The ability to apply this system to a wide range of *Medicago* species is important when considering the use of gene transfer techniques.

3 Embryo Development and Maturation

Once obtained, the globular embryos from callus in indirect systems or on the explant surface in direct somatic embryogenesis proceed through the next stages, i.e. development and maturation. The formation of tissues and organs in globular, torpedo and cotyledonary embryos is a process that includes many factors and is genotype specific. In most of the *Medicago*, the requirements of growth regulators during the induction, development and maturation are specific. A significant decrease or complete elimination of auxin is necessary for further normal embryo development and maturation (Denchev et al. 1991; Trinh et al. 1998; Iantcheva et al. 2001; Barbulova et al. 2002). Reduced cytokinin concentration is essential for proper embryo development in *M. suffruticosa* (Li and Demarly 1996; Chabaud et al. 1996; Iantcheva et al. 2001). Elimination of growth regulators for successful embryo development is essential in *M. truncatula* R 108 1 (Trinh et al. 1998). In the *M. falcata* system for direct somatic embryogenesis, removal of both auxin and cytokinin are necessary for embryo development; PEG (polyethylene glycol) and maltose lead to conversion of globular embryos to high number of vigorous torpedoes. This treatment of somatic embryos with an osmotic agent such as PEG resulted in a high rate of embryo development to the next stage. In *M. truncatula cv.* Jemalong and *cv.* R 108 1, the PEG in the culture medium (Iantcheva et al. 2001) also resulted in a high number of embryos in the torpedo stage but without a normally developed vascular system. Apparently, an increased osmolality of the culture medium does not improve further development of somatic embryos as in *M. falcata*. The stage of embryo maturation is critical for embryo development, and it is mostly characterized by the reserve accumulation which determines successful conversion of embryos into a vigorous plant. Alfalfa has been intensively investigated for reserve deposition in somatic embryos, and different compounds such as abscisic acid, amino acid and different types of carbohydrates have been monitored. This issue is still not fully solved and is one of the crucial steps which limits large-scale utilization of somatic embryogenesis for speeding and improving the breeding programme in this forage crop.

In alfalfa, ABA is found to regulate storage food accumulation and prevent precocious germination (Fujii et al. 1990; Denchev et al. 1991) and it also promotes desiccation tolerance in somatic embryos (Senaratna et al. 1989). The effect of ammonium ion alone or in combination with amino acids on alfalfa somatic embryogenesis is well documented (Walker and Sato 1981; Stuart and Strickland 1984; Lai and McKersie 1994; Barbulova et al. 2002). L-Proline emerges as the most stimulatory amino acid; the optimal level of L-proline that enhances embryo yield and quality is around 100 mM. In some cases, the synergistic interaction of proline and ammonium showed a positive effect on the embryo (Stuart et al. 1985). Successful application of 3 g/l proline in the medium for embryo development and maturation was earlier reported for commercial alfalfa cultivar (Barbulova et al. 2002). Proline is known to stimulate auxin-induced somatic embryogenesis and elongates alfalfa somatic embryos in a hormone-free medium. This may be due to improved cell signalling as proline is always associated with various signal transduction pathways in plants (Phang 1985).

Amino acids such as glutamine, serine and adenine are often added either alone (Stuart and Strickland 1984) or as a component of a mixture, such as ca-

Table 1 References for induction of somatic embryogenesis in perennial and annual *Medicago*

Species	Explant	Growth regulators for induction of SE	References
		Perennial *Medicago*	
Medicago sativa	MP, CS	2,4-D+Kin	Atanassov and Brown, 1984
	H, C	2,4-D+Kin	Brown and Atanassov, 1985
	L		Arcioni et al. 1989
	P		Finstad et al. 1993
			Shetty and McKersie, 1993
			Lecouteux et al. 1993
			Lai and McKersie, 1994
			Senaratna et al. 1995
			Horbowicz et al. 1995
	L, P, IN	NAA+IAA+Kin	Parrott and Bailey, 1993
	IE	BAP	Nincovic et al. 1995
	L, P	2,4-D+Kin+Lpro	Barbulova et al. 2002
	H	IAA+Z	Kim et al. 2004
Medicago falcata	L	2,4-D+Kin	Denchev et al. 1991
	L		Kuklin et al. 1994
	L		Shao et al. 2000
Medicago sativa (diploid)	L	2,4-D+BAP	Trinh et al. 1998
Medicago coerulea	L, MP	2,4-D+BAP	Arcioni et al. 1982
Medicago varia	S	2,4-D+Kin	Deak et al. 1986
Medicago lupulina	Ii	BAP	Li and Demarly, 1995
Medicago arborea	H, C, P, L	2,4-D+Kin/BAP/TDZ	Martin et al. 2000
Medicago marina	P	2,4-D+Kin	Walton and Brown, 1988
Medicago glutinosa	L, MP	2,4-D+Z	Arcioni et al. 1982
		Annual *Medicago*	
Medicago suffruticosa	L	2,4-D+BAP	Li and Demarly, 1996
Medicago truncatula	L	NAA+BAP	Nolan et al. 1989
	L, P	2,4-D+Z	Chabaud et al. 1996
	L	2,4-D+BAP	Trinh et al. 1998
	L	2,4-D+Z	Das Neves et al. 1999
	H, CB, PB	TDZ/BAP	Iantcheva et al. 1999
	L, P	2,4-D+Kin	Iantcheva et al. 2001
	R	2,4-D+Kin	Iantcheva et al. 2005
Medicago littoralis	H	2,4-D+BAP	Zafar et al. 1995
	H, CB, PB	TDZ/BAP	Iantcheva et al. 1999
Medicago murex	H, CB, PB	TDZ/BAP	Iantcheva et al. 1999

Table 1 Continued

Annual *Medicago*			
Species	Explant	Growth regulators for induction of SE	References
Medicago polymorpha	H	2,4-D+IAA	Scarpa et al. 1993
	H, CB, PB	TDZ/BAP	Iantcheva et al. 1999
Medicago scutelata	H, C, P, R	2,4-D+Kin	Walton and Brown, 1988
	L, P, R	2,4-D+Kin	Iantcheva et al. 2003
Medicago arabica	L, P, R	2,4-D+Kin	Iantcheva et al. 2003
Medicago orbicularis	L, P, R	2,4-D+Kin	Iantcheva et al. 2003
Medicago rugosa	H, C, P, R	2,4-D+Kin	Walton and Brown, 1988

Abbreviations: SE, somatic embryogenesis; MP, mesophyll protoplast; CS, cell suspension; H, hypocotyl; C, cotyledon; CB, cotyledon base; P, petiole; PB, petiole base; L, leaf; S, stem; R, root; IN, internode; Ii, immature inflorescence; IE, immature embryos; BAP, 6-benzylaminopurine; 2,4-D, dichlorophenoxyacetic acid; NAA, naphthaleneacetic acid; IAA, indole-3-acetic acid; Kin, kinetin; TDZ, thidiazuron; Z, zeatin; L, pro-L-proline; GA3, gibberellic acid

sein hydrolysate or yeast extract (Chabaud et al. 1996) or in combination with cytokinin (Iantcheva et al. 2001) for a high rate of embryo conversion. Accumulation of proteins at the maturation stage is a key step and is a prerequisite to high-vigour conversion of somatic embryos (Krochko et al. 1992; Lai and McKersie 1994).

Secondary embryo formation is mostly observed at the embryo maturation stage. If primary embryos fail to accomplish development to plants or recallus, secondary embryos appear on their surface as observed in *M. falcata* (Denchev et al. 1991), *M. sativa* (Barbulova et al. 2002) and *M. truncatula* (Chabaud et al. 1996; Iantcheva et al. 2001; Das Neves et al. 1999). A few of the secondary embryos develop into plants, the rest are arrested at the globular or torpedo stage or give rise to an additional round of embryos. Therefore, secondary embryogenesis may be useful in the clonal multiplication of alfalfa. Secondary embryo formation was originally described by Lupotto (1983) for alfalfa and reported for the other tetraploid alfalfa genotypes (Parrott and Bailey 1993; Ninkovic et al. 1995). Repetitive formation of embryos was observed when primary embryos were transferred on hormone-free medium. The capacity for production of the new cycle of embryos of these cultures remained stable for at least 2 years but was strongly dependent on the presence of sugars in the medium (Parrott and Bailey 1993). Repetitive de novo recycling of embryos was also established for different diploid *Medicago* (*M. truncatula*, *M. littoralis*, *M. murex* and *M. polymorpha*). If one regenerated cluster of embryos and secondary embryos is isolated and transferred again on TDZ embryo induction medium, the emergence of the new embryos is visible within 20 days of culture (Iantcheva et al. 1999). This recycling procedure opens up the possibility of scaling up embryo and plantlet formation, and maintains the embryogenic potential for an unlimited period. Such a cycling regeneration system is an advantage for gene transfer research, especially in the model plant *M. truncatula* (Iantcheva et al. 2005). Repetitive embryogenesis could be obtained from a single embryogenic cell developed in liquid culture medium. Separation of such a fraction composed of highly embryogenic cells into a fresh embryo induction medium led to new embryo formation. The whole regeneration period is shorter and the embryogenic potential may be kept for four to five passes (Iantcheva et al. 2005).

4
Embryo Conversion

Embryo conversion is the last stage in the process of somatic embryogenesis. Successful conversion and germination of somatic embryos is a consequence of a proper maturity in respect of desiccation, accumulation of reserves and proteins for future conversion of embryos to seedlings. In alfalfa, this stage showed an increased level of storage proteins and free amino acids (Horbowicz et al. 1995; Lai and McKersie 1994).

It seems that the exogenous application of ABA during the development and maturation stages resulted desiccation tolerance, followed by post-maturation quiescence which prevented precocious germination and enhanced the conversion rate (Senaratna et al. 1995; Kuklin et al. 1994). In the case of *M. falcata*, exogenous ABA application is effective against precocious germination and it also favours successful development of single embryos to plantlets. The presence of GA3 in the medium enhanced this process further (Denchev et al. 1991).

The conversion of somatic embryos to plants is sometimes genotype dependent. In *M. truncatula* the percentage of conversion in genotype R 108 1 was 20 times higher than that in *cv.* Jemalong (Iantcheva et al. 2001). Even after osmotic pre-treatment to primary explants, genotype dominated the conversion process (Iantcheva et al. 2005).

5
Phenotype of Regenerated Plants via Somatic Embryogenesis: Somaclonal Variation

Regenerated plants from annual and perennial *Medicago* produced via somatic embryogenesis in most cases displayed normal and vigorous growth in the greenhouse, and morphologically resembled their donor plants with flower and seed set (Matheson et al. 1990; Varga and Badea 1992; Arcioni et al. 1989; Nolan et al. 1989; Barbulova et al. 2002).

In *Medicago*, variation from tissue culture has, however, been observed. Hexaploid plants of *M. sativa* were obtained after tissue culture treatment from haploid (Latude-Data and Lucas 1983) or diploid (Reisch and Bingham 1981) donor plant material. Euploid and aneuploid alfalfa plantlets were regenerated via indirect somatic embryogenesis by Johnson et al. (1984). It is necessary to analyse regenerated plants in order to confirm their ploidy level and genome size. Larkin and Scowcroft (1981) proposed the general term "somaclonal variation" for the variation arising from tissue and cell culture. In *M. sativa*, somaclonal variation for qualitative genetic characters like disease resistance (Johnson et al. 1984; Latude-Data and Lucas 1983) and quantitative traits like forage yield (Johnson et al. 1984; Pfeifer and Bingham 1984) were previously reported. In Romania "Sigma" is the first cultivar from this forage crop created from in vitro regenerated somaclones via indirect somatic embryogenesis (Varga and Badea 1992). The same authors suggested the use of alfalfa somaclones in a breeding programme that could shorten the time for raising a new cultivar. In the paper of Arcioni et al. (1989) the authors' investigations on somaclonal variation do not provide novel phenotypes, absent in the donor cultivar. Among *Medicago* species, somaclonal variation is genotype specific and superior variants can be selected during the plant regeneration procedure. This issue needs further detailed studies, and methods such as DNA fingerprinting may be useful in this direction.

6
Conclusion

Somatic embryogenesis is the direct way to regenerate plants from single somatic cells, and opens up the possibility of understanding the process of cell cycle reprogramming from somatic to embryogenic type, cloning and characterization of genes involved in wounding, hormone activation, cell division, differentiation and developmental processes (Chugh and Khurana 2002). Considerable advances in the development of the somatic embryogenesis system in genera *Medicago* have been noted in the last 30 years. The development of a genome and proteome database of model annual *Medicago truncatula* species will serve as a genetically compatible model for alfalfa, which is tetraploid and perennial (Bell et al. 2001; Imin et al. 2004).

One of the important uses of somatic embryogenesis is to explore it as an approach to investigate the early events of zygotic embryogenesis in higher plants, because of the existing parallel events happening between the two processes (de Jong et al. 1993; Dodeman et al. 1997). The second important application of somatic embryogenesis is the mass propagation of commercially valuable genotypes—one of the most attractive uses of this morphogenic pathway. Because of the huge number of somatic embryo structures, easy scale-up is possible. Single-cell origin also permits synchronized, homogeneous and stable plant material; thus, somatic embryogenesis is the preferred method of regeneration rather than organogenesis (Merkle et al. 1990). Another use of somatic embryogenesis is in the generation of transgenic plants. Gene transfer into embryogenic cells may help in conventional plant breeding and crop improvement programmes.

References

Arcioni S, Davey MR, dos Santos AVP, Cocking C (1982) Somatic embryogenesis in tissues from mesophyll and cell suspension protoplasts of *Medicago coerulea* and *M. glutinosa*. Z Pflanzenphysiol 106:105–110

Arcioni S, Damiani F, Pezzotti M, Lupotto E (1990) Alfalfa, Lucerne (*Medicago* spp.). In: Bajaj YPS (ed) Biotechnology in agriculture and forestry, vol 10. Legumes and oil seed crops I. Springer, Berlin Heidelberg New York, pp 242–281

Arcioni S, Damiani F, Pupilli F, Pezzotti M (1989) Somatic embryogenesis and somaclonal variation in *Medicago sativa* L. J Genet Breed 43:223–230

Atanassov A, Brown DCW (1984) Plant regeneration from suspension culture and mesophyll protoplasts of *Medicago sativa* L. Plant Cell Tissue Organ Cult 3:149–162

Barbulova A, Iantcheva A, Zhiponova M, Vlahova M, Atanassov A (2002) Establishment of embryogenic potential of economically important Bulgarian alfalfa cultivars (*Medicago sativa* L.). Biotechnol Biotechnol Equip 16:55–63

Bell CJ, Dixon R, Farmer AD, Flores R, Inman J, Gonzales RA, Harrison MJ, Paiva NL, Scott AD, Weller JW, May GD (2001) The *Medicago* genome initiative: a model legume database. Nucleic Acids Res 29:114–117

Bingham ET, McCoy TJ, Walker KA (1988) Alfalfa tissue culture. In: Hanson A, Bames D, Hill R (eds) Alfalfa and alfalfa improvement. American Society of Agronomy, Madison, WI, pp 903–929

Brown DCW, Atanassov A (1985) Role of genetic background in somatic embryogenesis in *Medicago*. Plant Cell Tissue Organ Cult 4:111–122

Carman JG (1990) Embryogenic cell in plant tissue cultures: occurrence and behaviour. In Vitro Cell Dev Biol 26:746–753

Chabaud M, de Carvalho-Niebel F, Barker DC (2004) Efficient transformation of *Medicago truncatula* using hypervirulent *Agrobacterium tumefaciens* strain AGL 1. Plant Cell Rep 22:46–51

Chabaud M, Larsonneau C, Marmouget C, Huguet T (1996) Transformation of barrel medics (*Medicago truncatula* Gaetrn.) by *Agrobacterium tumefaciens* and regeneration via somatic embryogenesis of transgenic plants with the *MtENOD* 12 nodulin promoter fused to *gus* reporter gene. Plant Cell Rep 15:305–310

Chen THH, Marowitch J, Thompson BG (1987) Genotypic effect on somatic embryogenesis and plant regeneration from callus culture of alfalfa. Plant Cell Tissue Organ Cult 8:73–81

Chugh A, Khurana P (2002) Gene expression during somatic embryogenesis—recent advances. Curr Sci 83:715–730

Das Neves L, Duque S, Almeida J, Fevereiro S (1999) Repetitive somatic embryogenesis in *Medicago truncatula* ssp. *narbonensis* and *M. truncatula* Gaertn cv. Jemalong. Plant Cell Rep 18:398-405

Deak M, Kiss GB, Koncz C, Dudits D (1986) Transformation of *Medicago* by *Agrobacterium*-mediated gene transfer. Plant Cell Rep 5:97–100
De Jong AJ, Schmidt EDL, De Vries SC (1993) Early events in higher plant embryogenesis. Plant Mol Biol 22:367–377
Denchev P, Atanassov A (1988) Unconventional methods in lucerne breeding. In: Staszewski Z, Utrata A (eds) Inst. Extension Service, Poland, pp 17–21
Denchev P, Velcheva M, Atanassov A (1991) A new approach to direct somatic embryogenesis in *Medicago*. Plant Cell Rep 10:338–341
Dijak M, Simmonds DH (1988) Microtubule organization during early direct embryogenesis from mesophyll protoplasts of *Medicago sativa* L. Plant Sci 58:183–191
Ding Y-L, Aldao-Humble G, Ludlow E, Drayton M, Lin Y-H, Nagel J, Dupal M, Zhao G, Palaghy C, Kalla R, Emmerling M, Spangenberg G (2003) Efficient plant regeneration and *Agrobacterium*-mediated transformation in *Medicago* and *Trifolium* species. Plant Sci 165:1419–1427
Dodeman VL, Ducreux G, Kreis M (1997) Zygotic embryogenesis versus somatic embryogenesis. J Exp Bot 48:1493–1509
Dudits D, Gyorgyey J, Borge L, Bako L (1995) Molecular biology of somatic embryogenesis. In: Thorpe TA (ed) In vitro embryogenesis in plants. Kluwer, Dordrecht, pp 267–307
Finstad K, Brown DCW, Joy K (1993) Characterization of competence during induction of somatic embryogenesis in alfalfa tissue culture. Plant Cell Tissue Organ Cult 34:125–132
Fujii JA, Slade D, Olsen R, Ruzin SE, Redenbaugh K (1990) Alfalfa somatic embryo maturation and conversion into plants. Plant Sci 72:93–100
Gangopadhyay G, Basu S, Gupta S (1997) In vitro selection and physiological characterization of NaCl and manitol adapted callus line in *Brassica juncacea*. Plant Cell Tissue Organ Cult 50:161–169
Hoffmann B, Trinh TH, Leung J, Kondorosi A, Kondorosi E (1997) A new *Medicago truncatula* line with superior in vitro regeneration, transformation and symbiotic properties isolated through cell culture selection. Mol Plant Microbe Interact 10:307–315
Horbowicz M, Obendorf R, McKersie B, Viands D (1995) Soluble saccharides and cyclitols in alfalfa (*Medicago sativa* L.) somatic embryos, leaflets and mature seeds. Plant Sci 109:191–198
Hutchinson M, Murch S, Saxena KP (1996) Morphoregulatory role of thidiazuron: evidence of the involvement of endogenous auxin in thidiazuron-induced somatic embryogenesis of geranium (*Pelargonium horotorum* Bailey). J Plant Physiol 149:573–579
Iantcheva A, Bakalova E, Vlahova M, Kondorosi E, Elliott M, Atanassov A (1999) Regeneration of diploid annual medics via direct somatic embryogenesis promoted by thidiazuron and benzylaminopurine. Plant Cell Rep 18:904–910
Iantcheva A, Barbulova A, Vlahova M, Atanassov A (2004) Primary asymmetric division and embryo formation in a single cell suspension of *Medicago falcata*. Biotechnol Biotechnol Equip 18:27–33
Iantcheva A, Brown S, Atanassov A (2003) Flow cytometric analysis in diploid *Medicago* species from Algeria: relationship between genome size and competence for direct somatic embryo formation. Biotechnol Biotechnol Equip 17:44–49
Iantcheva A, Brown S, Vlahova M, Atanassov A (2004) Cell cycle plasticity in response to low temperature in root tips of tetraploid *Medicago*. Biotechnol Biotechnol Equip 18:39–46
Iantcheva A, Slavov S, Prinsen E, Vlahova M, van Onckelen H, Atanassov A (2005) Embryo induction and regeneration from root explants of *Medicago truncatula* after osmotic pre-treatment. Plant Cell Tissue Organ Cult 81:37–43

Iantcheva A, Vlahova M, Atanassov A (2005) Genetic transformation of *Medicago truncatula* using a system for direct somatic embryogenesis by TDZ. Biotechnol Biotechnol Equip 19:50–56

Iantcheva A, Vlahova M, Trinh TH, Brown S, Slater A, Elliott MC, Atanassov A (2001) Assessment of polysomaty, embryogenic potential, embryo formation and regeneration in liquid media for different species of diploid annual *Medicago*. Plant Sci 160:621–627

Iantcheva A, Trinh HT, Brown S, Atanassov A (2001) Early events of direct somatic embryo formation from single cells in *Medicago truncatula* observed using green fluorescent protein. Second meeting of the COST 843 WG3, quality enhancement of plant production through tissue culture, 19–21 April 2001, Carcavelos, Portugal, pp 1–2

Ibragimova SS, Smolenskaya SE (1997) Plant regeneration from seedlings apex in annual medics. Acta Agron Hung 45:109–116

Imin N, De Jong F, Mathesius U, van Noorden G, Saeed NA, Wang X-D, Rose RJ, Rolfe B (2004) Proteome reference maps of *Medicago truncatula* embryogenic cell cultures generated from single protoplasts. Proteomics 4:1883–1896

Ivanova A, Velcheva M, Denchev P, Atanassov A, van Onckelen HA (1994) Endogenous hormone levels during direct somatic embryogenesis in *Medicago falcata*. Physiol Plantarum 92:85–89

Jiménez VM (2001) Regulation of in vitro somatic embryogenesis with emphasis on the role of endogenous hormones. R Bras Fisiol Veg 13:196–223

Johnson LB, Stuteville DL, Schlarbaum SE, Skiner DZ (1984) Variation in phenotype and chromosome number in alfalfa protoclones regenerated from non-mutagenized calli. Crop Sci 24:948–951

Kim Y-S, Kim MY, Yang M-S (2004) Effect of plant growth regulators on plant regeneration through somatic embryogenesis of *Medicago sativa* L. Plant Biotechnol 6:87–90

Krochko J, Pramanik K, Bewley J (1992) Contrasting storage protein synthesis and mRNA accumulation during development of zygotic and somatic embryos of alfalfa (*Medicago sativa* L.). Plant Physiol 99:46–53

Kuklin A, Denchev P, Atanassov A, Scragg A (1994) Alfalfa embryo production in airlift vessels via direct somatic embryogenesis. Plant Cell Tissue Organ Cult 38:19–23

Lai F, McKersie BD (1994) Regulation of starch and protein accumulation in alfalfa (*Medicago sativa* L.) somatic embryos. Plant Sci 100:211–219

Larkin P, Scowcroft W (1981) Somaclonal variation—a novel source of variability from cell cultures for plant improvement. Theor Appl Genet 60:197–214

Latude-Data AO, Lucas JA (1983) Somaclonal variation and reaction to *Verticilium* wilt in *Medicago sativa* L. plants regenerated from protoplasts. Plant Sci Lett 32:205–211

Lecouteux C, Lai F, McKersie B (1993) Maturation of alfalfa (*Medicago sativa* L.) somatic embryos by abscisic acid, sucrose and chilling stress. Plant Sci 94:207–213

Li X-Q, Demarly Y (1995) Characterization of factors affecting regeneration frequency of *Medicago lupulina* L. Euphytica 86:143–148

Li X-Q, Demarly Y (1996) Somatic embryogenesis and plant regeneration in *Medicago suffruticosa*. Plant Cell Tissue Organ Cult 44:79-81

Lupotto E (1983) Propagation of an embryogenic culture of *Medicago sativa* L. Pflanzenphysiol 111:95–104

Maheswaran G, Williams EG (1984) Direct somatic embryoid formation on immature embryos of *Trifolium repens*, *T. pratense* and *Medicago sativa*, and rapid clonal propagation of *T. repens*. Ann Bot 54:201–211

Maheswaran G, Williams EG (1985) Origin and development of somatic embryoids formed directly on immature embryos of *Trifolium repens* in vitro. Ann Bot 56:618–630

Malik K, Saxena PK (1992) Regeneration in *Phaseolus vulgaris* L.: high frequency induction of direct shoot formation in intact seedlings by BAP and TDZ. Planta 186:384–388
Martin JP, Pintos B, Rebordinus I, Villalobos N, Guerra H, Martin L (2000) Embryogenic response in different *Medicago* arborea L. explants depending on cytokinin/auxin balances. J Plant Physiol 156:801–804
Matheson SL, Nowak J, MacLean N (1990) Selection of regenerative genotypes from highly productive cultivars of alfalfa. Euphytica 45:105–112
McKersie BD, Brown DCW (1996) Somatic embryogenesis and artificial seeds in forage legumes. Seed Sci Res 6:109–126
Meijer EGM, Brown DCW (1987) A novel system for rapid high frequency somatic embryogenesis in *Medicago sativa*. Physiol Plant 69:591–596
Merkle S, Parrott W, Williams E (1990) Application of somatic embryogenesis and embryo cloning. In: Bhojwani S (ed) Plant tissue culture: applications and limitations. Elsevier, Amsterdam, pp 67–101
Michaud E, Leman WF, Rumbaugh MD (1988) World distribution and historical development. In: Hanson A, Bames D, Hill R (eds) Alfalfa and alfalfa improvement. American Society of Agronomy, Madison, WI, pp 25–91
Mitten DH, Sato SJ, Skokut TA (1984) In vitro regenerative potential of alfalfa germplasm sources. Crop Sci 24:943–945
Murthy B, Murch S, Saxena PK (1995) Thidiazuron-induced somatic embryogenesis in intact seedlings of peanut (*Arachis hypogaea*): endogenous growth regulator levels and significance of cotyledons. Physiol Plant 94:268–276
Nagaradjan P, McKenzie JS, Walton PD (1986) Embryogenesis and plant regeneration of *Medicago* species in tissue culture. Plant Cell Rep 5:77–80
Ninkovic S, Miljus-Djukic J, Mirjana N (1995) Genetic transformation of alfalfa somatic embryos and their propagation through repetitive somatic embryogenesis. Plant Cell Tissue Organ Cult 42:255–260
Nolan KE, Irwanto R, Rose RJ (2003) Auxin up-regulates *Mt*SERK 1 expression in both *Medicago truncatula* root-forming and embryogenic cultures. Plant Physiol 133:218–230
Nolan KE, Rose RJ, Gorst JG (1989) Regeneration of *Medicago truncatula* from tissue culture: increased somatic embryogenesis using explant from regenerated plants. Plant Cell Rep 8:278–281
Parrott WA, Bailey MA (1993) Characterization of recurrent somatic embryogenesis of alfalfa on auxin-free medium. Plant Cell Tissue Organ Cult 32:69–76
Pfeifer TW, Bingham ET (1984) Comparison of alfalfa somaclonal sexual derivatives from the same genetic source. Theor Appl Genet 64:263–266
Phang J (1985) The regulatory functions of proline and pyrroline-5-carboxylic acid. Curr Top Cell Regul 25:92–132
Piccioni E, Valecchi G (1996) Effect of wounding and of rooting in the regenerative ability of micropropagated alfalfa tissues. Agr Med 126:311–315
Pintos B, Martin JP, Centeno ML, Villalobos N, Guerra H, Martin L (2002) Endogenous cytokinin levels in embryogenic and non-embryogenic calli of *Medicago arborea* L. Plant Sci 163:955–960
Reisch B, Duke SH, Bingham ET (1981) Selection and characterization of ethionine-resistant alfalfa (*Medicago sativa*) cell lines. Theor Appl Genet 59:89–94
Sanders J, Bingham ET (1972) Production of alfalfa plants from callus tissue. Crop Sci 12:804–808
Sanders JW, Bingham ET (1975) Growth regulator effect on bud initiation in callus culture of *Medicago sativa*. Am J Bot 62:850–855

Saxena PK, Malik K, Gill R (1992) Induction by TDZ of somatic embryogenesis in intact seedlings of peanut. Planta 187:421–424

Scarpa GM, Pupilli F, Damiani F, Arcioni S (1993) Plant regeneration from callus and protoplasts in *Medicago polymorpha.* Plant Cell Tissue Organ Cult 35:49–57

Seitz Kris MH, Bingham ET (1988) Interaction of highly regenerative genotypes of alfalfa (*Medicago sativa*) and tissue culture protocols. In Vitro Cell Dev Biol 24:1047-1052

Senaratna T, McKersie BD, Bowley SR (1989) Desiccation tolerance of alfalfa (*Medicago sativa* L.) somatic embryos. Influence of abscisic acid, stress pre-treatment and drying rates. Plant Sci 65:253–259

Senaratna T, Saxena PK, Rao MV, Afela J (1995) Significance of the zygotic seed coat on quiescence and desiccation tolerance of *Medicago sativa* L. somatic embryos. Plant Cell Rep 14:375–379

Shao C-Y, Russinova E, Iantcheva A, Atanassov A, McCormac A, Chen D-F, Elliott M, Slater A (2000) Rapid transformation and regeneration of alfalfa (*Medicago falcata* L.) via direct somatic embryogenesis. Plant Growth Regul 31:155–166

Shetty K, McKersie BD (1993) Proline, thioproline and potassium mediated stimulation of somatic embryogenesis in (*Medicago sativa* L.). Plant Sci 88:185–193

Stuart D, Strickland S (1984) Somatic embryogenesis from cell cultures of *Medicago sativa* L. The role of amino acid additions to the regeneration medium. Plant Sci Lett 34:165–174

Stuart D, Nelson J, Strickland SG, Nichol JW (1985) Factors affecting developmental processes in alfalfa cell cultures. In: Henke RR, Hughes RW, Constantin MP, Hollander A (eds) Tissue Culture in Forestry and Agriculture. Plenum Publ. Corp. New York, pp. 59–73

Trinh TH, Ratet P, Kondorosi E, Durand P, Kamaté K, Bauer P, Kondorosi A (1998) Rapid and efficient transformation of diploid *Medicago truncatula* and *Medicago sativa* ssp. *falcata* lines improved in somatic embryogenesis. Plant Cell Rep 17:345–355

Varga P, Badea EM (1992) In vitro plant regeneration methods in alfalfa breeding. Euphytica 59:119–123

Vergana R, Verbe F, Pitto L, Schiavo L, Terzi M (1990) Reversible variation in the methylation pattern of carrot DNA during somatic embryogenesis. Plant Cell Tissue Organ Cult 8:697–701

Walker K, Sato S (1981) Morphogenesis in callus tissue of *Medicago sativa*: the role of ammonium ion in somatic embryogenesis. Plant Cell Tissue Organ Cult 1:109–121

Walton PD, Brown DCW (1988) Screening of *Medicago* wild species for callus formation and the genetics of somatic embryogenesis. Genetics 67:95–100

Williams E, Maheswaran G (1986) Somatic embryogenesis: factors influencing coordinated behaviour of cells as an embryogenic group. Ann Bot 57:443–462

Zafar Y, Nenz E, Damiani F, Pupilli F, Arcioni S (1995) Plant regeneration from explant and protoplast derived callus of *Medicago littoralis.* Plant Cell Tissue Organ Cult 41:41–48

Plant Cell Monogr (2)
A. Mujib · J. Šamaj: Somatic Embryogenesis
DOI 10.1007/7089_038/Published online: 2 December 2005

Differential Gene Expression During Somatic Embryogenesis

P. Suprasanna (✉) · V. A. Bapat

Plant Cell Culture Technology Section,
Nuclear Agriculture and Biotechnology Division,
Bhabha Atomic Research Centre, Trombay, Mumbai 400 085, India
prasanna@magnum.barc.ernet.in, penna888@yahoo.com

Abstract Somatic embryogenesis is a complex developmental program in which somatic cells are induced for a commitment towards forming totipotent embryogenic cells capable of becoming complete plants. Plant somatic embryogenesis has been the choice model system to explore the earliest regulatory and morphogenetic events in the life of the plant. Somatic embryos are similar to zygotic embryos and undergo almost the same developmental stages except for the acquisition of embryogenic competence. Common regulatory mechanisms probably operate in the early stages in both types of embryogenesis and hence it is possible to investigate somatic embryogenesis either by analysis of the expression of genes (isolated and characterized in zygotic embryos) in somatic embryos or by analyzing the differential expression of genes in embryogenic and nonembryogenic tissues. Studies have been conducted either to identify the genes expressed and gene products that accumulate specifically during different stages of embryogenesis or to analyze the expression of a variety of genes that probably have some role in the embryogenic pathway. More often this has involved the comparison of somatic embryos, embryogenic callus or cells and embryos at an early stage. This review will cover the aspects outlined above and discuss current information.

1 Introduction

One of the most striking features of flexibility in plant development is the capability of several cell types, in addition to zygote, to initiate embryogenic development (Feher et al. 2003). Somatic embryogenesis is a complex developmental program in which somatic cells are induced for a commitment towards forming totipotent embryogenic cells capable of becoming complete plants. Plant somatic embryogenesis has been the choice model system to explore the earliest regulatory and morphogenetic events in the life of the plant (Zimmerman 1993; Rao 1996; von Arnold et al. 2002; Komamine et al. 2005). Somatic embryos are similar to zygotic embryos and undergo almost the same developmental stages (Dodeman et al. 1997) except for the acquisition of embryogenic competence. It is thought that common regulatory mechanisms probably operate in the early stages in both types of embryogenesis.

Plant development and differentiation are regulated directly or indirectly by changes in gene expression, especially during embryogenesis (Goldberg et al. 1989; Dong and Dunstan 2000). Initial research on zygotic embryogenesis was carried to estimate the number and distribution of distinct RNA species, to isolate and to characterize the seed protein genes, and to identify regulatory sequences and DNA-binding proteins that regulate expression of seed-specific genes (Goldberg et al. 1994; Meinke 1995). There are discrete developmental phases in somatic embryogenesis that are characterized by distinct biochemical and molecular events (Henry et al. 1994; Kawahara and Komamine 1995; Meinke 1995; Wilde et al. 1995; Dong and Dunstan 2000), suggesting that the number of genes specifically expressed during these events is rather limited (Komamine et al. 1992; Dodeman and Ducreux 1996; Schrader et al. 1997), and that changes in protein patterns are highly regulated posttranscriptionally, at the messenger RNA (mRNA) level (Wilde et al. 1995). Additionally, Dodeman and Ducreux (1996) indicated that changes in hormonal levels in tissue cultures may modify the synthesis of some somatic-embryogenesis-specific proteins. Dudits et al. (1995) opined that the gene expression is expected to be different during the processes of embryogenic commitment in primary explants or fully differentiated somatic cells from that acting in suspension cultures with proembryogenic structures, such as in the case of carrot. Figueroa et al. (2002) proposed that differential gene expression can modulate the embryogenic capacity of coffee cells and that the number of genes turned off in somatic cells to allow for the change from a somatic to an embryogenic state is higher than the number of genes that are turned on. Komamine et al. (2005) classified genes expressed during somatic embryogenesis into three categories: (1) genes involved in cell division, (2) genes involved in organ formation, and (3) genes specific for the process of somatic embryogenesis.

It is thus possible to investigate somatic embryogenesis either by the analysis of the expression of genes (already isolated and characterized in zygotic embryos) in somatic embryos or by analyzing the differential expression of genes in embryogenic and nonembryogenic tissues (reviewed by Chugh and Khurana 2002). Studies have been conducted either to identify the genes expressed and the gene products that accumulate specifically during different stages of embryogenesis or to analyze the expression of a variety of genes that probably have some role in the embryogenic pathway. More often, this has involved the comparison of somatic embryos, embryogenic callus or cells, and embryos at an early stage. In this chapter, we present an overview of studies on differential gene expression during different phases of somatic embryogenesis in higher plants. The information generated using different culture systems is opening up new approaches for understanding the embryogenic developmental pathway in higher plants. The reader may also refer to other reviews that have dealt with the subject of cellular and molecular aspects of somatic embryogenesis (Dudits et al. 1995; Chugh and Khurana 2002; Feher et al. 2003).

2
The Process of Somatic Embryogenesis

The process of somatic embryogenesis in culture starts with the induction of a somatic embryo when somatic cells become embryogenic. It has been recorded that either a single cell or a group of cells become embryogenic and these have been termed pre-embryo determined cells (Suprasanna and Rao 1997; Mordhorst et al. 1997; Feher et al. 2003). The shift from "somatic cells into embryogenic cells" is accompanied by the synthesis of RNA and DNA, a change in pH, an increase in the rate of oxygen uptake, elevated enzyme activity, mainly kinase, migration of nuclei towards the cell wall, changes in cytoskeleton, active conversion of ATP to ADP, and inactivation of cytosolic factors and maturation promotion factor. The cells destined to become embryogenic are isodiametric, rich in cytoplasm and starch, and have a callose deposition. Such cells are separated from the rest of the cells and during the process the plasmodesmata gets severed. It has also been observed that cells exude proteins into the culture medium that either promotes or inhibits ongoing embryogenic process and that these molecules act as signals (Schmidt et al. 1994). Several studies have been conducted to identify various genes responsible for the various stages of somatic embryogenesis and the genes have been grouped into five classes: class 1 genes consists of expressed genes which have functions required during normal plant growth, and thus are active throughout the entire plant, class 2 genes are embryo-specific genes of which the expression is restricted to the embryo proper and ceases prior to or at germination, class 3 genes are expressed during early embryogenesis, class 4 genes contain seed protein expressed during the expansion phase of the cotyledon and maturation, and class 5 genes are expressed throughout the embryo during the late embryogenesis to early germination.

During the last few years, there has been a tremendous surge in molecular biological research aimed at gaining insight into the somatic embryogenic pathway using culture systems of carrot, alfalfa, chicory and conifers. Mutants of *Arabidopsis thaliana* have been used to characterize the induction phase, i.e., the first stage during somatic embryogenesis. Most genes expressed differentially during somatic embryogenesis belong to the late embryo-abundant (*lea*) genes. Proposed functions for the products of this family of genes are the protections of cellular structures in mature embryos during seed desiccation and prevention of precocious germination of the zygotic embryos during seed development (Wilde et al. 1998; Dong and Dunstan 2000). Several genes expressed in carrot somatic embryos code for secreted extracellular proteins (Mordhorst et al. 1997). One gene product (EP1), with homology to *Brassica* S-locus glycoproteins, is present in nonembryogenic callus but not in somatic embryos themselves. Another gene that produces a lipid transfer protein (EP2) has been particularly useful as a marker for epidermal cell differentiation during embryogenesis. The precise role of these extracellular proteins remains to be established, but they may be involved in the regulation of cell expansion and the maintenance of biophysical features required for morphogenesis. Perhaps the most unexpected finding involves a secreted glycoprotein (EP3) that rescues a temperature-sensitive mutant of carrot (*ts11*) that fails to complete the transition from the globular to the heart stage of somatic embryogenesis (Meinke 1995). The *Dc3*, *Dc8*, *J4e* and *ECP31* genes represent another group of genes that are developmentally regulated in carrot suspension cultures, and are expressed at different moments during embryo development and localized in different cell groups within the proembryogenic masses and embryos (Wilde et al. 1998).

3 Genes Involved in the Cell Cycle

Cell cycle genes play a central role in somatic embryogenesis. Plant cyclin complentary DNAs (cDNAs) are expressed during carrot somatic embryogenesis (Hata et al. 1991). A cdc2 protein kinase cDNA (*cdc2MS*) from alfalfa shares 64% identity with the yeast and mammalian kinases. The transcript levels of *cdc2MS* were found to be higher in alfalfa shoots and auxin-induced suspension cultures (Hirt et al. 1991). Higashi et al. (1998) studied the nitrogen metabolism during zygotic and somatic embryogenesis in carrot. The expression pattern of three carrot cDNA clones coding for isoforms of glutamine synthetase (*CGS102*, *CGS103*, and *CGS201*) showed that transcript levels of *CGS102* and *CGS201* increased during the early stages of somatic embryogenesis and developing seeds, whereas *CGS103* was expressed only in the later stages of seed development and senescent leaves, and was absent in somatic embryos or young leaves. The expression of *CGS102* and *CGS201* decreased in the presence of medium supplemented with glutamine as a nitrogen source, indicating transcriptional regulation of glutamine synthetase activity, suggesting the involvement of a common regulatory system for nitrogen metabolism in somatic and zygotic embryogenesis (Higashi et al. 1998).

Changes in the expression of actin and tubulin genes have been demonstrated during embryogenesis as enhanced cell wall and membrane formation result in an increase in the expression of these genes as well (Cyr et al. 1987; Raghavan 1997). Kawahara et al. (1992) also found enhanced expression of two histone-coding genes, *H3-1* and *H3-11*, during alfalfa somatic embryogenesis in response to auxin treatment. A globular embryo-specific cDNA encoding for elongation factor-1α*CEM1* has been reported in the actively dividing cells (Sato et al. 1995). The encoded protein functions in the interaction of the aminoacyl transfer RNA with ribosomes during the synthesis of proteins for housekeeping chores in the cell. Another gene *CEM6* is specific to the preglobular and globular stages of carrot somatic embryo formation. *CEM6* encodes a glycine-rich protein and has a hydrophobic signal-sequence-like domain, possibly with a role in cell wall biogenesis during embryogenesis (Sato et al. 1995). DNA topoisomerase I is a key enzyme involved in various processes of DNA metabolism. Balestrazzi et al. (1996, 2001) observed that the levels of the *topI* (topoisomerase I) transcript increased during cell proliferation in 2,4-dichlorophenoxyacetic acid (2,4-D) induced carrot hypocotyls. The transcript levels rose with the proliferation of provascular cells and at the torpedo stage of somatic embryogenesis, showing the association of increased *topI* gene expression during cellular proliferative activities.

4 Triggering the Embryogenic Program—Stress and Hormones

In vitro culture conditions impose stress on the implanted plant cells as they are exposed to an artificial environment containing physical and chemical growth regimes. Stress promotes differentiation and is known to induce somatic embryogenesis. Alfalfa leaf protoplasts respond to different oxidative stress inducing compounds in the presence of exogenous auxins and cytokinins (Pasternak et al. 2002). Mitogen-activated protein kinase phosphorylation cascades may link oxidative stress responses to auxin signaling and cell cycle regulation (Hirt 2000; Feher et al. 2003). Among different plant growth regulators, auxins have been used as potent inducers of embryogenic response (Raghavan 1997). Exposure of an auxin to excised organs, cell cultures, and whole plants results in accumulation of mRNAs, thus leading to the isolation of corresponding cDNAs (Hagen et al. 1984, 1991; Abel and Theologis 1996; Guilfoyle 1999). Heat shock proteins (HSPs) have been shown to be expressed throughout somatic embryo development (Coca et al. 1994, Kitamiya et al. 2000). HSPs may serve as molecular chaperones with an assembly function during the developmental switch for the initiation of the embryogenic program. One of the HSPs, *Dchsp1*, is expressed throughout carrot somatic embryo development. *Dcarg-1* is an auxin-regulated gene, detected specifically during the early induction period (Kitamiya et al. 2000). It implies that auxin shock can induce a stress regime during which the embryogenic program is perceived. Auxin shock is also considered a stress signal and common elements can therefore be predicted to operate. The 3′ intergeneric element of an auxin-regulated gene cluster in soybean showed high homology to the sequence motif located 150 bp downstream of the stop codon soybean HSP gene 6834 (McClure et al. 1989). A small heat shock gene (*Mshsp 18*) is expressed in early, globular and heart- stage alfalfa embryos under normal cultural conditions. Following the induction of direct somatic embryogenesis, mRNA samples converted into cDNAs prior to RNA arbitrarily primed PCR (RAP-PCR) showed no significant homology of the clones (Fowler et al. 1998). One of the clones, A1.4, was characterized as a calnexin homolog with a potential chaperone function. Using the PCR-based cDNA subtraction approach, differentially expressed genes were identified in alfalfa (Russinova et al. 1998). A higher number of clones were obtained and were classified into expression categories: early (from the induction), medium (from 3 days after induction) and late (expression after 5–10 days). Sequencing of the clones revealed the presence of many transcription factors, kinases, and the phosphatase PP2C and auxin-inducible genes.

Alfalfa cells proceed from the G1 phase to the S phase in the cell cycle after high auxin shock, as shown by the expression of cell cycle related *cdk* and cyclin genes resulting in the formation of somatic embryos (Dudits et al. 1991). The exposure to auxin shock serves as a trigger, inducing cell division in the epidermal cells and promoting their further differentiation to somatic embryos. Thus, even a small pulse of auxin is sufficient for induction of competent cells to trigger embryogenesis. Differential screening of a cDNA library constructed from poly (A^+) RNA of 2,4-D-shocked cells revealed a set of genes with a characteristic expression pattern during different stages of embryogenesis (Dudits et al. 1991, 1995). Using differential display analysis, three partial cDNA clones (nos. 43, 87, 93) have been isolated from cell clusters during the earliest stage of carrot somatic embryogenesis (Yasuda et al. 2001). The transcripts of these clones preferentially accumulate in the embryogenic cell clusters formed after treatment with 2,4-D. The deduced amino acid sequence of the no. 43 and no. 93 cDNA clones showed homology with thaumatin-like protein and the precursor of the proline-rich Dc2.15 protein respectively (Yasuda et al. 2001).

Small auxin upregulated (*SAUR*) genes, *pJCW1* and *pJCW2*, are a class of auxin-induced genes with specificity to the embryogenic program. Auxin specifically induces accumulation of mRNAs hybridizing with these sequences (Hagen et al. 1984) and such probes can be useful for screening the embryogenic potential of different cell lines. The transcript levels of *pJCW1* and *pJCW2* declined in older alfalfa somatic embryo cultures, suggesting a change in the morphogenic program. Newly induced embryogenic callus lines generally produce competent embryos that convert readily into plantlets, while the older cultures fail to do so. This is attributable to the desensitization of auxin responsiveness leading to reduced embryogenic competence in callus lines following prolonged exposure to 2,4-D (Padmanabhan et al. 2001). Significant hypermethylation has been shown after 2,4-D application, whereas its removal caused rapid demethylation (LoSchiavo et al. 1989). A change in the methylation status is also seen when carrot embryogenic cells are treated with exogenous auxin, and in fact, an optimal level of methylation is a requisite for the normal development of somatic embryos as hypermethylation and hypomethylation both result in immediate and irreversible block of embryogenesis (LoSchiavo et al. 1989).

The role of abscisic acid (ABA) in embryo maturation and seed development has been demonstrated in detail (Quatrano 1986). The ABA-regulated gene expression program includes transcriptional as well as posttranscriptional events, such as transcript processing, mRNA stability, translational control, protein activity and turnover. Late embryogenesis abundant (LEA) proteins constitute an important component of the ABA-inducible systems. High levels of LEA transcripts accumulated during embryogenesis (Leal and Misra 1993). Several cDNAs of embryo-specific/embryogenic cell proteins have been isolated and characterized: *DcECP31* (Kiyosue et al. 1992), *DcECP40* (Kiyosue et al. 1993), *DcECP63* (Zhu et al. 1997) from carrot and *Arabidopsis AtECP31* (Yang et al. 1996), *AtECP63* (Yang et al. 1997). These LEA proteins showed specific, increased expression during the torpedo stage of somatic embryos. In sugarcane, Linacero et al. (2001) studied the accumulation of different transcripts (*lea* genes and barley hemoglobin gene) during somatic embryogenesis under the effect of ABA and desiccation stress. Only the *lea* genes were found to be dramatically increased in the embryogenic tissues treated with ABA. The *ECP* (extracellular protein) genes expressed during the embryogenic program also have ABA-responsive elements in their promoter regions containing a conserved motif (ACGT core motif). Promoter deletion analysis in *DcECP31* has revealed a – 250 bp upstream region for embryo-specific and ABA-inducible activity, while the distal (– 670 to – 390 bp) and proximal regions (– 140 to – 50 bp) are essential for the ABA-inducible expression (Ko et al. 2001). The molecular studies on the ABA-responsive–embryogenic program have highlighted that there are various factors involved in the hormone-induced signal transduction pathway.

5
Signal Transduction Cascade

A series of events associated with the molecular recognition of an environmental stimulus to a defined response constitute a signal cascade pathway and the phenomenon is described as signal transduction. Recognition of either hormone stimuli and/or a secondary messenger like calcium may set off various signal transduction cascades in the transition of single cells to somatic embryos. Protein kinases often undergo autophosphorylation for their activation and are involved in regulation of other successive transducer(s) in the signal transduction pathway. In alfalfa, three somatic embryo genes (*ASET1*, *ASET2*, and *ASET3*) had specificity in their expression to the early stages in embryogenic lines but not to nonembryogenic lines and mature embryos (Giroux and Pauls 1997). One of them, ASET2 protein, is predicted to encode several potential membrane-spanning domains and a potential phosphorylation site, making it a key candidate in the signaling pathway(s) (Giroux and Pauls 1997).

Calcium is a key regulator of various cellular and physiological processes of higher plants. Calmodulin (CaM) is an important protein involved in the calcium mediation signaling in plants. The CaM proteins are encoded by a multigene family in carrot and other plant species (Ling et al. 1991; Periera and Zielinski 1992). The role of calcium has been well investigated in carrot somatic embryogenesis and a threshold level of 200 μM was found to be essential for morphogenesis of undifferentiated cells into somatic embryos (Jansen et al. 1990; Overvoorde and Grimes 1994). Active calcium/CaM complexes have also been detected in the meristematic regions of heart- and torpedo-stage embryos, suggesting the regulatory role of activated CaM in embryonal regions showing rapid cell divisions (Overvoorde and Grimes 1994; Timmers et al. 1989). Elevated levels of CaM transcript were found to be associated with actively growing regions (Pereira and Zielinski 1992). Overvoorde and Grimes (1994) found that the quantity of CaM transcript increased somewhat in globular and heart-stage embryos compared with low levels in the undifferentiated callus. CaM is generally localized in the meristematic regions of developing embryos and also in the embryogenic cell cultures, supporting the view that CaM is important for embryogenesis. In our studies using sugarcane embryogenic cultures, CaM expression was examined, from the undifferentiated cells to embryogenic cultures and somatic embryo development stages. Expression of CaM was specific to the embryogenic stages compared with the nonembryogenic stage (Suprasanna et al. 2004). An increase in the CaM expression seems to be related to the stages where increased protein turnover in systems undergoing rapid cell division occurs and spatial regulation of CaM may be important for the regulation of the embryogenic program. Anil and Rao (2000) studied the possible involvement of Ca^{2+}-mediated signaling in the induction/regulation of somatic embryogenesis from proembryogenic cells of sandalwood. Blocking of the embryogenic process with an inhibitor reduced the embryogenic frequency, suggesting that blockage of the Ca^{2+} mediated signaling pathway involving sandalwood Ca^{2+}-dependant protein kinase (CDPK) and/or CaM causes the inhibition of embryogenesis.

Expression of CaM mRNA has also been seen to increase upon induction of somatic embryos and to remain constant thereafter. Genes coding for calcium-binding protein (*MsCa1*) also show an increase in the transcript levels after 2,4-D treatment and preferentially accumulate at early globular stages (Dudits et al. 1991). A cDNA encoding a typical protein kinase homologous to other plant kinases has been screened from the carrot somatic embryo cDNA library (Lindzen and Choi 1995). These somatic embryos expressed calcium-dependent related kinase (CRK) mRNA and the protein at a much higher level than the mature plant tissues. Two CDPKs of 55 and 60 kDa were identified in soluble protein extracts of embryogenic cultures of sandalwood (Anil et al. 2000). The proteins showed differential expression and were absent in plantlets regenerated from somatic embryos. The temporal expression of *swCDPK*s during the globular stage of somatic embryos and zygotic embryos, seed maturation (endosperm development), and germination indicates their involvement in the process of differentiation and development. *SwCDPK* is posttranslationally inactivated in zygotic embryos during seed dormancy and during precocious seed germination. In sandalwood, there is a fourfold increase in calcium levels during differentiation of proembryogenic masses into somatic embryos. Chelating agents arrest somatic embryo formation though the cells continue to proliferate, indicating the inhibition of calcium-mediated signaling pathways involving CDPKs and CRKs (Anil and Rao 2000). *MsCPK3* is a CaM-like protein kinase (CPK) from cultured alfalfa cells that encodes for a 553 amino acid polypeptide of 60.2 kDa (Daveletova et al. 2001). *MsCPK* gene expression increased during the early phase of somatic embryogenesis. Growth regulators like kinetin and ABA or NaCl treatment did not induce gene activity whereas heat shock was able to induce expression, suggesting the role of CPK in hormone and stress-activated reprogramming of embryogenic developmental pathways (Daveletova et al. 2001).

Somatic embryogenesis receptor kinase (*SERK*) is the only gene known to play a role in the acquisition of embryogenic competence in plants cells (Schmidt et al. 1997). *SERK* encodes for a protein having an N-terminal domain with five leucine-rich repeats (LRRs) acting as a protein-binding region. The SERK protein has the a proline-rich region between the extracellular LRR domain of *SERK* and the membrane-spanning region. This is a conserved feature of extensins (Schmidt et al. 1997). LRR sequence of *SERK* shows homology with the *Arabidopsis RLK5* (Walker 1994) and *Arabidopsis ERECTA* genes (Torii et al. 1996). *SERK* is also seen to be expressed from the induced-embryogenic cell stage to the globular stage of somatic embryos, but not in the nonembryogenic stages of embryogenic cultures. Thus, the gene can be useful as a molecular marker for distinguishing embryogenic competent and noncompetent cells. *SERK* promoter fused with the *LUC* reporter gene demonstrated that the elongating cells in carrot that express *SERK*s indeed have the ability to undergo somatic embryo formation. Shah et al. (2001a) studied the biochemical characterization of a transmembrane receptor kinase (from embryogenic carrot cell cultures) as a 40-kDa his-tag fusion protein in the baculovirus insect cell system. The kinase domain fusion protein showed in vitro autophosphorylation at serine and threonine residues. In *Arabidopsis*, Shah et al. (2001b) identified five members of the SERK family (*AtSERK1*, *AtSERK2*, *AtSERK3*, *AtSERK4*, and *AtSERK5*). *AtSERK1* had specific expression in the nucellus, the megaspore and the embryo sac besides in the stages of somatic embryogenesis. The seedling-derived callus that was overexpressing *AtSERK1* had 3–4 times higher embryogenic competence compared with the wild-type callus. This may indicate that the protein encoded by the *AtSERK1* gene can confer embryogenic competence in culture. The *SERK* gene also mediates acquisition of embryogenic competence in the egg cell during zygotic embryogenesis (Hecht et al. 2001). The embryogenic cells of *Dactylis glomerata* also express the *SERK* gene and whole mount in situ hybridization reveals that *SERK* is expressed differentially. The *SERK* gene was expressed in competent cells to the globular stage, but was not present in the clubbed-stage somatic embryos. In contrast to *Daucus carota*, the gene is also expressed in the shoot apical meristem region of the protoderm, coleoptile and coleorrhiza. The probe used for in situ hybridization was an expressed sequence tag cDNA clone R2976 from *Oryza sativa*. Interestingly, this partial cDNA clone is 70% identical to the *D. carota* and *SERK* cDNA sequence. At the amino acid level, they share 82% identity. The *Oryza* probe gave stronger signals than the *Daucus* probe and both exhibit a similar spatial expression pattern, thus indicating that the *SERK*-mediated embryo specific path is operational in grasses as well. Recently, two novel genes, *ZmSERK1* and *ZmSERK2* from maize (*Zea mays* L.), have been isolated using degenerate primers and PCR analysis (Baudino et al. 2001). These genes share all the unique features of the *SERK* family. Both genes are present as a single copy in the maize genome, and exhibit 70% identity among each other at the nucleotide level with an intron/exon structure similar to that of the other *SERK*s identified. The tissue-specific expression studies of these two genes have shown preferential expression of *ZmSERK1* in male and female reproductive tissues, with strongest expression in microspores, whereas *ZmSERK2* is uniformly expressed in all the tissues. Both genes are expressed in embryogenic as well as nonembryogenic cells.

A cDNA library constructed from cultured conifer tissue undergoing stage-1 embryo formation was screened against nonembryogenic tissues and six gene families were preferentially expressed during embryogenesis (Bishop-Hurley et al. 2003). The genes showed high mRNA transcript levels in embryogenic tissue compared with nonembryogenic tissue (roots, shoots, and needles or callus). The gene families identified included four putative extracellular proteins (germin, β-expansin, 21-kDa protein precursor, and cellulase), a cytochrome P450 enzyme, and a gene with unknown function (*PRE87*).

The search for markers of plant embryogenesis is an important aspect of modern plant breeding. Several physiological, biochemical, and molecular markers associated with embryogenic competence of cells have been reported, including isozymes and molecular markers. There are several candidate genes that could be used as molecular markers of single competent cells (Schmidt et al. 1997). One of these genes, the *SERK* gene, was found to mark single *Daucus* and *Dactylis* suspension cells that are competent to form somatic embryos (Schmidt et al. 1997; Somleva et al. 2000). Recently, Kitamiya et al. (2000) succeeded in isolating two genes that were induced after exposure of carrot hypocotyls to high concentrations of 2,4-D for 2 h, a treatment that initiated somatic embryogenesis directly on these explants. Expression analysis of the *CHI-GST1* gene (a cDNA encoding a glutathione S-transferase) by Northern blot indicated that the transcript accumulation is specific of the leaf developing somatic embryogenesis and is not observed in leaf tissue of the nonembryogenic cultivar (Galland et al. 2001). Similarly *EMB1* cDNA from carrot is expressed only in embryogenic tissues (globular and torpedo-stage embryos) and accumulates in the meristematic regions

(Wurtule et al. 1993). There are also genes that show specificity to the maturation stage; for example, the *Dc2.15* gene is maximally expressed at the heart and torpedo stages. The expression of the *Mat1* gene was found to be increased with desiccation and was missing upon rehydration (Liu et al. 1991). Lipid transfer proteins are very good "early" markers of somatic embryo induction in different systems (Schmidt et al. 1997; Sterk et al. 1991).

6 Conclusions

Somatic embryogenesis is a unique system to investigate the mechanisms that operate during the transition of a single somatic cell into an embryogenic entity with the potential of developing into a complete plant. Early research included molecular analysis of somatic embryogenesis that mostly relied on comparing genes and proteins being expressed in embryogenic and nonembryogenic cells as well as in the different stages of embryogenesis. Over the past few years, molecular understanding of this developmental program has been colossal based on experiments with different culture systems, especially carrot, alfalfa, chicory, and conifers. Isolation and identification of auxin-inducible genes and ABA-inducible genes have yielded clues to the hormonal control of gene expression during embryogenic development. Identification of genes such as *SERK* have generated great interest in inducing a switch in cell fate, and genes like *bbm, lec1*, and *lec2* can be used to induce embryogenic development. Future research in this area must center not only on isolating and characterizing large numbers of genes expressed during somatic embryo development, but also on deciphering the significance of these genes by demonstrating what happens when their function is disrupted. This is being attempted either by creating transgenic plants that express an antisense construct or by working with genes that have already been disrupted through loss-of-function mutations. It is expected that future research will also unravel many more intricacies, driving the developmental flexibility regulated by temporal and spatial patterns of gene expression during somatic embryogenesis.

References

Abel S, Theologis A (1996) Early genes and auxin action. Plant Physiol 111:9–17

Alexandrova KS, Conger BV (2002) Isolation of two somatic embryogenesis-related genes from orchardgrass (*Dactylis glomerata*). Plant Sci 162:301–307

Anil VS, Rao KS (2000) Calcium mediated signaling during sandalwood somatic embryogenesis. Role for exogenous calcium as second messenger. Plant Physiol 123:1301–1311

Anil VS, Harmon AC, Rao KS (2000) Spatio-temporal accumulation and activity of Calcium-dependent protein kinases during embryogenesis, seed development, and germination in sandalwood. Plant Physiol 122:1035–1043

Balestrazzi A, Toscano I, Bernacchia G, Luo M, Otte S, Carbonera D (1996) Cloning of a cDNA encoding DNA topoisomerase I in *Daucus carota* and expression analysis in relation to cell proliferation. Gene 183:183–190

Balestrazzi A, Bernacchia G, Pitto L, Luccarni G, Carbonera D (2001) Spatial expression of DNA topoisomerase I genes during cell proliferation in *Daucus carota.* Eur J Histochem 45:31–38

Baudino S, Hansen S, Brettschneider R, Hecht VF, Dresselhaus T, Lorz H, Dumas C, Rogowsky PM (2001) Molecular characterisation of two novel maize LRR receptor-like kinases, which belong to the SERK gene family. Planta 213:1–10

Bishop-Hurley SL, Gardner RC, Walter C (2003) Isolation and characterization of genes expressed during somatic embryo development in *Pinus radiata.* Plant Cell Tissue Organ Cult 74:267–281

Chugh A, Khurana P (2002) Gene expression during somatic embryogenesis—recent advances. Curr Sci 86:715–730

Coca MA, Almoguera C, Jordano J (1994) Expression of sunflower low-moelcular-weight heat-shock proteins during embryogenesis and persistence after germination: localization and possible functional implications. Plant Mol Biol 25:479–492

Cyr RJ, Bustos MM, Guiltinan MJ, Fosket DE (1987) Developmental modulation of tubulin protein and mRNA levels during somatic embryogenesis in cultured carrot cells. Planta 171:365–376

Daveletova S, Meszaros T, Miskolczi P, Obershall A, Torok K, Magyar Z, Dudits D, Deak M (2001) Auxin and heat shock activation of a novel member of the calmodulin like domain protein kinase gene family in cultured alfalfa cells. J Exp Bot 52:215–221

Dodeman VL, Ducreux G (1996) Total protein pattern expression during induction and development of carrot somatic embryos. Plant Sci 120:57–69

Dodeman VL, Ducreux G, Kreis M (1997) Zygotic embryogenesis versus somatic embryogenesis. J Exp Bot 48:1493–1509

Dodeman VL, Le Guilloux M, Ducreux G, de Vienne D (1998) Somatic and zygotic embryos of *Daucus carota* L. display different protein patterns until conversion to plants. Plant Cell Physiol 39:104–111

Dong J, Dunstan DI (2000) Expression of abundant mRNAs during somatic embryogenesis of white spruce. Planta 199:459–466

Dudits D, Bogre L, Gyorgyey J (1991) Molecular and cellular approaches to the analysis of plant embryo development from somatic cells in vitro. J Cell Sci 99:475–484

Dudits D, Gyorgyey J, Bogre L, Bako L (1995) Molecular biology of somatic embryogenesis. In: Thorpe TA (ed) In vitro embryogenesis in plants. Kluwer, Dordrecht, pp 267–308

Feher A, Pasternak TP, Dudits D (2003) Transition of somatic plant cells to an embryogenic state. Plant Cell Tissue Organ Cult 74:201–228

Figueroa FQ, Mendez-Zeel M, Teyer FS, Rojas-Herrera R, Loyola-Vargas VM (2002) Differential gene expression in embryogenic and non-embryogenic cell clusters from cell suspension cultures of *Coffea arabica.* J Plant Physiol 159:1267–1270

Fowler MR, Ong LM, Russinova E, Atanassov AI, Scott NW, Slater A, Eliott MC (1998) Early changes in gene expression during direct somatic embryogenesis in alfalfa revealed by RAP-PCR. J Exp Bot 49:249–253

Galland R, Randoux B, Vasseur J, Hilbert JL (2001) A glutathione S-transferase cDNA identified by mRNA differential display is upregulated during somatic embryogenesis in Cichorium. Biochim Biophys Acta 1522:212–216

Giroux RW, Pauls KP (1997) Characterization of somatic embryogenesis-related cDNAs from alfalfa (*Medicago sativa* L). Plant Mol Biol 33:393–404

Goldberg RB, Barker SJ, Perez-Grau L (1989) Regulation of gene expression during plant embryogenesis. Cell 56:149–160
Goldberg RB, de Paiva G, Yadegari R (1994) Plant embryogenesis: zygote to seed. Science 266:605–614
Guilfoyle TJ (1999) Biochemistry and Molecular Biology of Plant Hormones. In: Lib benga KR, Hall M, Hooykaas PJJ (eds) Biochemistry and molecular biology of plant hormones. Elsevier, Amsterdam, pp 423–459
Hagen G, Guilfoyle TJ (2002) Auxin-responsive gene expression: genes, promoters and regulatory factors. Plant Mol Biol 49:373–385
Hagen G, Kleinschmidt A, Guilfoyle TJ (1984) Auxin-regulated gene expression in intact soybean hypocotyl and excised hypocotyl sections. Planta 162:147–153
Hagen G, Martin G, Li Y, Guilfoyle TJ (1991) Auxin-induced expression of the soybean GH3 promoter in transgenic tobacco plants. Plant Mol Biol 17:567–579
Hata S, Kouchi H, Suzuka I, Ishii T (1991) Isolation and characterization of cDNA clones for plant cyclins. EMBO J 10:2681–2688
Hecht V, Vielle-Calzada J-P, Hartog MV, Schmidt ED, Boutilier K, Grossnikaulus U, de Vries SC (2001) The Arabidopsis *somatic embryogenesis receptor kinase-1* gene is expressed in developing ovules and embryos and enhances embryogenic competence in culture. Plant Physiol 127:803–816
Henry Y, Vain P, De Buyser J (1994) Genetic analysis of in vitro plant tissue culture responses and regeneration capacities. Euphytica 79:45–48
Higashi K, Shiota H, Kamada H (1998) Patterns of expression of the genes for glutamine synthetase isoforms during somatic and zygotic embryogenesis in carrot. Plant Cell Physiol 39:418–424
Hirt H (2000) Connecting oxidative stress, auxin and cell cycle regulation through a plant Mitogen activated protein kinase pathway. Proc Natl Acad Sci USA 97:2405–2407
Hirt H, Pay A, Gyorgyey J, Bako L, Nemet K, Bogre L, Schweyen RJ, Heberle-Bors E, Dudits D (1991) Complementation of a yeast cell cycle mutant by an alfalfa cDNA encoding a protein kinase homologous to p34cdc2. Proc Natl Acad Sci USA 88:1636–1640
Jansen MAK, Booij H, Schel JHN, de Vries SC (1990) Calcium increases the yield of somatic embryos in carrot embryogenic suspension cultures. Plant Cell Rep 9:221–223
Kawahara R, Komamine A (1995) Molecular basis of somatic embryogenesis. In: Bajaj YPS (ed) Biotechnology in agriculture and forestry, vol 30. Somatic embryogenesis and synthetic seed I. Springer, Berlin Heidelberg New York, pp 30–40
Kawahara R, Sunabori S, Fukuda H, Komamine A (1992) A gene expressed preferentially in the globular stage of somatic embryogenesis encodes elongation-factor 1α in carrot. Eur J Biochem 209:157–162
Kitamiya E, Suzuki S, Sano T, Nagata T (2000) Isolation of two genes that were induced upon initiation of somatic embryogenesis on carrot hypocotyls by high concentrations of 2,4-D. Plant Cell Rep 19:551–557
Kiyosue T, Shiota H, Higashi K, Kamada H, Shinozaki K (1998) A chromo box gene from carrot (*Daucus carota* L.): its cDNA structure and expression during somatic and zygotic embryogenesis. Biochem Biophysica Acta 1398:42–46
Kiyosue T, Yamaguchi-Shinozaki K, Shinozaki K, Higashi K, Satoh S, Kamada H, Harada H (1992) Isolation and characterization of a cDNA that encodes ECP31, an embryogenic-cell protein from carrot. Plant Mol Biol 19:239–249
Ko S, Thitamadee S, Yang H, Eun CH, Sage-Ono K, Higashi K, Satoh S, Kamada H (2001) Comparison and characterization of cis-regulatory regions in some embryo specific and ABA-responsive carrot genes, DcECPs. Plant Biotech 18(1):45–54

Komamine A, Kawahara R, Matsumoto M, Sunabori S, Toya T, Fujiwara A, Tsukuhara M, Smith J, Ito M, Fukuda H, Nomura K, Fujimura T (1992) Mechanisms of somatic embryogenesis in cell cultures: physiology, biochemistry, and molecular biology. In Vitro Cell Dev Biol Plant 28:11–14

Komamine A, Murata N, Nomura K (2005) Mechanisms of somatic embryogenesis in carrot suspension cultures—morphology, physiology, biochemistry, and molecular biology. In Vitro Cell Develop Biol-Plant 41:6–10

Leal I, Misra S (1993) Molecular cloning and characterization of a legumin-like storage protein cDNA of Douglas fir seeds. Plant Mol Biol 21:709–715

Linacero MR, Lopez-Bilbao MG, Vazquez AM (2001) Expression of different abscisic acid responsive genes during somatic embryogenesis in sugarcane (*Saccharum officianarum*). Protoplasma 217:199–204

Lindzen E, Choi JH (1995) A carrot cDNA encoding an atypical protein kinase homologous to plant calcium-dependent protein kinases. Plant Mol Biol 28:785–797

Ling V, Perera I, Zielinski RE (1991) Primary structures of Arabidopsis calmodulin isoforms deduced from the sequences of cDNA clones. Plant Physiol 96:1196–1202

Liu W, Hildebrand DF, Moore PJ, Collins GB (1991) Effects of exogenous auxins on expression of lipoxygenases in soybean tissue cultures. Plant Physiol 97:969–976

LoSchiavo F, Pitto L, Giulilano G, Torti G, Nuti Ronchi V, Marazziti D, Vergara R, Orselli S, Terzi M (1989) DNA methylation of embryogenic carrot cell cultures and its variation as caused by mutation, differentiation, hormones and hypomethylating drugs. Theor Appl Genet 77:325–331

Mayer U, Buttner G, Jurgens G (1993) Apical-basal pattern formation in the Arabidopsis embryo: studies on the role of the *gnom* gene. Development 117:149–162

McClure BA, Hagen G, Brown CS, Gee MA, Guilfoyle T (1989) Transcription, organization, and sequence of an auxin-regulated gene cluster in soybean. Plant Cell 1:229–239.

Meinke DW (1995) Molecular genetics of plant embryogenesis. Annu Rev Plant Physiol Plant Mol Biol 46:369–394

Mordhorst AP, Toonen MAJ, de Vries SC (1997) Plant embryogenesis. Crit Rev Plant Sci 16:535–576

Overvoorde PJ, Grimes HD (1994) The role of calcium and calmodulin in carrot embryogenesis. Plant Cell Physiol 35:135–144

Padmanabhan K, Cantliffe DJ, Koch KE (2001) Auxin-regulated gene expression and embryogenic competence in callus cultures of sweet potato, *Ipomoea batata* (L.) Lam. Plant Cell Rep 20:187–192

Pasternak T, Prinsen E, Ayaydin F, Miskoleczi P, Potters G, Asard H, Van Onckelen H, Dudtis D, Feher A (2002) The role of auxin, pH and stress in the activation of embryogenic cell division in the leaf protoplast-derived cells of alfalfa (*Medicago sativa* L.). Plant Physiol 129:1807–1819

Periera IY, Zielinski R (1992) Structure and expression of the Arabidopsis CaM-3 calmodulin gene. Plant Mol Biol 19:649–664

Quatrano RS (1987) The role of hormones during seed development. In: Davies PJ (eds) Plant hormones and their role in plant growth and development. Nijhoff, Dordrecht, pp 494–514

Raghavan V (1997) Molecular embryology of flowering plants. Cambridge University Press, New York, pp 467–499

Rao KS (1996) Embryogenesis in flowering plants: recent approaches and prospects. J Biosci 21:827–841

Russinova E, Slater A, Atanassov AI, Eliott MC (1995) Cloning of alfalfa cyclin sequences—a RACE-PCR approach. Cell Mol Biol 41:703–714

Sato S, Toya T, Kawahara R, Whittier RF, Fukuda H, Komamine A (1995) Isolation of a carrot gene expressed specifically during early-stage somatic embryogenesis. Plant Mol Biol 28:39–46

Schel JHN, Kieft H, Van Lammeren AAM (1984) Interactions between embryo and endosperm during early developmental stages of maize caryopses (*Zea mays*). Can J Bot 62:2842–2853

Schmidt EDL, De Jong AJ, de Vries SC (1994) Signal molecules involved in plant embryogenesis. Plant Mol Biol 26:1305–1313

Schmidt EDL, Guzzo F, Toonen MAJ, de Vries SC (1997) A leucine-rich repeat containing receptor-like kinase marks somatic plant cells competent to form embryos. Development 124:2049–2062

Schrader S, Kaldenhoff R, Richter G (1997) Expression of novel genes during somatic embryogenesis of suspension-cultured carrot cells (*Daucus carota*). J Plant Physiol 150:63–68

Shah K, Schmidt EDL, Vlak JM, de Vries SC (2001a) Expression of the *Daucus carota* somatic embryogenesis receptor kinase (DcSERK) protein in insect cells. Biochimie 83:415–421

Shah K, Gadella TW Jr, van Erp H, Hecht V, de Vries SC (2001b) Subcellular localization and oligomerization of the *Arabidopsis thaliana* somatic embryogenesis receptor kinase 1 protein. J Mol Biol 309:641–655

Somleva MN, Schmidt EDL, de Vries SC (2000) Embryogenic cells in *Dactylis glomerata* L. (Poaceae) explants identified by cell tracking and by SERK expression. Plant Cell Rep 19:718–726

Sopory SK, Munshi M (1998) Protein kinases and phosphatases and their role in cellular signalling in plants. Crit Rev Plant Sci 17:245–318

Sterk P, Booij H, Schellekens GA, van Kammen A, de Vries SC (1991) Cell-specific expression of the carrot EP2 lipid transfer protein gene. Plant Cell 3:907–912

Suprasanna P, Rao PS (1997) Somatic embryogenesis in crop plants. In: Ramana Rao TV, Kothari IL (eds) Plant structure and morphogenesis. Sardar Patel University, Vallabh Vidyanagar, India, pp 29–35

Suprasanna P, Desai NS, Nishanth G, Ghosh SB, Laxmi N, Bapat VA (2004) Differential gene expression in embryogenic, non-embryogenic and desiccation induced cultures of sugarcane. Sugar Technol 6(4):305–309

Timmers ACJ, de Vries SC, Schel JHN (1989) Distribution of membrane-bound calcium and activated camoldulin during somatic embryogenesis of carrot (*Daucus carota* L.). Protoplasma 153:24–29

Timmers ACJ, Reiss HD, Bohsung J, Traxel K, Schel JHN (1996) Localization of calcium during somatic embryogenesis of carrot (*Daucus carota* L.). Protoplasma 190:107–118

Toonen MAJ, Verhees JA, Schmidt EDL, van Kammen A, de Vries SC (1997b) AtLTP1 luciferase expression during carrot somatic embryogenesis. Plant J 12:1213–1221

Torii KU, Mitsukawa N, Oosumi T, Matsuura Y, Yokoyama R, Whittier RF, Komeda Y (1996) The Arabidopsis *ERECTA* gene encodes a putative receptor protein kinase with extracellular leucine-rich repeats. Plant Cell 8:735–740

von Arnold S, Izabela S, Bozhkov P, Julia D, Lada F (2002) Developmental pathways of somatic embryogenesis. Plant Cell Tissue Organ Cult 69:233–249

Walker JC (1994) Structure and function of the receptor-like protein kinases of higher plants. Plant Mol Biol 26:1599–1609

Walker JC, Key JL (1982) Isolation of cloned cDNAs to auxin-responsive poly(A)+ RNAs of elongating soybean hypocotyl. Proc Natl Acad Sci USA 79:7185–7189

Wilde HD, Seffens WS, Thomas TL (1995) Gene expression in somatic embryos. In: Bajaj YPS (ed) Somatic embryogenesis and synthetic seed I. Biotechnology in agriculture and forestry. vol 30. Springer, Berlin, Heidelberg New York, pp 41–52

Wilde HD, Nelson WS, Booij H, De Viries SC, Thomas TL (1998) Gene expression programs in embryogenic and non-embryogenic carrot cultures. Planta 176:205–211

Wurtle ES, Wang H, Durgerian S, Nikolau BJ, Ulrich TJ (1993) Characterization of a gene that is expressed early in somatic embryogenesis of *Daucus carota*. Plant Physiol 102:303–312

Yadegari R, de Paiva GR, Laux T, Koltunow AM, Apuya N, Zimmerman JL, Fischer RL, Goldberg RB (1994) Cell differentiation and morphogenesis are uncoupled in *Arabidopsis raspberry* embryos. Plant Cell 6:1713–1729

Yang H, Saitou T, Komeda Y, Harada H, Kamada H (1996) Late embryogenesis abundant protein in *Arabidopsis thaliana* homologous to carrot ECP31. Physiol Plant 98:661–666

Yang H, Saitou T, Komeda Y, Harada H, Kamada H (1997) *Arabidopsis thaliana ECP63* encoding a LEA protein is located in chromosome 4. Gene 184:83–88

Yasuda H, Nakajiima M, Ito T, Ohwada T, Masuda H (2001) Partial characterization of genes whose transcripts accumulate preferentially in cell clusters in the early stages of carrot somatic embryogenesis. Plant Mol Biol 45:705–712

Zhu C, Kamada H, Harada H, He M, Hao S (1997) Isolation and characterization of a cDNA encoded an embryogenic cell protein-63 related to embryogenesis from carrot. Acta Bot Sin 39:1091–1098

Zimmerman JL (1993) Somatic embryogenesis. Plant Cell 5:1411–1423

Zimmerman JL, Apuya N, Darwish K, O'Caroll C (1989) Novel regulation of heat shock genes during carrot somatic embryo development. Plant Cell 1:1137–1146

Plant Cell Monogr (2)
A. Mujib · J. Šamaj: Somatic Embryogenesis
DOI 10.1007/7089_031/Published online: 6 October 2005

Gametic Embryogenesis in *Triticum*: a Study of Some Critical Factors in Haploid (Microspore) Embryogenesis

L. Cistué[1] (✉) · K. J. Kasha[2]

[1]Departamento de Genética y Producción Vegetal, Estación Experimental de Aula Dei, C.S.I.C., Ap. 202, 50059 Zaragoza, Spain
lcistue@eead.csic.es

[2]Department of Plant Agriculture, University of Guelph, Guelph, ON N1G 2W1, Canada

Abstract This review of embryogenesis from microspores in two types of wheat examines the influences of various procedures used to induce and produce embryoids from the male gametes. The various factors that influence the quality of the microspores are critical and include growth of donor plants, pretreatments and culture and regeneration media. There are also many genetic and genotype factors, with a major problem being production of albino plants. Genotype differences in response and gametic selection during culture may be overcome through improved response in culture such as has been achieved in barley. Chromosome doubling to obtain uniform doubled haploid lines is being continually improved through pretreatments to cause nuclear fusion and through anti-microtubule agents in the induction media. Finally, the interesting aspects of cell and tissue changes that occur during the induction and development of the embryoids are briefly described and illustrated.

1 Introduction

Somatic embryogenesis is the process by which somatic cells differentiate into somatic embryos. Plants have great potential for cellular totipotency and it appears that any differentiated plant cell that retains its nucleus could have the ability to revert to the embryogenic condition and regenerate an entire plant (Reynolds 1997). Somatic embryo formation usually requires first a treatment of diploid cells with plant hormones, mostly auxin but also cytokinins, under specific culture conditions, and later auxin withdrawal to allow embryogenesis to continue (Toonen et al. 1994).

While this book deals extensively with somatic embryogenesis in several species of plants, this chapter describes the current status of male gametic embryogenesis focused mainly on *Triticum*. Reference will be made to other cereals, such as barley or maize, where certain aspects have been studied in more detail.

Most plant species exhibit a life cycle of alternating generations between a haploid gametophyte and a diploid sporophyte. In the angiosperm gametophyte phase, microspore mother cells in anther locules undergo meiosis and mitosis that result in the formation of male gametophytes known as pollen grains. Prompted by certain environmental cues, immature pollen (microspores) can be switched from gametophytic to sporophytic development, leading to the formation of "pseudo embryos" commonly known as embryoids (Reynolds 1997; Touraev et al. 2001; Zheng et al. 2001). Embryoids germinate to produce either haploid or doubled haploid (DH) plants. This phenomenon, an excellent example of plant cell totipotency, is defined as male gametic embryogenesis. Embryogenesis from microspores, either inside cultured anthers or as isolated microspore cultures, is induced by defined stress treatments. The stress converts a cell destined to produce a male gametophyte, i.e. a pollen grain, into an embryogenic cell that will develop into a sporophyte, i.e. an embryoid (Touraev et al. 1997; Zoriniants et al. 2005). These embryoids will usually have the gametic chromosome number and are described as haploid plants.

The study of plant embryogenesis via gametic embryogenesis is feasible and has the main advantage of starting from single, defined haploid cells which are available in large numbers as a fairly synchronous population and proceed to form embryoids. The frequency of embryos forming may only reach 1–2% of the microspore population but this is sufficient to produce large numbers of haploid plants. Microspore embryogenesis is not only a process of biotechnological importance but also a model system to study plant totipotency.

There are several advantages of microspore embryogenesis: (1) it is applicable to all pollen-producing plant species as demonstrated in a large number of crop species (Maluszynski and Kasha 2003); (2) there is a need to develop a good scientific understanding of microspore embryogenesis or other haploid production systems (Riley 1974); (3) the huge numbers of pollen grains in most species allows for a large number of DHs to be produced by microbiological techniques; (4) it is feasible to double the chromosome number during early stages of anther or microspore culture (Touraev et al. 1997; Kasha 2005). The disadvantages are the formation of albino plants (mostly in the Gramineae) and somaclonal variation, often a consequence of suboptimal tissue culture conditions.

First discovered by Bergner in *Datura stramonium* (Blakeslee et al. 1922), haploids are a very useful tool for plant breeders. Male gametic cells were first demonstrated to be totipotent in *D. innoxia* by Guha and Maheshwari (1964), and hundreds of haploid plantlets were induced from cultured anthers of tobacco (Nitsch and Nitsch 1969). DH plants are very useful for plant breeders for several reasons. Firstly, homozygous DH and thus uniform lines can be obtained in one generation via chromosome doubling of the haploid plants derived from an F_1 hybrid plant, whereas six or seven generations of inbreeding are often required for the production of fairly uniform diploids via conventional inbreeding methods. Hence DH production can be used to shorten the breeding cycle, save time, space and labour. Secondly, recessive characters are expressed in haploids from diploid species since there is only one set of chromosomes. This is particularly valuable in mutation breeding and mutagenic research. Thirdly, gametic genotypes are fully expressed at plant level in haploids and DH lines, so recombinant gametes are readily identified and selection of traits is easier (Hu 1997).

To be successfully used in a breeding programme, any particular DH system should usually meet certain criteria to make it cost effective relative to conventional breeding methods (Snape et al. 1986). These are (1) easy and consistent production of large numbers of DHs of all genotypes in the breeding programme, (2) DHs should be genetically normal and stable and (3) a DH population should contain a random sample of the parental gametes.

Since Ouyang et al. (1973) first obtained green plants from microspores of bread wheat (*Triticum aestivum*), great advances have been made (Tuvesson et al. 2000; Barnabás et al. 2001; Kasha et al. 2003)). The procedures developed have been improved to a level adequate for the application of these techniques in breeding programmes and a number of wheat cultivars have been produced (Thomas et al. 2003).

For durum wheat (*T. turgidum*) there is limited knowledge about androgenesis. Several authors reported the recalcitrant nature of this species to anther culture. Zhu et al. (1979) made the first reported attempt at androgenesis with durum wheat, obtaining only a few albino plantlets. Although several authors have reported the production of green plants during recent years, their frequencies are still low (Hadwiger and Heberle-Bors 1986; J'Aiti et al. 2000). The number of dividing microspores, the low number of microspores that develop into complete embryos and also the high frequency of albinism are the main restrictions. However, the most recent studies have produced improved results (Dogramaci Altuntepe et al. 2001; Jauhar 2003; Cistué et al. 2005).

In reviewing the state of the art in *T. aestivum* and *T. durum*, it is first necessary to consider many aspects involved in obtaining gametic embryogenesis and the genotypic or genetic factors before covering the pathways of embryogenesis that are being uncovered. This review will hopefully stimulate ideas and interactions for improving the different aspects of the gametic embryogenesis in both species of *Triticum* and allow for comparisons of gametic vs. somatic embryogenesis.

2
Critical Factors in Androgenesis

This chapter will provide a brief overview of factors and readers are referred to the publication edited by Maluszynski et al. (2003) for more details on many protocols.

2.1
Growing Donor Plants and Staging the Microspores

The first critical factor affecting androgenic response is the growth and vigour of the microspore donor plants. To ensure repeatable culture response one must optimize the environmental conditions for each species as well as for widely different types within a species, e.g. spring versus winter wheats (Kasha et al. 1990; Jähne and Lörz 1995). Obviously, biotic stresses such as diseases and insects should be avoided, as even the treatments for such pests in stages prior to spike collection can be deleterious. The main abiotic factors influencing donor plants are light, photoperiod and temperature, while good nutrition is also critical.

Variations in light intensity, quality and photoperiod are feasible but the control and coordination of all three are important for good results. Some laboratories grow the *T. aestivum* donor plants in growth chambers with day lengths of 16–17-h light (350–400 uE-2 s^{-1}) at approximately 20–22 °C and 8–7-h dark period at 17 °C (Kasha et al. 2003). Others have used higher temperatures of 27 °C for the day and 17 °C for the dark (Zheng et al. 2003). Good plants have been grown with natural light in a greenhouse if supplemental light is supplied (Tuvesson et al. 2003). The conditions may also be dependent upon the genotype and a compromise must be reached. Winter cultivars must have 6–8 weeks at 5–6 °C with low light for 8–10 h per day for vernalization. Immediately after vernalization an intermediate temperature of 16–18 °C may be required for a short time to avoid devernalization, resulting in poor pollen production. In the case of durum wheat, where the androgenic response is lower than for bread wheats, in most instances the donor plants have been grown in the greenhouse or in the field (Jauhar 2003). The only comparative study of growth conditions showed that plants grown in the field or the greenhouse produced a higher number of calli and green plantlets than those plants grown in growth chambers (Dogramaci-Altuntepe et al. 2001). However, Cistué et al. (2005) obtained good results with donor plants grown in growth chambers. Weekly or biweekly fertilization with soluble NPK 20 : 20 : 20 is used by most of the laboratories and this is important for all types of wheat.

Growth of donor plants must often be adjusted to the characteristics of the cultivars and to their climatic adaptations. German spring wheat cultivar seedlings were developed in a growth chamber with a 10-h photoperiod and a day/night temperature of 12/10 °C for improved tillering. Later, plants were transferred to a greenhouse with a 16-h photoperiod and a temperature of 16/18 °C for the day and 12/14 °C for the night (Foroughi-Wehr and Zeller 1990). However, Ghaemi et al. (1993) using Ethiopian lines, French commercial cultivars and drought-resistant cultivars from ICARDA used a 16-h photoperiod and temperatures of 25/15 °C (day/night). At Guelph, the winter wheats are grown at 20 °C, whereas spring wheats are grown at 22–24 °C with lights on.

The best time to collect the spikes is when the majority of microspores are at the mid to late uninucleate stage in *T. aestivum*. To check this stage, an anther is stained with acetocarmine or Alexander's stain (Kasha et al. 2001). Subsequently, additional tillers are collected when they reach a similar stage of morphological development. This morphological stage can vary from genotype to genotype. It is also desirable to check the percentage microspore viability by staining with fluorescein diacetate (Widholm 1972). Batches of plants with low viability of the microspores will obviously provide a poor response.

In durum wheat, microspores at the mid to late uninucleate stage have been used for anther culture (Otani and Shimada 1994), and at the late uninucleate or early binucleate stages for isolated microspore culture (J'Aiti et al. 2000).

2.2 Stress Pretreatment for Induction

To trigger androgenesis, i.e. switching the microspores from the gametophytic pathway to a sporophytic one, several and very different pretreatments have been applied, including physical, physiological and chemical treatments of excised spikes, anthers or microspores (Zoriniants et al. 2005). Some laboratories are using two of them in combination, for example low temperature and starvation (Kasha et al. 2003). While mannitol pretreatment was thought to act by starvation (Roberts-Oehlschlager and Dunwell 1990), it also appears to block cell wall formation that would disrupt the normal pathway of microspore development (Kasha et al. 2001; Shim et al. 2005). Nitrogen starvation, short days and high osmoticum treatment of donor plants are also found to increase embryos yields in anther cultures, but these treatments are not widely used for DH production or breeding programmes. The type and duration of pretreatment is very important in blocking gametophytic development and inducing embryogenic development.

Temperature shocks, both cold and hot, are used as stress pretreatments with good results. For *T. aestivum* Touraev et al. (1996) recommended culturing microspores at 33 °C for 4 days. Cold treatment has been a common system to induce embryogenic microspores. When tillers or spikes are given a cold shock at 4 °C for several days (Lazar et al. 1985; Armstrong et al. 1987; Tuvesson et al. 2000) prior to microspore culture, both response and "spontaneous" chromosome doubling increase.

For durum wheat Ghaemi et al. (1995) used a cold pretreatment by keeping spikes in the dark in flasks containing water at 4 °C for 7 days. Their results showed negative effects of cold pretreatment on the induction of embryoids but positive effects on plant regeneration. There is also evidence for an interaction between cold pretreatment and genotype (Mentewab and Sarrafi 1997). A treatment of 4 °C for 7 days was also used by Dogramaci-Altuntepe et al.

(2001). Saidi et al. (1997) kept spikes in the dark for 0 – 15 days at 3 °C. The 8-day pretreatment significantly improved the rate of embryogenesis.

Starvation is another method used to induce embryogenesis. Since Roberts-Oehlschlager and Dunwell (1990) first substituted mannitol to replace maltose for 4 days, as a pretreatment method in barley, this has often been a method of choice for wheat, because of its time-saving over longer cold treatments then in use. Kasha et al. (2003) used 0.4 M mannitol on collected spikes in large Petri dishes in a refrigerator (4 °C). Cistué et al. (2005) used 0.7 M mannitol solidified with agarose and kept for 4–6 days at 24 °C for both *T. aestivum* and durum wheat. Barnabás et al. (1991) used colchicine in the induction media for *T. aestivum* for up to 2 days and obtained improved induction of microspores as well as an increase in chromosome doubling.

Another inducer chemical formulation contains 0.001% (w/v) 2-hydroxynicotinic acid (2-HNA) (Liu et al. 2002). This system seems to be capable of initially inducing embryogenesis in 20–50% of microspores of some wheat genotypes (Zheng et al. 2003). This product is applied as a pretreatment to tillers or it can also be applied on isolated microspores.

2.3 Induction Media

Szarejko (2003) has summarized the components of the major media used in haploid production. Several media have been used to induce microspore divisions in both *T. aestivum* and *T. durum*, either after induction or before it in some instances. Research has been carried out not only on the basic media components but also on the growth regulators and their support system (solid, semisolid or liquid). Variations exist for mineral nutrients and vitamins, but basically microspores from bread wheat respond to the basal medium N6 or to the modified MS medium (Hu et al. 1995). Maltose has been shown to be a superior carbohydrate source for cereals (Hunter 1988) and is used in nearly all microspore cultures of cereals today.

The balance of exogenous hormones in the culture medium is crucial for both the yield and quality of embryoids. There is no general consensus on which kinds or amounts of hormones are the most effective but there is consensus about using the lowest possible concentrations. The most common auxins used are 2,4-dichlorophenoxyacetic acid (2,4-D) or phenylacetic acid; with kinetin or 6-benzylaminopurine as cytokinins (Hu and Kasha 1997; Zheng et al. 2001; Liu et al. 2002; Kasha et al. 2003). How plants are grown may greatly affect the endogenous hormone levels in donor plants and thus the levels required in culture. Stress to the donor plants can greatly alter the endogenous hormone levels.

For *T. durum*, the induction medium to use is still under development and much research is in progress. The first attempts to produce green plants from anther culture of durum wheat by Hadwiger and Heberle-Bors (1986) and

Cattaneo et al. (1991) used a modification of the Chinese potato-2 medium (PII). The frequency of embryogenesis varied from 0.0 to 28.9 embryoids per 100 anthers, and 1.6 green plants per 100 embryoids were regenerated from one out of the two cultivars assayed (Cattaneo et al. 1991). Much research followed with small modifications but no real progress in green plant production has occurred. Various comparisons of media were also made but with little progress in green plant production. Otani and Shimada (1995) compared four different liquid media, namely PII, C_{17}, W_{14} and MN_6, and found the C_{17} medium was the most effective for embryoid formation. Saidi et al. (1997) used anther culture to compare four solidified media, namely C_{17}, BPTG, PII and N_6. Again the C_{17} medium was consistently the best for embryo induction. Dogramaci-Altuntepe et al. (2001) compared the media BAC-1, BAD-1, BAD-3 and M-42. Anthers cultured on BAD-1 produced the highest mean number of calli, and although M-42 gave the lowest callus production, it gave the highest mean number of green plants regenerated. Jauhar (2003) also used BAD-1 induction medium for anther culture and then calli were transferred to BAD-3 medium for regeneration. Recently J'Aiti et al. (1999) demonstrated that co-culture of ovaries with anthers enhanced the response in three durum wheat genotypes.

In isolated microspore culture of bread wheat, the use of a preconditioned medium with ovaries is essential to increase the rate of divisions and embryos (Mejza et al. 1993; Hu and Kasha 1997; Devaux and Li 2001). Letarte et al. (2005) demonstrated that one of the important chemicals exuded from immature ovaries that enhances response in culture of wheat microspores is arabinogalactan proteins (AGP). Survival of the microspores is greatly enhanced by the application of AGP to the induction media for wheat and the potential for using a completely defined medium for wheats exists.

To our knowledge there are only two works on isolated microspore culture in durum wheat. In the first case, J'Aiti et al. (2000), the CHB3 medium preconditioned with ovaries was used. The percentage of pro-embryos was 8.4 per 100 microspores but only 2% of them developed into embryos. In the second recent study, Cistué et al. (2005) used cultivars and F_1 hybrids adapted to Spanish conditions. They cultured isolated microspores in a modified C_{17} medium preconditioned with ovaries and this was effective for obtaining both embryogenesis and green DHs from most genotypes. Perhaps a breakthrough for durum wheat microspore culture is close. A liquid medium is normally used, but a semisolid medium with Gelrite or agarose can also be used. Ficoll in the medium is another solution to prevent the microspores from sinking.

2.4 Regeneration Media

Although less attention has been focused on the regeneration medium for durum wheat, its composition is equally important for both bread and durum wheats, particularly if the embryos obtained in the induction media are sometimes not well developed.

For bread wheat the regeneration medium is less complicated because most of the embryoids produced are complete embryos with scutellum, coleoptile and coleorriza. Pauk et al. (2003), Barnabás (2003), Tuvesson et al. (2003) and Zheng et al. (2003) all used the 190-2 medium solidified with Gelrite, sucrose at 30 g l^{-1} and either without hormones or with small amounts (0.5 mg l^{-1}) of 1-naphthaleneacetic acid and kinetin. Kasha et al. (2003) used the modified medium MMS5 with low (30 g l^{-1}) maltose and Phytagel.

For durum wheat most of the authors transferred the embryoids or calli obtained from anther culture to a regeneration medium where they assayed many basal media and other components, such as growth regulators. Only a few examples will be mentioned since the production of green plants is rare and albinos plentiful. Foroughi-Wehr and Zeller (1990) used the Chinese regeneration medium (Chuang et al. 1978) with 50 mg l^{-1} glutamine, 30 g l^{-1} sucrose and solidified with agar. Cattaneo et al. (1991) cultured the embryoids on Murashige and Skoog (MS) medium containing 0.1 mg l^{-1} of 2,4-D and obtained 1.6 green plants per 100 embryoids plated. Ghaemi and Sarrafi (1994) transferred embryoids to the R_8 regeneration medium and the percentages of green and albino plants regenerated varied from 0.6 to 6.0 and from 4.0 to 14.2, respectively. Saidi et al. (1997) compared the regeneration of embryoids formed after 1–2 months by transferring them to R_9, C_{17} or N_6 media. The percentage of total plants regenerated was highest on C_{17} (20.1), followed by R_9 (15.7) and N_6 (10.7). In spite of this high percentage of regeneration, only two plants were green. Dogramaci-Altuntepe et al. (2001) transferred the embryos to a BAD-1-based medium modified with 30 g l^{-1} sucrose, 17.5 g l^{-1} glucose and solidified with agar. The average green plant production per 100 anthers, from three cultivars, was 0.29.

J'Aiti et al. (2000) used embryoids from culture of isolated microspores and compared different combinations of growth regulators (indole-3-acetic acid and kinetin) in MS regeneration medium but again green plant regeneration was low (0.55 green plants per 100 embryoids). The use of growth regulators actually reduced the percentage of green plants. Cistué et al. (2005) obtained good regeneration of embryoids cultured in the J25-8 regeneration medium (Jensen 1983), following the culture of isolated microspores. The high rate of total plant regeneration of the embryos derived from isolated microspores could also reflect the good quality of the embryos derived using this method.

3
Genotype and Genetic Considerations

3.1
Albino Plants

Microspore-derived haploid plants in cereals are often chlorophyll-deficient (albino). However, the three components, namely embryoid induction, plantlet regeneration and green vs. albino, are independently inherited, so the albino component can be studied. Genotype differences have been reported to account for 32.8% of the variation in percentage of green plant in studies of wheat anther culture (Holme et al. 1999). A quantitative trait locus explaining 31.5% of the genetic variation for green plant formation from the cross Ciano by Benoist was detected on chromosome 5BL (Torp et al. 2001). Studies with *T. aestivum* show significant genetic variability and relatively high heritability for androgenic traits (Moieni and Sarrafi 1995). However, Zheng (2003) observed a low heritability value of 0.68 for albino plant regeneration, indicating that it may be possible to reduce albinos by controlling the culture conditions or other environmental factors.

Besides the genetic factors, we have already discussed that response from anther culture is affected by growth conditions of the donor plants, culture media and that there are genotype x environment interactions. On the other hand, the longer the time that the microspores are in the induction culture medium, the higher will be the proportion of albino plantlets regenerated (Cistué et al. 1995). This is not unique to microspore cultures as it occurs in all types of plant tissue cultures and frequencies can be modified. For example, Liu et al. (2002) found that by adding 10% nutrient medium NPB98 to the microspore pretreatment with 2-HNA, green plant frequency was improved by 27%.

In contrast to previous studies (reviewed by Touraev et al. 2001), Ankele et al. (2004) proposed that deletions or rearrangements in the plastid genome DNA are not, most likely, the cause of the albino phenotypes. They observed that the RNA pattern of albino plants indicates the absence of the plastid-encoded polymerase. Control experiments of plastid gene translation revealed there were strongly reduced levels of plastid-encoded proteins, one of them an essential component of plastid ribosomes. This indicates that a complete deficiency of plastid protein synthesis in microspores could result in albino plants. On the other hand, if small DNA deletions occur in plastids in developing microspores, further rearrangements and large-scale deletions at hot-spot regions encoded in the plastid genome might follow them. However, it is difficult to conceive that a severe genetic defect such as a deletion of parts of the plastid genome can be overcome by manipulations of the conditions for microspore cultures, whereas an epigenetic mechanism, such as the translation deficiency by failure of ribosome assembly, provides a theoretical framework for interference in plastid function in order to prevent albino formation.

3.2 Genotype Response Dependence

As mentioned in Sects. 2.3 and 2.4, only a few durum wheat genotypes have given green plants until very recently. The presence of genotype interactions with stress pretreatments (Mentenwab and Sarrafi 1997), medium composition (Ghaemi et al. 1994) and cytoplasmic effects (Ghaemi et al. 1993) has been described. The location of nuclear genes has not been ascertained although some effort has been made to place genes on specific chromosomes or to identify markers linked to genes involved in the androgenic response (Zhang and Li 1984; Agache et al. 1989; Devaux et al. 1990; Ben Amer et al. 1995). According to Zhang and Li (1984) chromosomes 2A and 2D bear major genes for androgenesis, whereas chromosomes 2B, 4A, 5A and 5B carry minor genes acting to inhibit embryoid production. Ben Amer et al. (1997) mapped three quantitative trait loci for response in tissue culture on chromosome 2B of hexaploid wheat. However plantlet regeneration seems to be independent of callus response (Holme et al. 1999).

Several authors have attributed great importance to the presence of the D genome in bread wheat (AABBDD) in order to obtain a high androgenic response. The absence of this genome in durum wheat (AABB) may explain the low induction of embryogenesis and of green plant regeneration of this species (Ghaemi and Sarrafi 1994; Cattaneo and Vaccino 2001). Potential genes for controlling embryo induction are located on chromosomes 1D, 2D and 7D, genes for total plant regeneration on chromosomes 2D, 5D and 7D, and genes influencing green plants on chromosomes 2D and 7D (Galiba et al. 1986; Szakacs et al. 1988; Ghaemi et al. 1995; de Buyser et al. 1992). Henry and de Buyser (1985) and Agache et al. (1989) also reported a higher regeneration capacity of bread wheat cultivars carrying the 1B/1RS wheat-rye translocation. Cattaneo et al. (1991) also confirmed the beneficial effects of this translocation on the embryogenesis in durum wheat. Tersi et al. (2004) in a recent study also suggested that the presence of the D genome is necessary to obtain good anther culture in durum wheat. Finally, significant interactions exist between nuclear genes and cytoplasm types for all three components of the androgenic response (Ekiz and Konzak 1991). The influence of cytoplasm in androgenic response makes the choice of paternal vs. maternal parents in crosses an important step (Zheng 2003).

3.3 Gamete Selection During Culture

Gamete selection has been revealed in anther or isolated microspore culture of barley and many other species as associations between markers and culture response. The consensus is that the distortion results from allelic differences of genes related to survival in culture. Sayed et al. (2002) showed that the proportion of loci deviating from an expected monogenic segregation ratio in the DH population of barley was significantly higher (19/43 loci, 44%) than in the F_2 population, where some distortion was also detected (7/43 loci, 16%). The distributions were skewed towards those parent alleles that influenced better regeneration of green plantlets with the isolated microspore culture technique.

Another procedure to limit the extent of segregation distortion or the appearance of potentially undesired mutants is to avoid the selection of calli and utilize only complete embryoids in order to restrict somaclonal variation (Jain 2001; Mujib 2004). Related to this is the important aspect of improving the techniques used to develop the DH plants so that fewer negative genes can have an influence on the production of haploid plants. Hu and Kasha (1997) studied microspore-derived green DHs of the wheat cultivar Chris. By the third generation, the field-measured traits of microspore-derived DH lines were more similar to the parental check with 92 and 70% of DH lines having 1000-kernel weights and yields, respectively. This would suggest that a seed increase generation before examining quantitative traits could be beneficial. Ma et al. (1999) studied a spring wheat cross for the agronomic performance of lines derived from it by three techniques: anther culture (male gamete), maize pollination of wheat (female gamete) and single-seed descent. Thirty-three lines derived from each of those techniques were evaluated in six environments for six agronomic traits. The mean performance of the single-seed descent lines exceeded the performance of the anther culture lines for grain yield, kernel weight and plant height, with no apparent differences for grain protein content, test weight and heading date. Only differences in the means of plant height occurred between traits for the single-seed descent and maize pollination lines. However, to carry out meaningful comparisons larger population sizes are desired. Guzy-Wróbelska and Szarejko (2003) compared 76 DH lines derived using maize pollination with 122 anther derived lines and found that they all performed the same as the single seed descent population created from the same crosses. None or only very small differences in means and ranges were observed when the best 10% of the lines from all three methods were compared.

Comparisons among wheat DH lines produced by microspore culture using improved procedures (Kasha et al. 2003; Zheng et al. 2003) and lines obtained by the pedigree selection method should be made. These isolated microspore methods are capable of producing about 100–300 DH plants per wheat spike and if achieved across genotypes would be preferable to producing haploid wheats from wide crosses.

3.4
Chromosome Doubling

"Spontaneous" chromosome doubling during the first days of culture has been described in different cereals. In barley, a high rate of doubling is obtained, between 80 and 90% (Cistué et al. 2003), whereas in bread wheat this rate is lower (50–60%). In most other crop species it is low, requiring a separate step and time to produce the DH plants. The differences in the rate depend on the species, genotype, pretreatments used and the culture system. For example, nuclear fusion leads to chromosome doubling during mannitol pretreatment of barley (Kasha et al. 2001).

The numbers of haploid or DHs obtained spontaneously in durum wheat are less characterized because of the low numbers of green plantlets regenerated. Dogramaci-Altuntepe et al. (2001) cytologically studied 31 albino plants and seven were haploids ($n = 2x = 14$) and 24 were DHs ($2n = 4x = 28$). However, out of 102 green plants, 52 were DHs (tetraploid), 22 pentaploid, six hexaploid, one nonaploid, one decaploid and 20 aneuploid ($2n = 36$–39). Some of these plants showed translocations, dicentric chromosomes and also chromatid exchanges. It is important to note that the high rate of chromosomal abnormalities obtained could be due to the callus origin of these plants or the length of treatment or the time in culture might have been too long. However, Cistué et al. (2005) found 67% spontaneous DHs that were fertile plants. Therefore, perhaps the results will be similar to those for the bread wheat when microspore culture techniques work more efficiently, and embryos can be produced instead of calli.

Haploid plants in most species are usually treated with an agent for chromosome doubling. The most commonly used agent is colchicine as described by Jensen (1977, 1983) or Metz et al. (1988). The incorporation of this agent into the induction culture medium has proven to be successful in bread wheat (Barnabás et al. 1991; Zamani et al. 2000) as well as other species. The results of these experiments suggest that the effect of colchicine on cultured microspores is not limited to successful genome doubling; as it also functions as an inducing agent for microspore embryogenesis.

Although colchicine treatment is still the most effective way for doubling the chromosomes, its safety to handlers limits its application. A number of other anti-microtubule agents such as oryzalin, trifluralin, aminoprophos-methyl (AMP), 2-HNA and caffeine have been successfully used but are usually less effective than colchicine in producing DHs (Thomas et al. 1997; Kasha 2005).

4 Development of Embryos from Microspores

Section 2 summarized the many factors influencing embryogenesis from microspores and Sect. 3 covered the influences attributed to genotypes. This section will briefly describe our knowledge of the process of embryo development from microspores. This development is influenced by the treatments used as described in the Sect. 2 and therefore is generalized. Aionesei et al.

(2005) has reviewed this topic and the reader is referred to that review for more details.

The first clear evidence of induction is the disruption of the asymmetric first mitotic division (PMI) in the uninucleate microspore. Most induction treatments result in a symmetrical first division resulting in two large similar or symmetric nuclei that are typical of vegetative nuclei, but often without a cell wall between them (Fig. 1). This appears to be the result of the disruption of the asymmetric spindle formation that occurs perpendicular to the wall of the microspore in a nucleus that is located along the wall in the highly vacuolated microspore. The disruption results in the nucleus moving into the centre of the vacuolated nucleus with cytoplasmic streams attaching it to the cytoplasm lining the inside of the wall. This results in what has been described by Indrianto et al. (2001) as a "star-shaped" configuration (Fig. 2) that has been observed following many different types of pretreatment, but not often following a cold pretreatment (Kasha et al. 2001). When PMI occurs in such microspores, there is likely no restriction on spindle formation, so it is allowed to become symmetric, usually resulting in two large symmetric vegetative type nuclei with no cell wall between them.

The position of the cell wall formation during cell division is dictated by the actin microfilaments that are located inside the cell wall (Twell et al. 1998; Simmonds and Keller 1999). These filaments form a preprophase band along the cell wall that dictates where the new cell wall will form. In pathway development towards normal pollen formation, a thin wall is formed around the small densely staining generative nucleus after PMI. This produces a small generative cell with very few cytoplasmic organelles inside the large vacuolated microspore containing a large lightly staining vegetative nucleus. The generative nucleus usually continues on into the S phase and G_2 of the cell

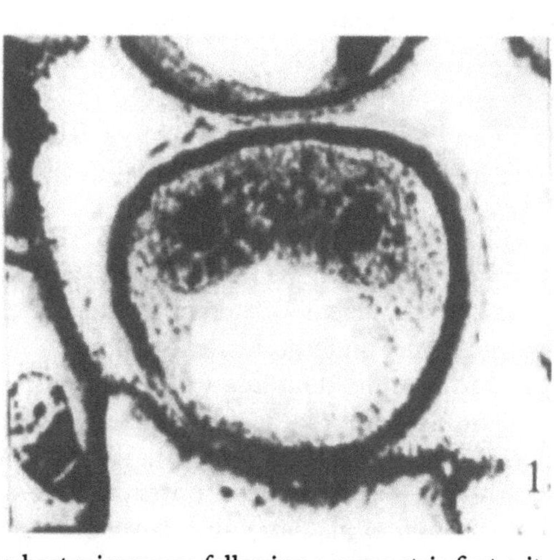

Fig. 1 A binucleate wheat microspore following a symmetric first mitotic division

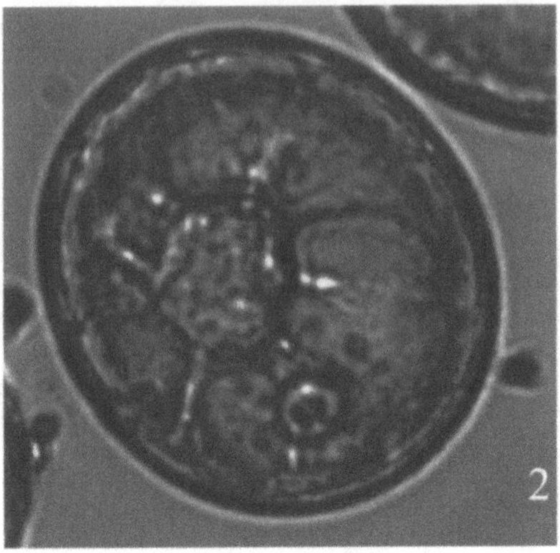

Fig. 2 A durum wheat uninucleate microspore in the "star" stage following an induction treatment

cycle. It may be arrested here or it may continue to divide (PMII) to form the two darkly staining sperm nuclei in cereals. This PMII does not occur immediately after PMI in all species of plants and could occur as late as during the developing pollen tube.

However, typical of induced microspores is the removal of the restriction on the vegetative nucleus to remain at the G_1 stage of the cell cycle. During normal pollen development it stays at the G_1 stage and is active in the development of the mature pollen and in developing the pollen tube. However, after induction of the microspore, it goes on to the S phase of the cycle and continues cycles of division. In some induced microspores, the generative nucleus is still distinct and it may divide again, but usually not more than two or three times. This continued generative division in induced microspores most likely occurs when microspores are pretreated at a very late uninucleate or an early binucleate stage so that it had possibly been programmed as a generative nucleus.

Following PMI with no wall formation, Shim et al. (2005) observed using time-lapse photography that the two nuclei are located along the wall and that one moves towards the other nucleus located near the pore in the microspore wall. Once together, the nuclei undergo nuclear fusion via membrane coalescence to result in a chromosome doubling of the nucleus. This nucleus can continue to divide, forming a cell that is multinucleate with cell wall formation occurring after 5–7 days in culture (Fig. 3). The evidence for this type of nuclear fusion in microspores has been observed in barley (Chen et al. 1984; Kasha et al. 2001; Shim et al. 2005) and *Zea mays* microspore cultures (Testillano et al. 2004).

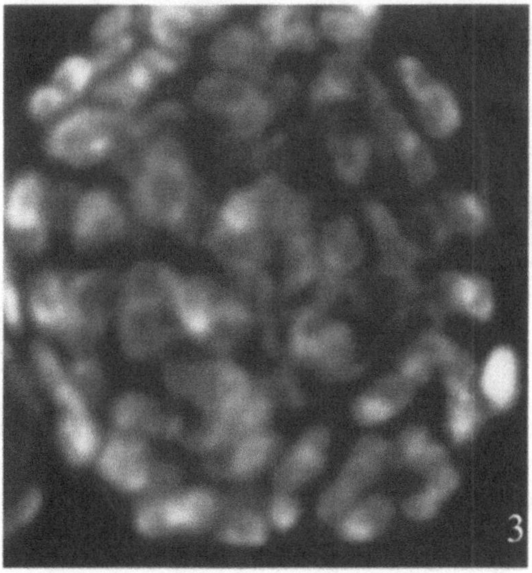

Fig. 3 A multicellular structure in wheat after 10 days in culture

Since the nuclei continue to divide after pretreatment, there is likely no disruption of the microtubules forming the spindle for mitosis but rather the actin filament assembly for development of the cell walls is disrupted, leading to the symmetric PMI. About 10 days in culture, the developing structures burst the microspore cell wall and become free in the culture medium. Maraschin et al. (2005) followed the development of uninucleate barley microspores through to microspore wall bursting and embryo development and noted differences along the way between those that formed embryos and those that did not. Those cells that were induced had a unique thin intine layer and an undifferentiated cytoplasm after the induction treatment. At the time of rupture of the microspore wall in the multicellular structure, there was a decrease in cell viability at the site of exine wall rupture, which was opposite to the position of the microspore pore. Thus, these proembryo structures appeared to have already become polarized and continued on to embryo development (Fig. 4). An earlier breakage of the cell wall and extrusion of a line of cells that develop much like a suspensor and developing embryo was described by Aionesei et al. (2005).

In other studies, differentiation of cell types was also observed to occur in the early multicellular structures from maize microspores (Testillano et al. 2002). One section had fairly small and cytoplasmic dense cells with straight-sided walls, while another sector was multinucleate and eventually formed large irregularly shaped cells with more than one nucleus and was much like the endosperm make-up of polyploid cells. Many questions arise from the observation of such structures, such as do they originate from the generative and vegetative nucleus, respectively. An earlier study in maize by Petrova et al.

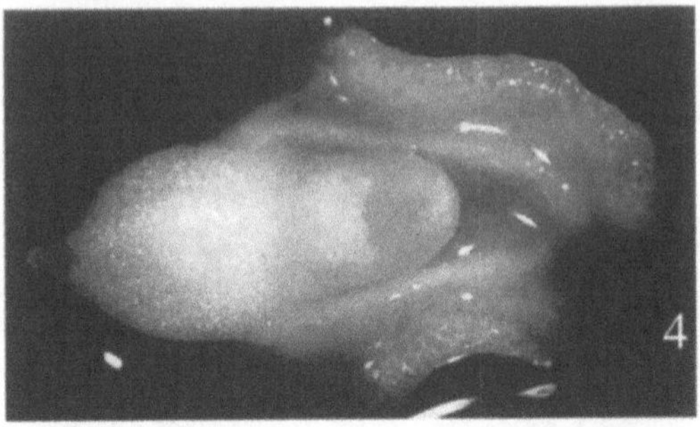

Fig. 4 A highly developed embryo from 30 days of wheat microspore culture

(1993) suggested that more than 60% of dividing microspores were derived from the generative nucleus. The study and development of such structures relative to different pretreatments needs to be examined further. Perhaps the type of development is determined by the species or the pretreatments used for induction.

5 Conclusions and Future Prospects

The protocols mentioned have proven to be very efficient for microspore culture of bread wheat. A great number of DHs can be obtained by microspore culture from most of the genotypes, making this technique competitive with the procedure of haploid production following intergeneric crossing with *Z. mays*.

However, these protocols have had a limited success in durum wheat. The best improvements in androgenesis of this species have been produced during the last year, but great effort is still needed concerning the study of the stress pretreatment and the induction media. The low rate of embryogenesis, the small size and poor quality of the embryos and the high rate of albinism (in many studies well over 90%) are the main limitations to utilizing the total potential of the technique. The efficiencies recently described are very promising, suggesting that the microspore culture technique could be the system to choose for haploid production in the future for durum wheat.

Much more detailed aspects related to biochemical, molecular, cytological, ultrastructural, stress and microspore development are discussed in the reviews edited by Palmer et al. (2005).

References

Agache S, Bachelier B, Buyser J, Henry Y, Snape S (1989) Genetic analysis of anther culture response in wheat using aneuploid, chromosome substitution and translocation lines. Theor Appl Genet 77:7–11

Aionesei T, Touraev A, Heberle-Bors E (2005) Pathways to microspore embryogenesis. In: Palmer CE, Keller WA, Kasha KJ (eds) Hapoids in crop improvement II. Biotechnology in agriculture and forestry, vol 56. Springer, Berlin Heidelberg New York, p 11–34

Amstrong TA, Metz SG, Mascia PN (1987) Two regeneration systems for the production of haploid plants from wheat anther culture. Plant Sci 51:231–237

Ankele E, Hofinger BJ, Heberle-Bors E, Pfosser M (2004) Alterations in the chloroplast genome of microspore-derived albino plants. In: COST action 851. Gametic cells and molecular breeding for crops improvement, Palermo, 11–13 November

Barnabás B (2003) Protocol for producing doubled haploid plants from anther culture of wheat (Triticum aestivum L.) In: Maluszynski M, Kasha KJ, Forster BP, Szarejko I (eds) Doubled haploid production in crop plants: a manual. Kluwer, Dordrecht, p 65–71

Barnabás B, Pfahler PL, Kovács G (1991) Direct effect of colchicine on the microspore embryogenesis to produce dihaploid plants in wheat (Triticum aestivum L.). Theor Appl Genet 81:675–678

Barnabás B, Szakács Ë, Karsai I, Bedö Z (2001) In vitro androgenesis of wheat: from fundamentals to practical application. Euphytica 119:211–216

Ben Amer IM, Worland AJ, Borner A (1995) Chromosomal location of genes affecting tissue-culture response in wheat. Plant Breed 114:84–85

Ben Amer IM, Korzun V, Worland AJ, Borner A (1997) Genetic mapping of QTL controlling tissue-culture response on chromosome 2B of wheat (Triticum aestivum L.) in relation to major genes and RFLP markers. Theor Appl Genet 94:1047–1052

Blakeslee AF, Belling J, Farnham ME, Bergner AD (1922) A haploid mutant in the Jimson weed, Datura stramonium. Science 55:1433

Cattaneo M, Vaccino P (2001) Dihaploid production for wheat breeding programmes: the Italian experience. In: Biotechnological approaches for utilization of gametic cells. COST 824 final meeting, Bled, Slovenia, 1–5 July 2000 p 53–59

Cattaneo M, Qiao YM, Pogna NE (1991) Embryoid induction and green plant regeneration from cultured anthers in a durum wheat line homozygous for the 1BL/1RS translocation. J Genet Breed 45:369–372

Chen CC, Howarth MJ, Peterson RL, Kasha KJ (1984) Ultrastructure of androgenic microspores of barley during the early stages of anther culture. Can J Genet Cytol 26:484–491

Chuang CC, Ouyang JW, Chia H, Chou SM, Ching CK (1978) A set of potato media for wheat anther culture. In: Proceedings of the symposium on plant tissue culture. Science, Beijing p 51–56

Cistué, L, Ziauddin A, Simion E, Kasha KJ (1995) Effects of culture conditions on isolated microspore response of barley cultivar Igri. Plant Cell Tiss Org Cult 42:163–169

Cistué, L, Vallés MP, Echávarri B, Sanz JM, Castillo A (2003) Barley anther culture. In: Maluszynski M, Kasha KJ, Forster BP, Szarejko I (eds) Doubled haploid production in crop plants: a manual. Kluwer, Dordrecht, p 29–33

Cistué L, Soriano M, Castillo AM, Vallés MP, Sanz JM, Echávarri B (2005) Production of doubled haploids in durum wheat (Triticum turgidum) through isolated microspore culture. Plant Cell Rep (in press)

De Buyser J, Hachemi-Rachedi S, Lemee ML, Sejourne S, Marcotte JL, Henry J (1992) Aneuploid analysis of anther response in wheat. Plant Breed 109:339–342

Devaux P, Li H (2001) Enhancement of microspore culture efficiency of recalcitrant barley genotypes. Plant Cell Rep 20:475–481

Devaux P, Lonnet P, Jean R (1990) Significant higher proportions of 1BL-1RS wheat-rye translocation lines among doubled haploid progenies derived from F1 hybrids between translocated and non-translocated wheat cultivars. In: Nijkamp HJJ, van der Plas LHW, van Aartrijk J (eds) Progress in plant cellular and molecular biology. Kluwer, Dordrecht, p 422–426

Dogramaci-Altuntepe M, Peterson TS, Jauhar PP (2001) Anther culture-derived regenerants of durum wheat and their cytological characterization. J Hered 192:56–64

Ekiz H, Konzak CF (1991) Nuclear and cytoplasmic control of anther culture response in wheat: III. Common wheat crosses. Crop Sci 31:1432–1436

Foroughi-Wehr B, Zeller FJ (1990) In vitro microspore reaction of different German wheat cultivars. Theor Appl Genet 79:77–80

Galiba G, Kovacs G, Sutka J (1986) Substitution analysis of plant regeneration in callus culture of wheat. Plant Breed 97:261–263

Ghaemi M, Sarrafi A, Alibert G (1993) Influence of genotype and culture conditions on the production of embryos from anthers of tetraploid wheat (Triticum turgidum). Euphytica 65:81–85

Ghaemi M, Sarrafi A, Alibert G (1994) The effects of silver nitrate, colchicine, cupric sulfate and genotype on the production of embryoids from anthers of tetraploid wheat (Triticum turgidum). Plant Cell Tiss Org Cult 36:355–359

Ghaemi M, Sarrafi A, Alibert G (1995) Influence of genotype, media composition, cold pre-treatment and their interactions on androgenesis in durum wheat (Triticum turgidum). Cereal Res Commun 23:215–227

Ghaemi M, Sarrafi A, Morris R (1995) Reciprocal substitution analysis of embryo induction and plant regeneration from anther culture in wheat (Triticum aestivum L.). Genome 38:158–165

Guha S, Maheshwari SC (1964) In vitro production of embryos from anthers of Datura. Nature 204:497

Guzy-Wróbelska J, Szarejko I (2003) Molecular and agronomic evaluation of wheat doubled haploid lines obtained through maize pollination and anther culture methods. Plant Breed 122:305–313

Hadwiger MA, Heberle-Bors E (1986) Pollen plant production in Triticum turgidum ssp durum In: Horn W, Jensen CJ, Odenbach W, Schieder O (eds) Genetic manipulation in plant breeding. de Gruyter, Berlin, p 303–305

Henry Y, De Buyser J (1985) Effect of the 1B/1R translocation on anther culture ability in wheat (Triticum aestivum L.). Plant Cell Rep 4:307–310

Holme IB, Olesen A, Hansen NJP, Andersen SB (1999) Anther and isolated microspore culture response of wheat lines from northwestern and eastern Europe. Plant Breed 118:111–117

Hu H (1997) In vitro induced haploids in wheat. In: Jain SM, Sopory SK, Veilleux RE (eds) In vitro haploid production in higher plants, vol 4. Kluwer, Dordrecht, p 73–97

Hu TC, Kasha KJ (1997) Performance of isolated microspore-derived doubled haploids of wheat (Triticum aestivum L.). Can J Plant Sci 77:549–554

Hu TC, Kasha KJ (1997) Improvement of isolated microspore culture of wheat (Triticum aestivum L.) through ovary co-culture. Plant Cell Rep 16:520–525

Hu TC, Ziauddin A, Simion E, Kasha KJ (1995) Isolated microspore culture of wheat (Triticum aestivum L.) in a defined media: I. Effects of pretreatment, isolation methods, and hormones. In Vitro Cell Dev Biol 31:79–83

Hunter CP (1988) Plant regeneration from microspores of barley (Hordeum vulgare). PhD thesis, Wye College, University of London, London, p 235

Indrianto A, Barinova I, Touraev A, Heberle-Bors E (2001) Tracking individual wheat microspores in vitro: identification of embryogenic microspores and body axis formation in the embryo. Planta 212:163–174

J'Aiti F, Benlhabib O, Sharma HC, El Jaafari S, El Hadrami I (1999) Genotypic variation in anther culture and effect of ovary coculture in durum wheat. Plant Cell Tiss Org Cult 59:71–76

J'Aiti F, El Jaafari S, El Hadrami I, Nachit M, Baum M, De Buyser J, Picard E (2000) Problématique de la culture de microspores chez le blé dur Triticum turgidum L. var. Durum (2000). In: Durum wheat improvement in the Mediterranean region: new challenges. Instituto Agronómico Mediterráneo Zaragoza, 12–14 April 2000 p 149–152

Jähne A, Lörz H (1995) Cereal microspore culture. Plant Sci 109:1–12

Jain SM (2001) Tissue culture-derived variation in crop improvement. Euphytica 118:153–166

Jauhar PP (2003) Haploid and doubled haploid production in durum wheat by anther culture. In: Maluszynski M, Kasha KJ, Forster BP, Szarejko I (eds) Doubled haploid production in crop plants: a manual. Kluwer, Dordrecht, p 167–172

Jensen CJ (1977) Monoploid production by chromosome elimination. In: Reinert J, Bajaj YPS (eds) Applied and fundamental aspects of plant cell tissue organ culture. Springer, Berlin Heidelberg New York, p 299–340

Jensen CJ (1983) Producing haploid plants by chromosome elimination. In: Cell and tissue culture techniques for cereal crop improvement. Science, Beijing, p 55–79

Kasha KJ (2005) Chromosome doubling and recovery of doubled haploid plants. In: Palmer CE, Keller WA, Kasha KJ (eds) Haploids in crop improvement II. Biotechnology in agriculture and forestry, vol 56. Springer, Berlin Heidelberg New York, p 123–152

Kasha KJ, Ziauddin A, Cho U-H (1990) Haploids in cereal improvement: anther and microspore culture. In: Gustafson JP (ed) Gene manipulation in plant improvement II. Plenum, New York, p 213–235

Kasha KJ, Simion E, Oro R, Yao QA, Hu TC, Carlson AR (2001) An improved in vitro technique for isolated microspore culture of barley. Euphytica 120:379–385

Kasha KJ, Hu TC, Oro R, Simion E, Shim YS (2001) Nuclear fusion leads to chromosome doubling during mannitol pre-treatment of barley (Hordeum vulgare L.) microspores. J Exp Bot 52:1227–1238

Kasha KJ, Simion E, Miner M, Letarte J, Hu TC (2003) Haploid wheat isolated microspore culture protocol. In: Maluszynski M, Kasha KJ, Forster BP, Szarejko I (eds) Doubled haploid production in crop plants: a manual. Kluwer, Dordrecht, p 77–82

Lazar MD, Schaeffer GW, Baenziger PS (1985) The physical environment in relation to high frequency callus and plantlet development in anther culture of wheat (Triticum aestivum L.) cv. Chris. J Plant Physiol 121:103–109

Letarte J, Simion E, Miner M, Kasha KJ (2005) Arabinogalactans and arabinogalactan-proteins induce embryogenesis in wheat (Triticum aestivum L.) microspore culture. Plant Cell Rep (in press)

Liu W, Zheng MY, Konzak CF (2002) Improving green plant production via isolated microspore culture in bread wheat (Triticum aestivum L.). Plant Cell Rep 20:821–824

Liu W, Zheng Y, Polle E, Konzak CF (2002) Highly efficient doubled-haploid production in wheat (Triticum aestivum L.) via induced microspore embryogenesis. Crop Sci 42:686–692

Ma H, Busch RH, Riera-Lizarazu O, Rines HW, Dill-Macky R (1999) Agronomic performance of lines derived from anther culture, maize pollination and single-seed descent in a spring wheat cross. Theor Appl Genet 99:432–436
Maluszynski M, Kasha KJ, Forster BP, Szarejko I (eds) (2003) Doubled haploid production in crop plants: a manual. Kluwer, Dordrecht, p 428
Maraschin S de F, Vennik M, Lamers GEM, Spaink HP, Wang M (2005) Time-lapse tracking of barley androgenesis reveals position-determined cell death within proembryos. Planta 220:531–540
Mejza SJ, Morgant V, DiBona D, Wong JR (1993) Plant regeneration from isolated microspores of Triticum aestivum. Plant Cell Rep 12:149–153
Mentewab A, Sarrafi A (1997) Influence of genotype and cold pre-treatment on the production of embryoids and their regeneration in tetraploid and hexaploid wheats. J Genet Breed 51:59–62
Mentewab AB, Lazarevich SV, Souvre A, Sarrafi A (1997) Androgenic ability of different Triticum species. J Genet Breed 51:109–113
Metz SG, Sharma HC, Amstrong TA, Mascia PN (1988) Chromosome doubling and aneuploidy in anther-derived plants from two winter wheat lines. Genome 30:177–181
Moieni A, Sarrafi A (1995) Genetic analysis for haploid regeneration responses of hexaploid-wheat anther cultures. Plant Breed 114:247–249
Mujib A (2004) In vitro variability in tissue culture: a fresh look. In: A Mujib, M-J Cho, S Predieri, S Banerjee (eds) In vitro application in crop improvement. Science, Enfield, NH, USA p 261–273
Nitsch JP, Nitsch C (1969) Haploid plants from pollen grains. Science 163:85–87
Otani M, Shimada T (1994) Pollen embryo formation and plant regeneration from cultured anthers of tetraploid wheat. J Genet Breed 48:103–106
Otani M, Shimada T (1995) Effect of synthetic medium on microspore-derived embryoid formation of tetraploid wheat species. Cereal Res Commun 23:345–350
Ouyang JW, Hu H, Chuang CC, Tseng CC (1973) Induction of pollen plants from anther of Triticum aestivum L. cultured in vitro. Sci Sin 16:79–95
Palmer CE, Keller WA, Kasha KJ (eds)(2005) Haploids in crop improvement II. Biotechnology in agriculture and forestry, vol 56. Springer, Berlin Heidelberg New York p 318
Pauk J, Mihály R, Puolimatka M (2003) Protocol of wheat (Triticum aestivum L.) anther culture. In: Maluszynski M, Kasha KJ, Forster BP, Szarejko I (eds) Doubled haploid production in crop plants: a manual. Kluwer, Dordrecht p 59–64
Petrova A, de Ruijter N, van Lammeren A, Schel JH (1993) Structural observations during androgenic microspore culture of the 4c1 genotype of Zea mays L. Euphytica 65:61–69
Reynolds TL (1997) Pollen embryogenesis. Plant Mol Biol 33:1–10
Riley R (1974) The status of haploid research. In: Kasha KJ (ed) Haploids in higher plants; advances and potential. Proceedings of the 1st international symposium. University of Guelph, p 3–9
Roberts-Oehlschlager SL, Dunwell JM (1990) Barley anther culture: Pre-treatment on mannitol stimulates production of microspore-derived embryos. Plant Cell Tiss Org Cult 20:235–240
Saidi N, Cherkaoui S, Chlyah A, Chlyah H (1997) Embryo formation and regeneration in Triticum turgidum ssp. durum anther culture. Plant Cell Tiss Org Cult 51:27–33
Sayed H, Kayyal H, Ramsey L, Ceccarelli S, Baum M (2002) Segregation distortion in doubled haploid lines of barley (Hordeum vulgare L.) detected by simple sequence repeat (SSR) markers. Euphytica 225:265–272

Shim YS, Kasha KJ, Simion E, Letarte J (2005) The relationship between induction of embryogenesis and chromosome doubling in microspore cultures. Protoplasma (in press)

Simmonds DH, Keller WA (1999) Significance of preprophase bands of microtubules in the induction of microspore embryogenesis of Brassica napus. Planta 208:383–391

Snape JW, Simpson E, Parker BB (1986) Criteria for the selection and use of doubled haploid systems in cereal breeding programmes. In: Horn W, Jensen CJ, Odenbach W, Schieder O (eds) Genetic manipulation in plant breeding. de Gruyter, Berlin, p 217–229

Szakacs E, Kovacs G, Paulk J, Barnabás B (1988) Substitution analysis of callus induction and plant regeneration from anther culture in wheat (Triticum aestivum L.). Plant Cell Rep 7:127–129

Szarejko I (2003) Major media composition. In: Maluszynski M, Kasha KJ, Forster BP, Szarejko I (eds) Doubled haploid production in crop plants: a manual. Kluwer, Dordrecht, p 405–414

Tersi M, Xynias IN, Gouli-Vavdinoudi E, Roupakias DG (2004) Effect of the D genome on anther culture response in wheat: preliminary results. In: COST action 851. Gametic cells and molecular breeding for crops improvement, Palermo, 11–13 November

Testillano P, Georgiev S, Mogensen HL, Coronado MJ, Dumas C, Risueno MC, Matthys-Rochon E (2004) Spontaneous chromosome doubling results from nuclear fusion during in vitro maize induce microspore embryogenesis. Chromosoma 112:342–349

Testillano P, Ramirez C, Domenech J, Coronado MJ, Vergne P, Matthys-Rochon E, Risueno MC (2002) Young microspore-derived maize embryos show two domains with defined features also present in zygotic embryogenesis. Int J Dev Biol 1:498–503

Thomas J, Chen Q, Howes N (1997) Chromosome doubling of haploids of common wheat with caffeine. Genome 40:552–558

Thomas WTB, Forster BP, Gertsson B (2003) Doubled haploids in breeding. In: Maluszynski M, Kasha KJ, Forster BP, Szarejko I (eds) Doubled haploid production in crop plants: a manual. Kluwer, Dordrecht, p 337–350

Toonen MAJ, Hendriks T, Schmidt EDL, Verhoeven HA, van Kammen A, de Vries SC (1994) Description of somatic-embryo-forming single cells in carrot suspension cultures employing video cell tracking. Planta 194:565–572

Torp AM, Hansen AL, Andersen SB (2001) Chromosomal regions associated with green plant regeneration in wheat (Triticum aestivum L.) anther culture. Euphytica 119:377–387

Touraev A, Indrianto A, Wratschko I, Vicente O, Heberle-Bors E (1996) Efficient microspore embryogenesis in wheat (Triticum aestivum L.) induced by starvation at high temperature. Sex Plant Reprod 9:209–215

Touraev A, Vicente O, Heberle-Bors E (1997) Initiation of microspore embryogenesis by stress. Trends Plant Sci 2:297–302

Touraev A, Prosser M, Heberle-Bers E (2001) The microspore: a haploid multipurpose cell. Adv Bot Res 35:53–109

Tuvesson S, Ljungberg A, Johansson N, Karlsson KE, Suijs LW, Josset JP (2000) Large-scale production of wheat and triticale double haploids through the use of a single-anther culture method. Plant Breed 119:455–459

Tuvesson S, von Post R, Ljungberg A (2003) Wheat anther culture. In: Maluszynski M, Kasha KJ, Forster BP, Szarejko I (eds) Doubled haploid production in crop plants: a manual. Kluwer, Dordrecht, p 71–76

Twell D, Park SK, Lalanne E (1998) Asymmetric division and cell-fate determination in developing pollen. Trends Plant Sci 3:305–310

Widholm JM (1972) The use of fluorescein diacetate and phenosafranine for determining viability of cultured plant cells. Stain Technol 47:189–194

Zamani I, Kovács G, Gouli-Vavdinoudi E, Roupakias DG, Barnabás B (2000) Regeneration of fertile doubled haploid plants from colchicines-supplemented media in wheat anther culture. Plant Breed 119:461–465

Zhang YL, Li DS (1984) Anther culture of monosomics in Triticum aestivum. Hereditas 6:7–10

Zheng MY (2003) Microspore culture in wheat (Triticum aestivum)—doubled haploid production via induced embryogenesis. Plant Cell Tiss Org Cult 73:213–230

Zheng MY, Liu W, Weng Y, Polle E, Konzak CF (2001) Culture of freshly isolated wheat (Triticum aestivum L.) microspores treated with inducer chemicals. Plant Cell Rep 20:685–690

Zheng MY, Liu W, Weng Y, Polle E, Konzak CF (2003) Production of doubled haploids in wheat (Triticum aestivum L.) through microspore embryogenesis triggered by inducer chemicals. In: Maluszynski M, Kasha KJ, Forster BP, Szarejko I (eds) Doubled haploid production in crop plants: a manual. Kluwer, Dordrecht, p 83–94

Zhu ZQ, Wang JJ, Sun JS (1979) Induction of albino plantlets from Triticum durum Desf. Pollen and preliminary observation of their ploidy. Acta Bot Sin 21:295–297

Zoriniants S, Tashpulatov AS, Heberle-Bors E, Touraev A (2005) The role of stress in the induction of haploid embryogenesis. In: Palmer CE, Keller WA, Kasha KJ (eds) Haploids in crop improvement II. Biotechnology in agriculture and forestry, vol 56. Springer, Berlin Heidelberg New York, p 35–52

Plant Cell Monogr (2)
A. Mujib · J. Šamaj: Somatic Embryogenesis
DOI 10.1007/7089_020/Published online: 30 November 2005

In Vitro Culture of *Arabidopsis* Embryos

Michael Sauer · Jiří Friml (✉)

Zentrum für Molekularbiologie der Pflanzen, Universität Tübingen,
Auf der Morgenstelle 3, 72076 Tübingen, Germany
jiri.friml@zmbp.uni-tuebingen.de

Abstract *Arabidopsis thaliana* is currently the most important model organism for basic molecular plant research. It is also a favourable model for developmental biology, as its embryogenesis follows a nearly invariant pattern of cell divisions and cell type specifications. Study of embryogenesis can involve genetic, physiological or biochemical approaches, but is always limited by the inaccessibility of the embryos which develop deep inside maternal tissue. Thus, for developmental studies, there is an increasing demand for methods which allow embryogenesis under artificial conditions, providing better accessibility to experimental manipulation. In this chapter, we address theoretical aspects of embryo culture, give some thoughts on which embryo culture system is suited best for which application and finally discuss three current methods which have been successfully used in *Arabidopsis* embryo culture.

1 Introduction

The culture of plant embryos outside of their normal environment within maternal tissue is a general approach for multiple purposes in basic biology research, agriculture and horticulture. Not surprisingly, the translation into a specific method for embryo development ex planta varies greatly with the intended application. We first want to give some general theoretical thoughts about different reasons for embryo culture and the requirements of the culture method resulting from them. We will then focus on embryo culture in the genetic model organism *Arabidopsis thaliana*, and in particular on a relatively recent method for the ex planta culture of its embryos.

2 Different Applications for Embryo Culture

In agriculture and horticulture, both production and research, one common aim is to generate genetically identical offspring from a potentially limited source of original plant material with desirable traits. This is often done by partitioning of vegetative tissue, therefore omitting fertilization and embryogenesis. This vegetative reproduction relies on a plant's amazing capability of regeneration, which has to do with its capacity for extensive postembryonic development, including the formation of new organs. In many plant species, asexual reproduction is even the predominant form of reproduction under natural circumstances. However, for many applications in crops, it is desirable to manipulate embryos rather than parts of vegetative tissue. The origin of the embryos in this case should not be zygotic, as one wishes to avoid recombinatory events to establish genetically identical offspring from a specific individual plant. A special method to generate multiple clones from vegetative source tissue is somatic embryogenesis. It differs from simple partitioning and regenerating vegetative tissue in that it artificially triggers the embryogenetic programme of the plant, yielding mature embryos without the need for gametophytic structures or a fertilization event. As it is an extremely widespread method within the whole plant research field we will dedicate a section to somatic embryogenesis in *Arabidopsis*.

One application which is equally present in research and commercial applications is the rescue of embryos of certain mutants or hybrids, where under normal conditions the zygotic embryos would be aborted. For example, in cases where fruit or seed development is impaired and would result in death of the embryo, embryo culture can overcome this problem. For this application, it is generally not important if embryo development closely resembles embryo development in planta, as the focus lies on regenerating the plant. The priority here is robustness of the method to reliably yield regenerated plants, paired with ease of use.

For developmental biology research, embryogenesis is the key event in an organism's life cycle. Although in plants postembryonic development is much more pronounced than in animals, as most habitually relevant features are established after embryogenesis, embryogenesis is still a central developmental process, because it generates the axes and symmetries along which all subsequent steps of development will occur. The developing embryo of most plants is a tiny, delicate structure and it is therefore clear that embryogenesis takes place in a well-protected environment, deeply embedded in maternal gametophytic and vegetative tissues and organs. Consequently, this process is not readily accessible for experimental analysis, especially when the experimental setup requires the manipulation of embryos in vivo. Methods aiming to overcome these problems must provide the embryo with an environment which is suitable for its development, preferentially one in which embryogenesis occurs identically to the development in planta, while still being accessible to experimental manipulation. The generation of large numbers of developing embryos is not a primary objective, but if a method allows large sample numbers, statistical analyses become possible.

The examples just given are by far not all possible situations in which the culture of embryos is desired, but they illustrate that each application has its own, special profile of requirements and solutions. We will proceed by showing different methods to culture *Arabidopsis* embryos in vitro and as *Arabidopsis* is mainly a model organism for basic research, we will discuss them with special attention to research applications.

3
Culture of *Arabidopsis* Embryos

3.1
Arabidopsis Embryogenesis in Vivo

Before we speak about specific methods for culturing *Arabidopsis* embryos, we think it is necessary to describe how *Arabidopsis* embryos develop under natural conditions.

Arabidopsis is a member of the Brassicaceae; its fruit is a silique with two valves, separated by a dividing wall, the replum. In each valve, about 20–30 immature seeds, the ovules, contain the female gametophyte and the endosperm. The ovules are connected to the plant by the funiculus, which provides the developing seeds with water, nutrients and potentially signals from the plant. (For a detailed description of *Arabidopsis* morphology, see Bowman 1994). After fertilization, the embryo develops through a highly uniform series of cell divisions and cell fate changes, which are also reflected by molecular markers, leading to morphologically distinct intermediate stages (Friml et al. 2003; Vroemen et al. 1996). The initial zygotic division is asymmetric: it gives rise to a smaller apical and a larger basal cell. The basal cell and its lineage divide anticlinally and form the suspensor, an extraembryonic file of cells which connects the embryo to the maternal tissue. After the 32-cell stage, the uppermost suspensor cell becomes incorporated into the embryo as the hypophysis cell, from which most of the future root meristem is derived. All other embryonic parts are derived from the apical cell, which divides periclinally and anticlinally to first form an eight-cell stage proembryo, then establishes the protoderm (at the 16-cell stage) and finally gives rise to the round globular embryo. At this stage, a new axis of bilateral symmetry is established, with the positions of the cotyledon primordia becoming apparent at the triangular stage. At the heart stage, the patterning is complete and during later development this pattern is merely elaborated upon by elongation growth (Jürgens and Mayer 1994). Embryogenesis is complete when the embryos have reached the bent cotyledon stage, after that seed maturation sets in. The whole process of embryogenesis from zygote to bent cotyledon stage takes about 8 days at 25 °C under continuous light (Mansfield et al. 1991) From the developmental perspective the most relevant period from fertilization to the triangular or heart stage takes only 3 days.

3.2 Requirements for a Research-Oriented *Arabidopsis* Culture System

When thinking of an optimal system for research applications, especially developmental research, the following requirements come up:

1. The system should simulate the in planta situation in a way that embryogenesis occurs identical to the natural situation. This is a problematic requirement, as it is not easy to assess whether embryogenesis is taking the natural course in all accounts. Microscopic analysis does not give a full report on the status of the embryo and there are only a limited number of molecular markers for cell fates. Cell wall composition, cytosolic composition, physiological responses and so on are difficult to assess and thus there remains some uncertainty about the faithfulness of a culture system's ability to simulate the in planta situation, although embryos might morphologically appear unaltered.
2. The system must be experimentally accessible. It should allow manipulation of the embryo's surroundings to study the influence of potential signals, such as light, temperature, nutrients, hormones, or chemical compounds affecting physiological processes. At the same time, the embryos should be analysable by common techniques, especially microscopy, preferentially without interfering with the development to allow for time-lapse studies over extended periods.
3. The system should be easy to use. One problem with *Arabidopsis* embryos is that they are very small: right after the first zygotic division, the embryo is about 20 μm in length, at the globular stage about 40 μm and at the heart stage, where all symmetry axes are already established, it still measures only 100 μm (Bowman 1994). This makes direct handling of young-stage embryos almost impossible, and it is still time-consuming for older embryos.
4. The system should be efficient, so that large numbers of developing embryos can be generated. This is especially necessary for studies in which statistical analysis must be performed, for example when weak phenotypic defects are analysed, or when subsequent steps (like staining procedures) will further reduce the number of embryos available for analysis.

We will now describe and discuss three different methods which have proven useful in current *Arabidopsis* research.

3.3 Somatic Embryogenesis in *Arabidopsis*

Plant embryogenesis can be recapitulated to some extent from somatic cells by a process known as somatic embryogenesis. Vegetative cells of virtually all tissue types can produce somatic embryos and even immature pollen grains can be induced to develop into androgenetic embryos (Custers et al. 1994). Somatic embryogenesis occurs spontaneously very rarely; more often it is induced in vitro in plant tissue cultures by application of exogenous auxins, stress treatment or in specific mutant backgrounds (Mordhorst et al. 1998). In culture, only a small proportion of single cells, so-called competent cells, can be induced to change their developmental programme and recapitulate embryogenesis (Toonen et al. 1997). The mechanism of this developmental switch is largely unknown (there is some although not conclusive knowledge about embryo-competent cells—this will be covered by other chapters) in either *Arabidopsis* or other species.

In carrot (*Daucus carota*) tissue culture, the expression of *SOMATIC EMBRYOGENESIS RECEPTOR KINASE* (*DcSERK*) specifically marks the competent cells destined to develop into somatic embryos. *SERK* encodes a Leu-rich repeat transmembrane receptor-like kinase (RLK) (Schmidt et al. 1997). Numerous examples also from *Arabidopsis* show that RLKs appear to have a prominent role in cellular signalling in plants (Morris and Walker 2003). DcSERK's closest homologue in *Arabidopsis* (*AtSERK1*) is expressed in developing ovules, early embryos and importantly in cultured cells competent for embryogenesis. Moreover, ectopic expression of *AtSERK1* confers sustained embryogenic competence to seedlings under in vitro conditions (Hecht et al. 2001). Interestingly, expression of SERK1 is induced by auxin (Nolan et al. 2003), providing a possible mechanism for how auxin increases embryo competence of in vitro cultured cells. Although the exact nature of the signal activating SERK or the other components of this pathway remains elusive, it seems clear that *AtSERK1* plays an important role in determining embryogenic competence. Another gene which is prominently expressed during the initiation of embryo development is *BABY BOOM* (*BBM*) coding for an APETALA2-like transcription factor. *BBM* has been isolated as being expressed at early stages of *Brassica napus* microspore embryo cultures. Overexpression of *BBM* in *35S::BBM Arabidopsis* transgenic seedlings conferred spontaneous formation of somatic embryos and/or cotyledon-like structures on aerial organs (Boutilier et al. 2002). Other transcription factors having similar effects on induction of somatic embryogenesis are *LEAFY COTYLEDON1* (*LEC1*) and *LEAFY COTYLEDON2* (*LEC2*) (Lotan et al. 1998; Stone et al. 2001). There are also loss-of-function mutants showing defects in both embryo identity and seed maturation processes (West et al. 1994). *lec* mutant embryos exhibit morphological characteristics otherwise confined to postembryonic development and are desiccation-intolerant. Ectopic expression of either gene promotes somatic embryo formation on vegetative tissues of the plant. The transformation of postgerminative plant organs into embryo-like structures is also induced by mutations in another *Arabidopsis* gene, *PICKLE* (*PKL*). The roots of *pkl* seedlings show embryonic characteristics and form somatic embryos when excised and placed on minimal tissue culture medium (Ogas et al. 1997). *PKL* encodes a chromatin remodelling factor with possible transcriptional repressor function (Ogas et al. 1999). Thus, at the moment, a number of proteins, mainly transcriptional regulators, have been identified as functionally associated with the competence of cells to produce somatic embryos. Most of them are also expressed in the course of zygotic embryogenesis in *Arabidopsis* and seem to be involved in regulation of this process.

This raises an important question to which extent somatic embryogenesis follows the same developmental programme as zygotic embryogenesis in terms of mechanism and factors involved. It is difficult to compare early embryo development as all somatic embryogenesis methods suffer from the absence of well-defined early stages. Somatic *Arabidopsis* embryos (derived from zygotic embryos) resembling the four-cell stage were described (Luo and Koop 1997); however, the cell division pattern becomes very irregular subsequently. Also somatic embryos derived from leaf protoplast exhibited some resemblance to normal zygotic embryos, but they were less regular and naturally lacked a suspensor. During following stages (after the globular stage) somatic and zygotic embryos show pronounced morphological similarities as well as the same sequence of developmental stages, suggesting a common developmental programme. The issue about the involvement of the same molecular factors in regulating both zygotic and somatic embryogenesis has also been addressed using genetic approaches. Three *Arabidopsis* mutants lacking an embryonic shoot apical meristem (SAM), *shoot meristemless* (*stm*), *wuschel* (*wus*) and *zwille/pinhead* (*zll/pnh*), were used to establish embryogenic cell cultures. Somatic embryos derived from all three mutants showed the same mutant phenotypes as their zygotic equivalents. These results show that a functional SAM is not essential for induction of somatic embryogenesis. They also suggest that the developmental programmes of somatic and zygotic embryos require the same regulatory genetic pathways (Mordhorst et al. 2002). Nevertheless, lack of clearly defined early stages and early embryo markers in somatic embryos makes it difficult to evaluate to what extent the same developmental programme is followed during the most important early patterning events.

That fact brings about the limitations of somatic embryogenesis for studying the normal course of embryogenesis: the comparability of developmental programmes is not entirely clarified, culture media usually contain high levels of phytohormones, which also strongly affect normal zygotic embryogenesis (Liu et al. 1993; Hadfi et al. 1998; Friml et al. 2003) and the mutants used, such as the *primordia timing* (*pt*) (Mordhorst et al. 1998) generally also show aberrations in their zygotic embryonic development. In light of these facts, the feasibility of somatic embryo culture for the study of early zygotic embryogenesis is limited.

3.4
Embryo Culture Based on Isolated Embryos

The culture of isolated zygotic embryos is the most direct approach, as embryogenesis will occur without any cues from surrounding maternal tissue. It has been successfully carried out in *Brassica* from the globular stage on (Liu et al. 1993; Hadfi et al. 1998) to study the effects of auxin or auxin transport inhibitors. Both studies reveal a connection between auxin transport and proper embryogenesis. Liu et al. (1993) showed that inhibition of auxin transport at the globular stage leads to improper placement or insufficient separation of cotyledon primordia, resulting in cotyledon fusion. In contrast, cotyledon fusion did not occur in embryos treated at the heart stage or later, where the bilateral symmetry of the cotyledon primordia is already established. Hadfi et al. (1998) also found younger embryos (at the globular stage) to be more sensitive than transition-stage embryos when treated with either auxin or auxin transport inhibitors. At early stages, auxin treatment led to severe phenotypes showing embryos developed into ball-shaped structures with no recognizable axes. These studies show that axis formation is a process which depends on auxin transport and that already-established symmetries are maintained even if auxin homoeostasis is altered.

In *Arabidopsis*, however, reports on culture of excised embryos are scarce and none of them deal with stages earlier than the heart stage, at which the main patterning events have already taken place. Harding et al. (2003) cultured bent cotyledon-stage embryos constitutively expressing AGL15, a MADS domain protein on a medium similar to normal seed germination medium. The transgenic embryos produced secondary somatic embryos and remained embryogenic for long periods of time, suggesting that AGL15 is involved in the promotion and maintenance of embryo identity. In a recent study, Blilou et al. (2005) used isolated embryo culture to analyse the postembryonic phenotype of an otherwise lethal combination of mutations in the auxin transport components PIN1, PIN3, PIN4 and PIN7. The quadruple mutants lack roots and show ectopic shoot-like structures at the apex, while wild-type embryos develop normal root and shoot structures. This provides additional evidence for the involvement of polar auxin transport in root and shoot pole specification. In both of these studies, more or less mature embryos were cultured. For earlier stages, there is one report by Eastmond et al. (2002) where heart-stage embryos were isolated and successfully cultured on a B5-based medium supplemented with glutamine. The response of *tps1* mutant embryos, deficient for trehalose-6-phosphate synthase 1, which is essential for embryo maturation, was tested when cultured on different sucrose concentrations. Reduced sucrose levels led to a partial rescue, implicating that TPS1 is involved in the regulation of sugar metabolism or transport in embryo development.

It seems that in *Arabidopsis* isolating methods are especially limited to later stages of embryo development owing to the small size of early embryos. The study by Eastmond et al. (2002) shows that isolation and culture of heart-shaped embryos is possible and actually the only way of creating an environment in which concentrations of solutes are not buffered by surrounding maternal tissue. Embryogenesis seems at least superficially to follow the natural course, and the experimental accessibility of embryos in this culture system is very high, also for microscopy. However, it is not a method for generating large sample numbers, nor it is easy to use. But for some applications, where the embryo must be separated from all maternal influences, there appears to be no other alternative.

3.5
Culture of Embryos Inside Their Ovules

The culture of embryos inside their ovules is a widespread method in the plant field, although until recently, there were no studies with a focus on development in *Arabidopsis*. For example, there are protocols to rescue specific hybrids or mutants (e.g. for grapes, Cain et al. 1983; for wheat, Kumlehn et al. 1997), or applications where the development of the ovule itself is analysed (e.g. cotton fibre development, Beasley 1971).

Recently, the establishment of an ovule-based culture system for the study of *Arabidopsis* embryogenesis was reported (Sauer and Friml 2004). Its applicability also to the earliest stages of embryonic development was verified. The aberrations in development induced by the culture are acceptably low and the method is compatible with various subsequent analytical methods, such as fluorescence microscopy or immunocytochemistry. This method was used for administration of physiologically active drugs or hormones and was instrumental in elucidating the role of auxin and its spatial and temporal distribution during apical–basal axis formation (Friml et al. 2003). Here for the first time consequences of treatments with auxin and auxin transport inhibitors were shown in *Arabidopsis* preglobular embryos. Auxin activity was indirectly monitored using the synthetic auxin responsive promoter DR5rev::green fluorescent protein (GFP), and its activity was analysed in embryos excised from cultured ovules with fluorescence microscopy. Treatments with auxins led to an accumulation of DR5rev::GFP fluorescence in the early preglobular proembryo, whereas auxin transport inhibitor treatments led to accumulation in the suspensor, indicating a flow of auxin towards the proembryo at early stages, which is reversed later on (Fig. 1). The same approach was used to examine the role of auxin and polar auxin transport in the establishment of bilateral symmetry and organ primordia (Benkova et al. 2003). The DR5rev::GFP expression pattern which normally marks the cotyledon primordia was enhanced and more diffuse in embryos from ovules cultured in the presence of the polar auxin transport inhibitors. A similar

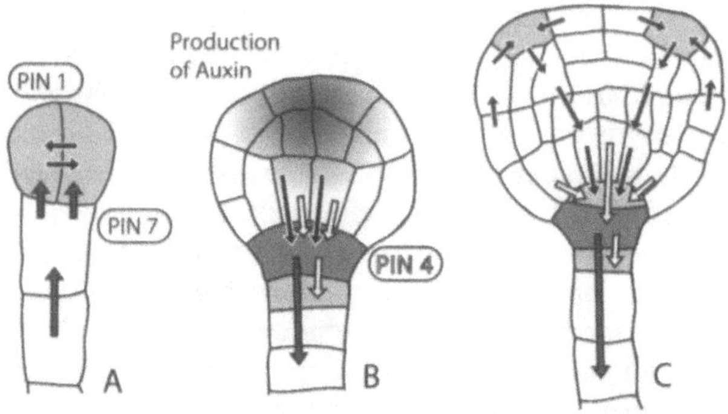

Fig. 1 Auxin flux during the establishment of the apical–basal axis and bilateral symmetry in *Arabidopsis* embryogenesis. **a** In the suspensor of preglobular embryos, PIN7 mediates transport into the embryo proper, while PIN1 does not facilite auxin translocation. This results in auxin accumulation in the embryo proper. **b** At the globular stage, PIN7 polarity in the suspensor is reversed, PIN1 and PIN4 start to be polarly localized and mediate a flow of auxin out of the embryo with a new auxin maximum at the basal embryo pole. Treatments with transport inhibitors suggest a new site of auxin production in the apical part of the proembryo. **c** At the transition between globular and heart stages, PIN1 is localized in protodermis cells with a polarity pointing towards the incipient cotyledon primordia. There, auxin accumulates, probably facilitating primordium development. In inner-tissue layers, PIN1 is basally localized in the elongating vasculature precursor cells. Auxin accumulates in the hypophysis, where it presumably plays an important role in root meristem development

study of Weijers et al. (2005) shows that PIN auxin transport components are not only required for the establishment of auxin gradients but also for their maintenance. Auxin levels in the embryo were increased either by exogenous application of auxin to the culture medium or by expression of a bacterial gene for auxin synthesis from tryptophan, *iaaM*, under the *RPS5a* promoter active in the proembryo. Increased auxin levels in the embryo lead to no apparent defects in embryogenesis in wild-type background, however, in the *pin4* mutant background, significantly increased rates of embryos with pattern defects were found, such as fused cotyledons. Interestingly, auxin synthesis by the *iaaM* protein could be strongly enhanced when tryptophan was added to the medium, as demonstrated by increased *DR5rev::GFP* expression. Taken together, the three studies show that axis formation in *Arabidopsis* embryogenesis relies on auxin gradients actively established and maintained by members of the PIN protein family. At the same time they validate the use of the ovule-based in vitro culture system for developmental studies.

Ovule culture is a more direct approach than somatic embryogenesis, because it does not artificially induce embryogenesis and also makes the very early stages accessible to analysis, which is difficult in somatic embryogenesis, but it is less direct than the culture of isolated zygotic embryos. The embryos are surrounded and buffered by several layers of maternal tissue and are connected to the endosperm by the suspensor; thus, they still receive cues from maternal tissue and are to some extent protected from any chemical treatment administered in the medium. Nevertheless, the results with this method have shown that this barrier is not absolute: all soluble substances tested so far have been shown to permeate through the integuments to an extent which enables visible effects on embryo development.

The integument also limits access to the embryo for microscopic analysis, as it is not transparent enough to allow high-resolution imaging. For studies aiming to monitor the living embryo noninvasively, the ovule culture method is unsuitable, although advances in multiphoton fluorescence microscopy may provide a solution for this problem.

4
Conclusions and Future Prospects

In this chapter we have described three methods for *Arabidopsis* embryogenesis in vitro. None of them are without limitations, but each of them allows deeper insights into *Arabidopsis* embryogenesis than classical genetic, physiological or biochemical approaches could give alone. The strength of embryo culture lies in the ability to combine some of these approaches while the embryo is developing, and possibly also to directly monitor the effects on the living embryo.

From our perspective, the ultimate goal in terms of culture methods would be a system which allows for live real-time imaging of very early stages of development and external manipulation of environment, while still enabling normal embryogenesis. Although recent advances with new culture methods or with multiphoton fluorescence microscopy seem promising, we are still far away from an artificial ovule.

Until then, researchers will have to carefully consider the advantages and limits of each culture protocol and chose the method best suited for their specific application.

References

Beasley CA (1971) In vitro culture of fertilized cotton ovules. BioScience 21:906–907

Benková E, Michniewicz M, Sauer M, Teichmann T, Seifertova D, Jürgens G, Friml J (2003) Local, efflux-dependent auxin gradients as a common module for plant organ formation. Cell 115:591–602

Blilou I, Xu J, Wildwater M, Willemsen V, Paponov I, Friml J, Heidstra R, Aida M, Palme K, Scheres B (2005) The PIN auxin efflux facilitator network controls growth and patterning in *Arabidopsis* roots. Nature 433:39–44

Boutilier K, Offringa R, Sharma VK, Kieft H, Ouellet T, Zhang L, Hattori J, Liu CM, van Lammeren AA, Miki BL, Custers JB, van Lookeren Campagne MM (2002) Ectopic expression of BABY BOOM triggers a conversion from vegetative to embryonic growth. Plant Cell 14:1737–49

Bowman J (ed) (1994) *Arabidopsis*: an atlas of morphology and development. Springer, Berlin Heidelberg New York

Cain D, Emershad R, Tarailo R (1983) In-ovulo embryo culture and seedling development of seeded and seedless grape (Vitis vinefera L). Vitis 22:9–14

Custers JBM, Cordewener JHG, Nöllen Y, Dons JJM, Van Lookeren Campagne MM (1994) Temperature controls both gametophytic and sporophytic development in microspore cultures of Brassica napus. Plant Cell Rep 13:267–271

Eastmond PJ, van Dijken AJ, Spielman M, Kerr A, Tissier AF, Dickinson HG, Jones JD, Smeekens SC, Graham IA (2002) Trehalose-6-phosphate synthase 1, which catalyses the first step in trehalose synthesis, is essential for *Arabidopsis* embryo maturation. Plant J 29:225

Friml J, Vieten A, Sauer M, Weijers D, Hamann T, Schwarz T, Offringa R, Jürgens G (2003) Efflux-dependent auxin gradients establish the apical-basal axis of *Arabidopsis*. Nature 426:147–153

Hadfi K, Speth V, Neuhaus G (1998) Auxin induced developmental patterns in Brassica juncea embryos. Development 125:879–887

Harding EW, Tang W, Nichols KW, Fernandez DE, Perry SE (2003) Expression and maintenance of embryogenic potential is enhanced through constitutive expression of AGAMOUS-like 15. Plant Physiol 133:653–663

Hecht V, Vielle-Calzada JP, Hartog MV, Schmidt ED, Boutilier K, Grossniklaus U, de Vries SC (2001) The *Arabidopsis* SOMATIC EMBRYOGENESIS RECEPTOR KINASE 1 gene is expressed in developing ovules and embryos and enhances embryogenic competence in culture. Plant Physiol 127:803–816

Jürgens G, Mayer U (1994) *Arabidopsis*. In: Bard JBL (ed) Embryos colour atlas of development. Wolfe, London pp 7–21

Kumlehn J, Schieder O, Lörz H (1997) In vitro development of wheat (Triticum aestivum L) from zygote to plant via ovule culture. Plant Cell Reports 16:663–667

Liu C, Xu Z, Chua N (1993) Auxin polar transport is essential for the establishment of bilateral symmetry during early plant embryogenesis. Plant Cell 5:621–630

Lotan T, Ohto M, Yee KM, West MAL, Lo R, Kwong RW, Yamagishi K, Fischer RL, Goldberg RB, Harada JJ (1998) *Arabidopsis* LEAFY COTYLEDON1 is sufficient to induce embryo development in vegetative cells. Cell 93:1195–1205

Luo Y, Koop H (1997) Somatic embryogenesis in cultured immature zygotic embryos and leaf protoplasts of *Arabidopsis thaliana*. Planta 202:387–396

Mansfield SG, Briarty LG (1991) Early embryogenesis in *Arabidopsis thaliana*. II. The developing embryo. Can J Bot 69:461–476

Mordhorst A, Hartog M, El Tamer M, Laux T, de Vries S (2002) Somatic embryogenesis from *Arabidopsis* shoot apical meristem mutants. Planta 214:829–836

Mordhorst A, Voerman K, Hartog M, Meijer E, van Went J, Koornneef M, de Vries S (1998) Somatic embryogenesis in *Arabidopsis thaliana* is facilitated by mutations in genes repressing meristematic cell divisions. Genetics 149:549–563

Morris ER, Walker JC (2003) Receptor-like protein kinases: the keys to response. Curr Opin Plant Biol 6:339–342

New DAT (1990) Whole embryo culture, teratogenesis and the estimation of teratologic risk. Teratology 42:635–642

Nolan KE, Irwanto RR, Rose RJ (2003) Auxin up-regulates MtSERK1 expression in both Medicago truncatula root-forming and embryogenic cultures. Plant Physiol 133:218–230

Ogas J, Cheng J-C, Sung RZ, Somerville C (1997) Cellular differentiation regulated by gibberellin in the *Arabidopsis thaliana* pickle mutant. Science 277:91–94

Ogas J, Kaufmann S, Henderson J, Somerville C (1999) PICKLE is a CHD3 chromatin-remodeling factor that regulates the transition from embryonic to vegetative development in *Arabidopsis*. Proc Natl Acad Sci USA 96:13839–13844

Sauer M, Friml J (2004) In vitro culture of *Arabidopsis* embryos within their ovules. Plant J 40:835–843

Schmidt ED, Guzzo F, Toonen MA, de Vries SC (1997) A leucine-rich repeat containing receptor-like kinase marks somatic plant cells competent to form embryos. Development 124:2049–2062

Stone SL, Kwong LW, Yee KM, Pelletier J, Lepiniec L, Fischer RL, Goldberg RB, Harada JJ (2001) LEAFY COTYLEDON2 encodes a B3 domain transcription factor that induces embryo development. Proc Natl Acad Sci USA 98:11806–11811

Toonen MA, Verhees JA, Schmidt ED, van Kammen A, de Vries SC (1997) AtLTP1 luciferase expression during carrot somatic embryogenesis. Plant J 12:1213–1221

Vroemen C, Langeveld S, Mayer U, Ripper G, Jürgens G, Van Kammen A, de Vries S (1996) Pattern formation in the *Arabidopsis* embryo revealed by position-specific lipid transfer protein gene expression. Plant Cell 8:783–791

Weijers D, Sauer M, Meurette O, Friml J, Ljung K, Sandberg G, Hooykaas P, Offringa R (2005)Maintenance of embryonic auxin distribution for apical-basal patterning by PIN-FORMED-dependent auxin transport in Arabidopsis. Plant Cell 17:2517–2526

West MAL, Yee KM, Danao J, Zimmerman JL, Fischer RL, Goldberg RB, Harada JJ (1994) LEAFY COTYLEDON1 is an essential regulator of late embryogenesis and cotyledon identity in *Arabidopsis*. Plant Cell 6:1731–1745

Subject Index

Kolibri 360 GmbH
Ferdinand-Jühlke-Straße 7
99095 Erfurt, Deutschland
produktsicherheit@kolibri360.de